STRUCTURED FLUIDS

Structured Fluids

Polymers, Colloids, Surfactants

T. WITTEN

The University of Chicago,
Chicago, Illinois, USA

with

P. PINCUS

University of California,
Santa Barbara, USA

OXFORD

UNIVERSITY PRESS

OXFORD
UNIVERSITY PRESS

Great Clarendon Street, Oxford OX2 6DP

Oxford University Press is a department of the University of Oxford.
It furthers the University's objective of excellence in research, scholarship,
and education by publishing worldwide in

Oxford New York

Auckland Cape Town Dar es Salaam Hong Kong Karachi
Kuala Lumpur Madrid Melbourne Mexico City Nairobi New Delhi
Shanghai Taipei Toronto

With offices in

Argentina Austria Brazil Chile Czech Republic France Greece
Guatemala Hungary Italy Japan Poland Portugal
Singapore South Korea Switzerland Thailand Turkey Ukraine Vietnam

Oxford is a registered trade mark of Oxford University Press
in the UK and in certain other countries

Published in the United States
by Oxford University Press Inc., New York

First published 2004
First published in paperback 2010
Reprinted 2012

British library catalogue in Publication Data
Data available
Library of Congress Cataloging-in-Publication Data
Data available

Typeset by Newgen Imaging Systems (P) Ltd., Chennai, India
Printed in Great Britain
on acid-free paper by
CPI Group (UK) Ltd, Croydon, CR0 4YY

ISBN 978-0-19-958382-9

10 9 8 7 6 5 4 3 2

For Molly

Preface

This book is about fluids containing polyatomic structures such as polymer molecules or colloidal grains. Such structured fluids have come to be known as soft condensed matter. The book takes a scaling approach, following de Gennes's important monograph *Scaling Concepts in Polymer Physics*. My purpose is to provide a unified, pedagogical introduction to soft-matter phenomena that embodies this scaling approach. The book focuses on how to account in the simplest way for the distinctive length, time, and energy scales that characterize each phenomenon. This approach allows unity and simplicity, but it provides only an initial glimpse into the rich phenomena and science to be found in soft condensed matter. Notably, I omit an important aspect of soft matter: the broken symmetry of the liquid crystalline state and the distinctive responses and structures that arise from it. This realm of soft matter has benefited greatly from the modern tools of differential geometry and field theory, as discussed elegantly in the advanced text *Principles of Condensed Matter Physics* by Paul Chaikin and Tom Lubensky (Cambridge, 1995).

The book began life in the late eighties as a joint venture with Phil Pincus of the University of California, Santa Barbara. The chapters developed over several cycles of teaching a course on structured fluids at Chicago. The intended audience is the advanced undergraduate in physical science or engineering who is comfortable with elementary physics and has seen elementary statistical mechanics. The needed knowledge of physics is at the level of e.g., D. Halliday and R. Resnick's *Fundamentals of Physics* (New York, Wiley 2000). I assume the student has seen statistical physics at the level of Reif's book, the second item of the list below. A working knowledge of Fourier transformation will also be helpful.

Important conceptual points are illustrated by problems interspersed through the text. The serious student should work these problems since the logical development remains incomplete without them. Almost all of the problems have been assigned, graded, and refined at least once. The book also includes a number of suggested experimental projects using household materials. These are intended to show concretely the principles discussed. They also give the student a sense of how well the idealizations treated in the book apply to real liquids. Students in the structured fluids course were required to do one of these projects or some other project of their own devising. The projects are meant to open a line of inquiry; only some have been tested, and the student may well find that great modifications are needed to make them work. To guide the student to further information, the book includes references to more detailed work. I have tried to suggest sources in each major subject area. In some cases I have cited primary journal articles, but these references are far from being thorough or balanced. Corrections and additions to the text will be available via the Oxford University Press web site, http://www.oup.com/.

In teaching the course, I found that the text contains too much material to be covered in a one-quarter term with 27 class hours. It could probably be covered rapidly in a one-semester course.

In learning from this book, there are several useful reference works to keep in mind. The first and most important is de Gennes's monograph on polymer physics, Item 1 on the list below. Polymers serve as the paradigmatic system in the book. Most of the ideas have been developed using polymers as an example, then applied to other structured fluids. For further depth or clarification of these ideas, de Gennes's book is the best single source of which we are aware. The subject depends heavily on ideas of statistical physics. The book develops all the needed ideas within the text and aims to keep their number to a minimum. Still, more explanation or depth on statistical physics may be helpful; a good source is Reif's book cited in Item 2. For the discussion of colloids and surfactants, Item 3 is useful. This practical book by Jacob Israelachvili is about liquids generally but contains much information about colloids and surfactants. For collective properties of surfactant assemblies, Sam Safran's book (Item 4) is useful.

1. P.-G. de Gennes, *Scaling Concepts in Polymer Physics* (Ithaca NY: Cornell, 1979).
2. F. Reif, *Fundamentals of Statistical and Thermal Physics* (New York: McGraw-Hill, 1965).
3. Jacob N. Israelachvili, *Intermolecular and Surface Forces*, 2nd ed. (London: San Diego, CA: Academic Press, 1991)
4. Samuel A. Safran, *Statistical Thermodynamics of Surfaces, Interfaces, and Membranes* (Reading, MA: Addison-Wesley Publication, 1994).

The reader should know about other classic and recent works that cover the same material. The polymer physics material in Chapters 3 and 4 is well covered by other textbooks and monographs. Two of the best known are Paul J. Flory's *Statistical Mechanics of Chain Molecules* (New York, Interscience Publishers, 1969) and Charles Tanford's *Physical Chemistry of Macromolecules* (New York, Wiley 1961). There is also an important property handbook: *Polymer Handbook*, 4th edition, by J. Brandrup, E. H. Immergut and E. A. Grulke (New York, Wiley, 1999). An important monograph written from a statistical physicist's point of view is *Polymers in Solution* by G. Jannink and J. Des Cloizeaux (Oxford, Clarendon Press, 1992). Another monograph gives an authoritative treatment of polymer motions. It is *The Theory of Polymer Dynamics* by Masao Doi and S. F. Edwards (New York, Oxford University Press, 1986). Two recent textbooks offer a pedagogical treatment of polymers: *Statistical physics of Macromolecules* by A. Yu. Grossberg and A. R. Khokhlov (New York, AIP Press, 1994), and *Polymer Physics* by M. Rubinstein, R. H. Colby (New York, Oxford University Press, 2003). A recent text covering a variety of structured fluids is Ronald G. Larson's *The Structure and Rheology of Complex Fluids* (New York, Oxford University Press, 1999). It is more advanced and more comprehensive than this book. Richard A. L. Jones's new text *Soft Condensed Matter* (Oxford, 2002), is similar to the present book in scope and level.

By now the efforts of several people have improved the text greatly. The book owes much to the generations of students who took the course, read early drafts, and forced me to be more clear. Arlette Baljon, Alfred Liu, Keith Bradley, Joe Plewa, and other students weeded out errors and confusing statements in the text. Jung-ren Huang improved every chapter by his close reading and thoughtful suggestions. He also compiled the data for the big table of semi-dilute universal ratios in that chapter. Olgica Bakajin did the simulation to make the dilation symmetry figure in the appendix of Chapter 3. Sidney Nagel and other colleagues at the James Franck Institute took a lively interest in the course, and the book benefited from their discussions and support. Oxford editor Bob Rodgers provided early encouragement and obtained several very useful reviews of an early draft of the book. Seven other reviewers looked at the completed book last year and many improvements were made in response to their wise suggestions. Phil Pincus inspired this entire project and wrote the first draft of Chapter 5. He also spent many hours over several collaborative visits to bring the book along. The book would not exist without the support and encouragement of my wife Molly.

Chicago   T. A. Witten
May 2003

Contents

Overview

1

1.1 Introduction

As physical scientists, we are concerned with the behavior of matter in all its forms. We want to know what matter does and why. This is our goal in studying the primordial universe, the tenuous interstellar medium, the gaseous atmosphere of the Earth, the ionized plasma of the Sun's corona, the mundane liquids and solids of our human surroundings, and the exotic dense matter within a molecule, an atomic nucleus, or a proton. This book is about a tiny subset of this vast range of forms of matter: structured fluids. Structured fluids are liquids, i.e., condensed matter in which the atoms are adjacent but freely mobile on a local scale. These liquids contain connected polyatomic structures, such as solid grains or large molecules[†]. Though this is a small sampling of all forms of matter, it is a significant one. These materials are important to our basic goal in studying matter: to explore the limits of what matter is capable of doing. Though they form a narrow range of materials, structured fluids show a wide range of behavior. Indeed, they show behavior not seen in other forms of matter, however exotic.

> [†] By "polyatomic structure" we mean any assembly of many atoms that retains its integrity over experimental time scales. This includes macromolecules and also more weakly-bound structures.

We may glimpse this distinctive behavior by recalling a few structured fluids from everyday experience: pancake syrup, egg white, Silly Putty, corn-starch-and-water paste, or toothpaste. Though each of these gooey materials resembles a liquid, all respond to forces in a distinctive way that distinguishes them qualitatively from simple fluids and from one another. They owe their distinctive properties to connected structures much larger than an atom, but much smaller than a macroscopic, classical body. They belong to an intermediate, *mesoscopic* size regime. These structures give raw egg white its springy consistency. They cause flowing corn-starch-and-water paste to shatter like a brittle solid when struck. And they cause toothpaste to flow out of its tube as a plug. These structures individually confer distinctive properties to a liquid. More importantly, they interact to create new forms of cooperative behavior and *self-organization*, as we shall illustrate below. Some of these structured fluids have an additional significance: they are important constituents of living cells or are important technologically.

It was not possible until recently to study the molecular basis of structured-fluid behavior. Now however experimental probes such as electron microscopy and dynamic scattering allow us to study structured fluids with a resolution suited to the distinctive mesoscopic structures in them. In addition present-day sophistication in chemical synthesis has made it possible to make fluids whose polyatomic constituents are precisely known. Such fluids serve as models for the types of behavior possible in structured fluids.

In the following chapters we shall explore how the special properties of structured fluids arise from their polyatomic structures. Our goal is to show how a wide variety of important properties can be understood and quantitatively predicted using a few simple principles. But before embarking on this study, it seems good to give you a foretaste of the variety of behavior that structured fluids can show in the laboratory. The next section shows a few striking examples. The following section describes the broad classes of structured fluid.

1.2   A gallery of structured fluids

1.2.1   Self-organization

Figure 1.1 shows the striking spatial self-organization that structured fluids can give to matter. The figure is an electron micrograph of a fluid of polyatomic molecules called block copolymers. Each polymer is a long, flexible chain hydrocarbon molecule. Roughly one-third of each polymer chain is a repeated sequence of monomers called isoprene. The other two-thirds of each polymer are made of another type of monomer called styrene. Once prepared and equilibrated, the fluid is frozen and cut into slices tens

Fig. 1.1
Transmission electron micrograph of the bi-continuous copolymer domain structure, reproduced from [1]. © 1987 American Chemical Society. Repeat distance is about 100 nm. The minority species in black forms two disjoint domains, labeled 1 and 2. The sketch at right, after [2], shows the three-dimensional structure inferred from such micrographs and from x-ray diffraction studies. © American Physical Society. The vertical dashed line gives the line of sight for the micrograph. The thickness of the two minority domains is reduced to make the illustration clearer. Domains 1 and 2 are shown with different shadings to distinguish them. The magnified sketch at bottom shows the orientation of individual polymers in the domains.

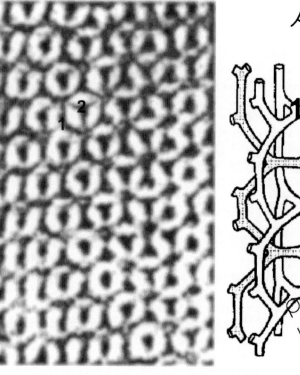

of nanometers thick. The two types of monomers can be made to scatter electrons differently, so that when a beam of electrons passes through a slice, regions denser in isoprene cast shadows; it is these shadows that form the micrograph. Evidently the polymers have assembled themselves into a periodic pattern. It is strikingly different from conventional crystals in both shape and size. Half of the isoprene forms a network of connected nodes in the so-called "gyroid" structure [3]. The other half forms an identical network, displaced from the first by half a unit cell. Each spatial unit cell is several tens of nanometers across and contains several hundred polymer molecules. The same structure has now been seen in several types of block copolymer fluids. How is it that these molecules choose to organize themselves spontaneously into such a domain structure? What features of the polymer molecule control the size and shape of the domains? In the chapters to follow we shall develop the concepts needed to address such questions.

Figure 1.2 shows another form of strong, mesoscopic organization, caused by small surfactant molecules. It is an electron micrograph of a liquid containing water, oil, and surfactants. Surfactants have the property of segregating themselves at water–oil interfaces; they are roughly the size of a water or an oil molecule. To make the micrograph, the liquid was rapidly frozen, and the frozen liquid was fractured. Staining techniques were used to reveal the topography of the fracture surface. We see that the fracture surface is made up of rounded regions a few tens of nanometers in size. Many studies of such mixtures support the view that these are regions of unmixed water or oil. The surfactant has induced the limited mixing of the

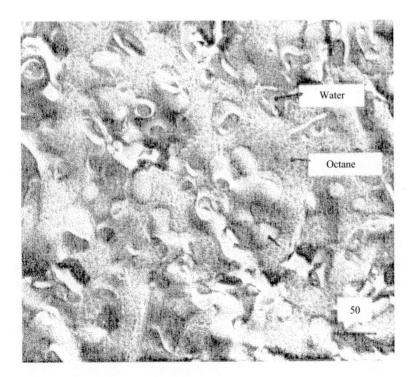

Fig. 1.2
Freeze-fracture micrograph of a water–oil–surfactant microemulsion reprinted with permission from [4]. © 1988 American Chemical Society. The smooth hills and valleys are thought to represent the boundaries of oil- and water-containing regions.

Fig. 1.3
Labyrinth pattern of a ferrofluid drop
(courtesy of R. E. Rosensweig) [5]. A drop
of ferrofluid was put between two
horizontal glass plates about a millimeter
apart. The plates were then placed between
the poles of a magnet. As the (vertical)
field increases from zero, the circular
droplet of fluid develops a fringe of spikes,
which elongate and branch to fill the area
between the plates with the pattern shown.
The thickness of the channels is
comparable to the spacing of the plates.

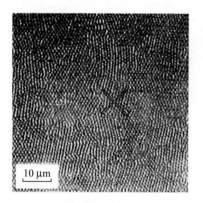

Fig. 1.4
An emulsion of water, decane oil, and
surfactant in which the oil droplets were
purified to have uniform size and then
concentrated [6], © American Physical
Society, courtesy J. Bibette. The droplet
diameter is 0.93 μm. The long parallel
rows of particles and the crosshatched
texture indicate an incipient periodic
lattice.

oil and water into these mesoscopic domains. In later chapters we shall be interested in how small surfactant molecules can induce oil and water to mix at this coarse scale. We shall consider what features of the molecules govern the size of the domains and discuss how the evident randomness of the interfaces may be characterized quantitatively. Such questions are of broad significance, because surfactant-mediated interfaces between liquids are ubiquitous. In everyday life and in technology surfactants at interfaces are used to control the mixing of oily and watery liquids. And living cells use these surfactant interfaces as a fundamental structural building block.

Mesoscopic structures can create *macroscopic* self-organization. The spatial scale of the stripes in Fig. 1.3 is a few millimeters. This is a photograph of a black droplet of "ferrofluid" sandwiched between two glass plates a few millimeters apart. The plates and fluid are in a magnetic field of several hundred Gauss, pointing out of the figure. As the field increases from zero the ferrofluid droplet goes from a circular shape to the labyrinthine shape seen here. Evidently this fluid interacts strongly with magnetic fields. It consists of 10-nm bits of the magnetic mineral magnetite dispersed in kerosene. In this fluid the polyatomic structures of interest are these bits of dispersed solid. Solids dispersed in this "colloidal" form often have strong interactions with external forces such as the magnetic field used here. As we proceed, we shall want to account for these strong interactions and explore their limits.

Just as colloidal particles interact strongly with external fields, they can interact strongly with each other, leading to further forms of self-organization. They can *repel* each other so strongly that they form a crystalline arrangement like that of Fig. 1.4. Figure 1.5 shows what can happen when dispersed colloidal particles are suddenly made to *attract* each other strongly. The particles assemble into a complex branched aggregate. The figure is a transmission electron micrograph of colloidal silica particles in water that have aggregated and then settled on a carbon film. A drop of the water was placed on a carbon film, dried, and then placed in the microscope. The aggregate looks different from a bulk precipitate: it is wispy and tenuous. Aggregates like this are used commercially to toughen rubber in tires and to thicken liquids like ice cream. In the chapters to follow, we shall consider what makes aggregates take this form and we shall explore ways to characterize its wispiness quantitatively. We shall also investigate how such wispy objects interact with the fluid and with other aggregates near them.

1.2.2 Rheology

The self-organizing behavior of structured fluids illustrated above is only part of their distinctive behavior. Further striking behavior is revealed when the equilibrium of a structured fluid is perturbed. A classic example is shown in on the right side of Fig. 1.6, a photograph of a polymer solution in a beaker. A rotating rod has been inserted into the center. The polymer liquid climbs this rod like bread dough in a mixer. A simple liquid behaves oppositely; the rotating liquid drops in the center and rises up the outer wall because of centrifugal force. In the sequel we shall identify the forces that cause

Fig. 1.5
Electron micrograph of aggregated
3.5-nm-radius particles of colloidal silica
from Lin *et al.* [7], courtesy of D. Weitz,
reprinted by permission from *Nature*
© 1989 Macmillan Publishers Ltd.
Aggregate was formed in water under
diffusion-limited conditions. The fraction
of the volume occupied by particles is
expected to vary as the sphere radius to
the −1.3 power whenever aggregation is
diffusion limited. This "universal" scaling
has been tested for a variety of materials.

Fig. 1.6
Rod-climbing liquid, reproduced from [8]
reproduced by permission of John Wiley &
Sons, Inc. (© 1977 John Wiley & Sons).
The control fluid in the left has the same
viscosity as the polymer solution on the
right, but is depressed in the center rather
than climbing the rod.

this rod-climbing behavior. We shall investigate what rate of stirring is
required to produce this effect. A less classical effect is sketched in Fig. 1.7.
This sketch depicts another polymer liquid. But the polymers in this liquid
contain surfactants on some of their monomers. When the liquid is properly
formulated, the act of quickly inverting its container produces a qualitative
change. The quiescent liquid has the consistency of motor oil; the disturbed
liquid is like half-cooled Jello. Phenomena like this show the potentialities
of these liquids for surprising behavior. We can only begin to sketch its
origins in what follows.

1.2.3 Scaling

When discussing the properties of a structured fluid like those shown
above, it is useful to imagine that the mesoscopic structures in the fluid
are *indefinitely* large relative to their small-molecule constituents. Thus

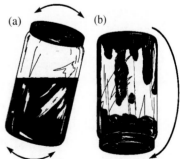

Fig. 1.7
Shear-induced gelation in an associating
polymer solution. (a) Moderate-viscosity
state in gentle flow. (b) Gel state after
rapidly inverting the jar.

various measured properties can be expected to show a simple asymptotic dependence on the structures' size. A fundamental question of this type concerns the viscosity of a polymer liquid. It is natural to expect that this viscosity may grow as a power of the molecular weight of the polymers. Our focus will be to ask whether this asymptotic dependence is indeed a power law and to predict the exponent. We shall keep our focus on such fundamental scaling issues and avoid more detailed predictions. In this way we can isolate the basic respects in which our structured fluids are distinctive. We can also account quantitatively for a broad range of liquids from a unified point of view.

1.3   Types of structured fluids

Having illustrated these examples of structured fluid behavior, we now describe the major classes of structured fluids in more detail, with some important examples for each. We begin with the simplest type, colloids.

1.3.1   Colloids

Colloids are fluids containing compact, polyatomic particles suspended in a liquid solvent. A familiar example is black ink, which is made from colloidal carbon. Colloidal particles give distinctive physical properties of fluids. Thus the colloidal form of the carbon in black ink is what makes it absorb light efficiently. These distinctive physical properties are incidental to many applications. Instead the colloidal particles are often present to give the fluid some *chemical* property of interest. Examples are the cells in blood, manufactured particles which remove molecules from a liquid by adsorbing them, the light-sensitive grains in photographic film, or the pH-buffer particles in detergent motor oil. In other cases, such as paint and rubber cement, the colloidal particles produce the desired property not in the fluid but in the solid that forms when the solvent dries. Powder processing of ceramics utilizes the fluid properties of suspensions of ceramic powders to optimize packing in order to achieve low defect bulk ceramics for structural materials.

There are colloids that are important for the *physical* properties they impart to fluids. One such colloid is the dispersion of 10-nm particles in the ferrofluid of Fig. 1.3. In moderate magnetic fields the magnetic energy of the particles is strong enough to alter the fluid's energy significantly. In a similar way, colloidal particles, because of their bulk, respond strongly to electric fields or to flow. In *emulsions* and *foams* the dispersed particles are liquid or gaseous droplets. These can be concentrated so much that the dispersed particles push against each other and give the dispersion the solid-like consistency of mayonnaise or whipped cream. Figure 1.4 shows an example of a concentrated emulsion.

The interaction energy of two colloidal particles in a given solvent is also magnified because of their bulk. Consequently, small changes in the solvent can have a large effect on the interaction energy. This makes it possible to change the interaction between two colloidal particles abruptly from an effective hard-core repulsion to an attraction whose strength is many times the thermal energy $k_B T$[†]. With such an attraction the particles must stick together when they encounter each other. The particles flocculate or precipitate. This effect is exploited to sense small changes in a solution or to determine the presence or absence of a biological antigen—as in certain early pregnancy tests.

† Here T is the absolute temperature and k_B is the Boltzmann constant. The next chapter reviews the significance of the thermal energy $k_B T$.

1.3.2 Aggregates

The enhancement of particle–particle interaction between colloids makes possible a form of self-organization not seen in simple molecular liquids— namely colloidal aggregates like that of Fig. 1.5. Such aggregates form when the attraction between two particles in contact is so strong that they must stick together permanently. They cannot slide or roll around in order to maximize the amount of contact. The result is the tenuous structure shown in the figure. It is qualitatively unlike the dense-particle phase that forms when small molecules precipitate from solution. The average particle density within a radius r of a given particle decreases as r increases. Thus the average density in an arbitrarily large aggregate becomes arbitrarily small: the aggregates are "fractal" structures [9]. The origins and the consequences of fractal structure will be a large concern in the chapters to follow.

Aggregated colloids show the enhancement effects discussed above for dispersed colloids. In addition they have properties arising from their fractal structure. Even though a tenuous aggregate occupies an arbitrarily small fraction of the volume it pervades^{††}, it transmits forces efficiently throughout the pervaded volume. An aggregate in a shear flow screens the surrounding solvent: the fluid is obliged to flow around rather than through the aggregate. Since this screening inhibits flow, it enhances viscosity. Thus the tenuous property of these aggregates makes them particularly effective in increasing viscosity. Their fractal structure confers many other distinctive properties to be explored in the course.

†† The pervaded volume of an object is the spherical region of space that contains the object.

1.3.3 Polymers

Another way of producing a tenuous, polyatomic structure is to link small molecules together into a flexible chain to form a polymer, as shown in Fig. 1.8. The successive bonds between the monomers making up a flexible chain have some randomness in their relative directions. Thus the directional correlation between bonds more than a few bond lengths apart becomes negligible; accordingly, a long polymer has the statistical properties of a random walk. In some situations the statistical properties of a long polymer are instead those of a self-repelling random walk. Like colloidal aggregates, such polymers have the spatial scaling properties of a fractal: the average monomer density of a self-repelling polymer of size R scales as $R^{-4/3}$. This power for a simple random walk is -1. Chains whose pervaded-volume fraction is a tenth of a percent or less can be readily produced. In a solution of such polymers, there is room for a thousand chains in the volume pervaded by one chain.

The flexibility of a tenuous polymer gives it properties that a tenuous aggregate does not have. Unlike an aggregate, a polymer is not quenched

Fig. 1.8
Detail of a large polystyrene molecule as it might appear in a good solvent such as toluene. Each sphere represents a carbon atom and one or two small hydrogen atoms attached to it. The chemical bonds are superimposed on one repeating unit or monomer, and on a section of the chain backbone. The backbone bonds may rotate freely. A few successive backbone repeat units are labeled 0, 1, 2, ... The vector $\vec{a}$ for a four-monomer segment is shown. This structure was generated by a Monte Carlo computer simulation [Simulation by M. Mondello, H.-J. Yang and R. J. Roe at the University of Cincinnati using the Cray Y-MP at the Ohio Supercomputer Center, circa 1990, private communication], which simulates the random rotations of the bonds as they might occur in solution. The simulated molecule has about 1/20 the mass of those in a typical polystyrene foam cup.

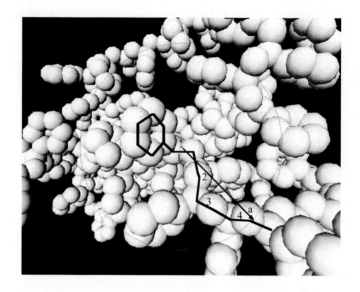

into some particular configuration but is free to explore the ensemble of random bond directions. The randomness in each chain's configuration thus amounts to thermodynamic entropy, which may serve as a reservoir for heat and work. The chain may be dramatically deformed by mild perturbations without permanent effects. Thus the spontaneous thermal fluctuations in the end-to-end distance of a flexible polymer are about as large as the average end-to-end distance. Consequently, externally imposed energies as small as the thermal energy $k_B T$ are sufficient to distort the shape of a polymer by factors of order unity.

Unlike rigid colloidal aggregates, polymers may be concentrated to volume fractions up to unity. In this solvent-free limit called the melt state, chains interpenetrate and entangle strongly. Each chain interacts directly with hundreds of others and the forces thus communicated can produce large, reversible deformation in each chain. The deformation of these random chains is what produces the restoring force in a rubber band or a plate of Jello. When a polymer liquid is abruptly deformed, it too responds elastically over short times, like a rubber band. But over longer times the chains disentangle and forget their initial distortions. The time scale for disentanglement can easily be as long as seconds. By mixing polymers of different lengths and architectures, one may produce liquids that behave like a tough rubber on short time scales, like a weak rubber on longer time scales and like a flowing latex at yet longer times [10]. This power to control the storage of energy over time allows one to adapt polymer solutions to the needs of a many-step manufacturing process, such as the assembly of a car tire. Similarly, many everyday structured fluids such as pancake syrup, shampoo, and paint are deliberately thickened to keep them from flowing too much during application. The energy stored in a polymer when it is deformed has striking effects on the flow properties of the liquid. Elongating a chain produces spring tension along its length. This "normal stress" combines with other applied forces to accelerate each small volume of the fluid. The elastic energy stored in the polymers can easily exceed the kinetic energy in the flowing fluid. This produces nonintuitive flow properties like the rod-climbing behavior of Fig. 1.6 [8]. One such effect of commercial importance is turbulent drag reduction. Trace amounts of a polymer can substantially reduce the power required to push a turbulent fluid through a pipe, in spite of the small increase in viscosity due to the polymer [11].

The deformability of a polymer has dramatic consequences when electrically charged species are attached along the polymer chain at a given density. The electrostatic repulsion in a long enough chain is sufficient to stretch the chain from a random-walk configuration to that of a rigid rod, for which the end-to-end distance is proportional to the molecular weight. Polyelectrolytes, as such polymers are called, can be controlled in a fashion not possible for neutral polymers. When the interaction along the chain is screened, by the addition of some salt or the presence of other chains, the electrostatic repulsion is reduced and the polyelectrolyte chain shrinks in size. This alters fluid properties such as the viscosity.

Another important consequence of the deformability of polymers is seen in their behavior near an interface. Even when the binding energy of a monomer to a surface or interface is much smaller than the thermal energy $k_B T$, the total binding energy of a polymer chain made up of such monomers may be several times the thermal energy. Then the polymer chain can increase its binding by flattening itself closer to the surface, with little cost in deformation energy. The distinctive features of the adsorbed state will be considered later.

1.3.4 Surfactant assemblies

Self-assembled surfactants like those at the interfaces in Fig. 1.2 are a further major category of structured fluid. A surfactant molecule is amphiphilic: it incorporates parts that if not connected would be strongly immiscible. A common example is hexadecyl trimethyl ammonium bromide (CTAB)—a hydrophobic 16-carbon chain with a bulky ammonium ion and a bromide counterion at one end. Isolated surfactant molecules cannot exist in either oil or water (except in extremely low concentrations). Instead, these molecules assemble themselves into "micelles". A micelle is a configuration in which the molecules' immiscible parts clump together, thus minimizing their contact with the solvent. In water for example, the hydrophobic hydrocarbon tails clump together, while the polar heads point outward towards the water. In an oil the situation is inverted: the inverted micelles have their ionic parts clumped together.

Micelles typically have diameters on the order of a few nanometers; they constitute a colloid-like polyatomic species in their own right. Furthermore a micelle-containing fluid shows interfacial properties unlike anything discussed above. Any water–oil interface clearly forms a very favorable environment for a surfactant; at the interface the molecule's hydrophilic end can be in water and the hydrophobic tail can be in oil. Accordingly surfactants assemble readily at these interfaces. The assembly's energy is lower than that of a random dispersion of surfactants in either oil or water. The lowering of the energy amounts to a lowering of the interfacial free energy or interfacial tension. Surfactants thus make water and oil more nearly miscible.

In certain conditions a surfactant can reduce the interfacial tension virtually to zero, allowing the interfacial area between oil and water to grow spontaneously. The result is a microemulsion like that shown in Fig. 1.2—a thermodynamically stable mixture of oil, water, and surfactant that is full of fluctuating oil–water interfaces.

1.3.5 Association

The broad range of behavior discussed above is augmented even further when we consider "association" of mesoscopic structures. By association we mean a temporary joining together of the structures. The structures thus joined can transmit forces and thus alter mechanical properties strongly. But still these junctions are weak enough to break and reform over the time of an experiment. Thus the associations alter themselves in response to the local stress or flow in the liquid. We have already met the most prevalent form of association: the joining together of surfactant molecules to make a micelle fits the definition of association. Because these micelles are temporary, the number and the average size of the micelles relaxes over time when the micellar solution is subjected to e.g., a temperature jump. Association behavior is common in colloidal aggregates. For example, silica aggregates in water may be made to associate with one another via hydrogen bonds. The result is a type of gel network whose links may be broken and readily re-formed. Such a "network fluid" holds its shape in quiescent conditions but flows under sufficiently strong shear. Upon removal of this shear, the network is reestablished and the flow stops. These associating aggregates are used to keep paint from running before it dries.

One can make polymers associate by attaching immiscible chemical groups sparsely along the chains. Like surfactants, these groups assemble themselves into micelles and form temporary crosslinks between the polymers. Such polymer solutions can show a new form of response to shear, namely reversible, shear-induced gelation. This is the phenomenon sketched in Fig. 1.7. Another form of polymer association occurs in the block copolymers of Fig. 1.1. Block copolymers, being amphiphilic, have properties similar to those of surfactants. But the polymers' entanglement and deformability cause their resulting micellar microdomains to have some distinctive properties not seen in surfactants. We discuss these in Chapter 7. An example of commercial importance is Kraton, a rubbery polymer tipped at each end with a small section of an immiscible, glass-forming polymer. The ends of the Kraton polymer congregate into spherical micelles, each containing many chain ends. The midsections of such a copolymer, being attached at both ends, cannot disentangle themselves, and a strong, rubbery material results. But when the material is heated above about 100° C the spherical micelles at the ends melt and become more miscible, so that the material flows. It can then be molded and processed. Such polymers, called thermoplastic elastomers, are used e.g., in adhesive coatings and for the elastic stripes painted on disposable diapers.

1.4 The chapters to follow

Now that we have surveyed the varieties of structured fluids, we are ready to try to understand them in the chapters to follow. Chapter 2 deals with fundamental principles that underlie the treatment of all the systems to be treated later. The first step is to review the principles of statistical physics that form the basis of all the phenomena to follow. Next, we discuss the orders of magnitude of the microscopic phenomena on which structured-fluid properties are based. That is, we consider the energy, length, and time scales characteristic of simple liquids. From these we infer the order of magnitude of the viscosity of typical liquids. Finally, we discuss the experimental techniques—some new, some old—by which structured fluids are studied. We shall want to keep these concrete tests of our ideas in mind as we try to make predictions about the fluids studied later.

In the rest of the book we discuss the classes of structured fluid in turn. In Chapters 3 and 4 we start by treating polymers. We discuss the characteristic length, energy, and time scales of single polymers and of strongly interacting polymer solutions. In so doing we lay the groundwork for quantitative understanding of the distinctive responses of polymer liquids: their elasticity and their high viscosity. In Chapter 5 we turn to colloidal fluids. Many aspects of these fluids prove to be directly understandable using ideas developed for polymers. But there are also new aspects. The chief one is to understand why large colloidal particles are difficult to disperse in a solvent, and how these difficulties may be overcome. Chapter 6 discusses the interaction between fluids and interfaces: liquid–liquid, liquid–gas, and liquid–solid. We discuss surface and interfacial tensions, statics and dynamics of wetting, and other related topics including adhesion and forces between colloidal particles induced by adsorption and depletion layers. In Chapter 7 we consider surfactant assemblies. Here our central question is how amphiphilic molecules are able to promote the mixing of immiscible liquids. In understanding this puzzle, we are led to explore the statistical properties of the fluctuating surfactant-coated interfaces.

References

1. H. Hasegawa, H. Tanaka, K. Yamazaki, and T. Hashimoto, *Macromolecules* **20** 1651 (1987).
2. M. W. Matsen, *Phys. Rev. Lett.* **80** 4470 (1998).
3. D. A. Hajduk, P. E. Harper, S. M. Gruner, and C. C. Honeker, *Macromolecules* **27** 4063 (1994); M. W. Matsen and F. S. Bates, *Macromolecules* **29** 1091 (1996).
4. W. Jahn and R. Strey, *J. Phys. Chem.* **92** 2294 (1988).
5. See e.g. R. E. Rosensweig, in *Physics of Complex and Supermolecular Fluids*, ed. S. A. Safran and N. A. Clark (New York: Wiley-Interscience, 1987), p. 699.
6. J. Bibette, D. Roux, and F. Nallet, *Phys. Rev. Lett.* **65** 2470 (1990).
7. See e.g. M. Y. Lin, H. N Lindsay, D. A. Weitz, R. C. Ball, and R. Klein, *Nature* **339** 360 (1989); R. Jullien and R. Botet, *Aggregation and Fractal Aggregates* World Scientific (1987).

8. R. B. Bird, O. Hassager, R. C. Armstrong, and C. F. Curtiss, *Dynamics of Polymeric Liquids* (New York: Wiley, 1977).

9. See L. M. Sander, *Sci. Am.* **255** 94 (1987); B. B. Mandelbrot, *The Fractal Geometry of Nature* (San Fransisco: Freeman, 1982).

10. For a review, see D. S. Pearson, *Rubber Chem. Technol.* **60** 439 (1987).

11. R. B. Bird and C. F. Curtiss, "Fascinating Polymeric Liquids," *Phys. Today* January, (1984) p. 36.

Fundamentals

2

This chapter assembles some underlying concepts that we will need throughout our study of structured fluids. The first section reviews the main needed ideas from statistical mechanics. For systems in equilibrium, we recall how probabilities of states are determined, and the connection between probability and the work to alter a system's state. We also describe ways to estimate the time for a system to reach equilibrium. The second section is about the orders of magnitude common to the liquid environment of structured fluids. In this section we justify the magnitude of a simple liquid's viscosity. The final section surveys the experimental probes that determine the useful questions we can ask about structured fluids.

2.1 Statistical physics

2.1.1 Thermal equilibrium

All the structured fluids mentioned in the last chapter are intimately controlled by the laws of thermodynamics. The conditions of thermal equilibrium impose the dominant limitations on how these fluids can behave. The bulk of the understanding achieved to date about these systems comes from the laws of thermodynamics and statistical mechanics. In this section we recall a few principles of statistical physics that will be important for the rest of the course. We will discuss what determines the probability of random states in our structured fluid. We'll recall the meaning of temperature and the ability of a statistical system to do work.

The structured fluids we'll meet in the course contain many forms of randomness, from the random bond angles of a polymer chain to the random positions of the molecules of a simple liquid. But for this discussion let us think of a simple and familiar example: a dilute monatomic gas. Each atom of the gas has a random position and momentum. Nevertheless, not all positions and momenta are equally likely.

To understand the relative likelihood of these random variables, we define the notion of a configuration or microstate of the gas. The configuration is the set of all the positions $\{\vec{r}_i\}$ and momenta $\{\vec{p}_i\}$ of all the N atoms. It will save us from some awkwardness later to imagine that these positions and momenta are discretely spaced at very fine intervals. This is natural since the coordinates and momenta can only be known to limited precision in any case. Thus all the randomness of our system is specified by telling the relative probabilities $f(\{r_i\}, \{p_i\})$. In an arbitrary system, we shall denote a given configuration by the simple label c; any configuration c has a relative probability $f(c)$. The absolute probability $p(c)$ is just obtained by normalizing the $f(c)$: $p(c) = f(c)/(\sum_c f(c))$. Since f is only a relative probability, multiplying it by a constant (independent of c) does not change its meaning at all. Any two f's whose ratio is independent of c are clearly equivalent.

In principle a system such as our gas can be prepared in any specific configuration c or with some initial probability distribution $f_0(c)$. But in general it does not stay in its initial distribution. In our gas for example the coordinates and momenta of the particles are constantly changing because of the motions of the atoms and collisions between them. Thus their relative probabilities in general change from what they were initially as time goes on, until a final steady distribution $f(c)$ is reached. This final steady state is called the equilibrium state.

A gas isolated in a container is a closed mechanical system. If it is initially in a configuration c, it can evolve into many others as time goes on. But there are many configurations it cannot evolve into. An immediate reason for this is the conservation of energy. Since the total energy E in the gas is conserved, the system is confined to configurations c' with the same energy as the original one c_0: $E(c') = E(c_0)$. That is, $f(c') = 0$, except if $E(c') = E(c_0)$. For many systems this is the only restriction on the configurations c', however; all the configurations with the proper energy are accessible. The hypothesis of statistical mechanics is that all these accessible states in a closed and isolated system are *equally* probable. This relaxation to this uniform probability distribution can be proven to occur under a broad range of conditions; we shall assume that it occurs for the systems we will study. That is, we will assume our system attains **thermodynamic equilibrium**[†].

The systems we will study consist of many **degrees of freedom**. In our gas, each cartesian coordinate or momentum of each atom is a degree of freedom; in our polymer molecule each bond angle of each monomer is a degree of freedom. We specify the configuration c of a system by specifying the values of all its degrees of freedom.

This behavior of an isolated system leads to an interesting behavior of *parts* of the system. In our gas system, for example, we can infer the probabilities of a single atom's coordinates and momenta. A subsystem, such as a single atom, is a subset of the degrees of freedom of the overall system. A single atom of mass m in a dilute gas has a fairly well-defined energy ϵ that depends on the configuration of its own degrees of freedom: $\epsilon = \frac{1}{2}p^2/m$. This energy is not conserved, since the atom occasionally transfers energy by collisions. During a collision, the energy of an atom is not simply the kinetic energy formula given above. Still, with increasing dilution, this formula becomes more and more accurate.

In situations like this, we may infer the equilibrium distribution for subsystems like our atom, which are a small part of a large, isolated system with fixed total energy E_t. The subsystem has its own set of configurations c, each with some relative probability $f(c)$. We consider subsystems which have a well-defined energy $E(c)$; this energy is assumed to be little affected by anything beyond the subsystem, as is the case for a dilute-gas atom. The rest of the system beyond the subsystem has a configuration labeled by d; this large set of degrees of freedom outside the subsystem of interest is called the **reservoir**. Thus to specify the configuration of the whole system, we must specify both c and d. First we consider two subsystem configurations c and c' with the same energy: $E(c') = E(c)$. The relative probability $f(c)$ is given by the total number of reservoir configurations d that are compatible with that c. The only constraint on these d's is that their total energy must be the fixed total energy E_t less the subsystem energy $E(c)$. All reservoir configurations with energy $E_t - E(c)$ are equally allowed. This same set of reservoir configurations is equally allowed for the c' configuration. The number of overall configurations with sub-configuration c' is the same as the number with sub-configuration c; thus $f(c') = f(c)$. We are led to the conclusion that configurations of a given subsystem with the same energy have equal probability. Thus, $f(c)$ can only depend on c through its energy: $f(c) = f(E(c))$.

To determine how f depends on E we consider two separate subsystems within a given larger reservoir (e.g. two atoms in our gas). The first subsystem has configurations denoted c_1 with energy $E(c_1)$; the second one has configurations denoted c_2 and energy $E(c_2)$. (The two subsystems need not be the same size. For example, subsystem 1 could contain three atoms, and subsystem 2 could contain 75 other atoms.) The only way our separate subsystems have any connection is that they belong to the same larger reservoir. Thus if one has more energy there is in principle less energy available for the other. But the total energy E_t is so much larger than the energy of either subsystem that this effect is negligible. In the limit of a large reservoir, the effect of one subsystem on the other must become negligibly small. We now consider the subsystem made up of subsystems 1 and 2 combined. Since the two have negligible effect on each other, the probability $f_{1+2}(c_1, c_2)$ of the combined system must be simply the probability that subsystem 1 has configuration c_1 times the probability that subsystem 2 has configuration c_2: $f_{1+2}(c_1, c_2) = f_1(c_1) \times f_2(c_2)$. As we have seen, these probabilities depend only on the respective energies: The energy $E(c_1)$ for the first subsystem, $E(c_2)$ for the second subsystem and the total energy of the two for the combined system: $E(c_1, c_2) = E(c_1) + E(c_2)$. The relative probabilities f must satisfy $f_{1+2}(E_1 + E_2) = f_1(E_1) \times f_2(E_2)$.

One way to satisfy this requirement is $f(E) = (\text{constant})e^{-E/e_0}$, where e_0 is some constant. This $f(E)$ in fact gives the equilibrium probability of a configuration c of any small subsystem whose energy is $E(c)$. This probability distribution is called the Boltzmann distribution; deriving it is a major topic in statistical mechanics courses. By assuming this form one readily verifies that $f_{1+2}(E_1 + E_2) = f_1(E_1) \times f_2(E_2)$; moreover, the constant e_0 must be the same for subsystems 1, 2, and $(1+2)^{\dagger}$. Thus any two (small) subsystems of the same overall isolated system have the same value for the coefficient e_0. Problem 2.1 gives another way of understanding how the Boltzmann distribution arises.

2.1. *Justifying the Boltzmann distribution* It was stated above that the probability that a system in thermal equilibrium is in a configuration c of energy $E(c)$ varies as $\exp(-E(c)/e_0)$. This is true for any conserved quantity in a small subsystem of a closed random system. To illustrate this point, consider the following

$\dagger$ To see that f_{1+2} has the same value of the constant e_0 as f_1, we first suppose otherwise that $f_{1+2} = (\text{constant})e^{-E_{1+2}/e_0'}$ with $e_0' \neq e_0$. Now we calculate the relative probability for some configuration $\tilde{c}_1$ of subsystem 1 using the $f_{1+2}(c_1, c_2)$ of the combined system. Evidently we must sum the probabilities of all combined configurations for which the simple subsystem is in configuration $\tilde{c}_1$. Thus our desired probability $f(\tilde{c}_1) = \sum_{c_2} f(\tilde{c}_1, c_2)$. Using our Boltzmann formula for $f_{1+2}(c_1, c_2)$ this means

$$f_1(\tilde{c}_1) = \exp(-E(\tilde{c}_1)/e_0')$$
$$\times \sum_{c_2} \exp(-E(c_2)/e_0').$$

The sum factor is a mere constant independent of $\tilde{c}_1$ and has no significance in our relative probability. Comparing this $f_1(\tilde{c}_1)$ with our original formula, $\exp(-E(\tilde{c}_1)/e_0)$, we see that the two expressions will be the same only if $e_0 = e_0'$.

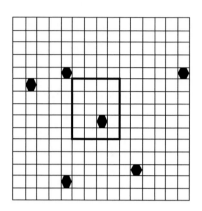

Museum-of-Science-and-Industry example. Our system is a 1000×1000 grid of $U = 1,000,000$ cells. We designate a section of $V = 10,000$ cells as our subsystem. (The figure shows a typical configuration in a scaled-down version. The inner rectangle is the subsystem.) Into the system a fixed total of $N = 1,000$ identical balls are distributed at random. (The fixed number N of balls corresponds to a fixed total energy. Just as different pieces of energy in a system do not have an individual identity, neither do our balls. These cells are allowed to contain more than one ball.) A specific k-ball configuration c_k of the subsystem is then specified by stating the number of balls in each of its V cells. We define $p(c_k)$ as the probability of a configuration c_k of the subsystem with k balls. It is the fraction of all system configurations in which the subsystem has the configuration c_k. We define the configuration of the outer part of the system (which has $N - k$ balls) as d_{N-k}. Then $p(c_k) = (constant)$ $[\sum_{\{d_{N-k}\}} 1]$: the probability of a given *single* subsystem configuration is given by the number of system configurations d_{N-k} compatible with it. We denote this number as z_{N-k}. (Clearly any configuration with k balls has the same probability as any other.) For simplicity in what follows, you may assume that $k \ll N$ (subsystem is small) and $N \ll U$ (system is dilute). (a) Given an m-ball configuration of the outer part of the system (the "reservoir"), how many $m + 1$ ball configurations can be made from it by adding one ball? Denote this number as $b(m)$. (b) Given an $m + 1$ ball configuration of the reservoir, now many different m-ball configurations could have led to this one by adding one ball? Is $z_{m+1} = z_m \times b(m)$? (c) Show that the ratio $p(c_{k+1})/p(c_k)(= z_{N-k-1}/z_{N-k})$ is independent of k. (This can be done without calculating the (horrendous) $z_{N-k} \equiv [\sum_{d_{N-k}} 1]$ explicitly.) What is this ratio? (d) Show that $p(c_k)$ has the form $p(k = 0) \exp(-k/constant)$, and find the constant.

See the end of the chapter for the solution.

The coefficient e_0 is clearly an important parameter in the distribution. Clearly, a configuration c' with energy $E(c')$ more than a few times e_0 greater than $E(c)$ has a relatively insignificant probability. By analyzing how e_0 affects thermodynamic properties of the system, one can infer that e_0 is essentially the absolute temperature T: $e_0 = k_B T$, where k_B is the Boltzmann constant mentioned Chapter 1. One way to make the connection between e_0 and T is along the lines of Problem 2.3. The Boltzmann constant k_B is merely a conversion factor between conventional temperature units (defined before the connection between temperature and energy was known) and conventional energy units. We may avoid this constant by simply quoting temperatures directly in energy units. For our purposes in this course, the temperature will nearly always be room temperature—300 K. This temperature in conventional energy units is 1/40 eV or 0.6 kcal/mol.

These conclusions were first drawn by Ludwig Boltzmann in the nineteenth century. We may summarize them as follows. Any subsystem which interacts weakly with a very large "reservoir" system necessarily has at equilibrium a "temperature" T, characteristic of the reservoir. The relative probability of a configuration c of that subsystem has the form $f(c) = \exp(-E(c)/T)$. We shall refer often to this **Boltzmann principle** henceforth.

2.2. *Sinking colloids*   It is possible to make colloidal particles in water out of metallic iron, whose density is 10 times that of water. If such particles were large enough, they would settle under gravity to the bottom of their container. (a) For what mass of particle would the mean height be 1 cm in water solution?

Express your answer in atomic mass units $(6.023 \times (10^{23})^{-1}$ g). (b) What would the particle radius be in this case?

An important part of statistical mechanics is the relationship between the temperature of a subsystem and its ability to do work. To recall this relationship, we consider a specific example: an atom in a container in a uniform gravitational field. This field exerts a constant force mg on the atom and gives it a potential energy mgz where z is the vertical distance above (say) the bottom of the container. According to Boltzmann's principle, we may take this atom as our subsystem. Its relative probability to be at height z is then given by $f(z) = \exp(-mgz/T)$[†]. From this one may immediately find the mean height

$$\langle z \rangle = \int_0^\infty dz\, z\, p(z) = \int_0^\infty \frac{dz f(z) z}{\int_0^\infty dz f(z)}.$$

Simple integration yields $\langle z \rangle = T/(mg)$.

The mean height is evidently influenced by the temperature and the gravitational field g. By increasing g slightly, we would do work $-mg\,\delta z$ on the system. This fluctuating work is difficult to give experimental meaning, but the average work $\delta W \equiv -mg\delta\langle z \rangle$ is well defined.

Thus the work ΔW done by an external agent to change g from g_1 to g_2 is $\int_{g_1}^{g_2} \delta W$. Expressing $\delta\langle z \rangle$ as $(d\langle z \rangle/dg)\delta g$, or $-(T/mg)\delta g/g$, we conclude

$$\Delta W = \int_{g_1}^{g_2} mg\left(\frac{T}{mg}\right)\frac{dg}{g} = T\log\left(\frac{g_2}{g_1}\right) = T\log\left(\frac{\langle z_1 \rangle}{\langle z_2 \rangle}\right). \quad (2.1)$$

The point of this example is to remind you that (a) a system in thermal equilibrium exerts forces on its surroundings due to its thermal fluctuations, (b) it can be made to do work, and (c) the amount of work can be calculated using the Boltzmann probabilities $f(c)$. Since the work is evidently recoverable (by returning g to its initial value), the work done on the system represents a form of stored energy.

2.1.2 Probability and work

The connection between probability and work is an important aspect of statistical mechanics. Below, we describe how the work done by a general change in a system in thermal equilibrium can be expressed in terms of the Boltzmann probabilities $f(c)$. We first discuss the work associated with changing a degree of freedom from one definite value to another. This work is related to the probability that the degree of freedom has various values when it is allowed to fluctuate. Next, we relate this work to the work associated with altering the system's energy, as in the gravity example above. Finally, we show the use of these ideas via an example, the lattice gas.

The probabilities that influence work done on the system are more involved than the simple Boltzmann probabilities $f(c)$. In itself $f(c)$ gives the relative probability of a *particular* configuration c. Often we need to know a probability that involves *many* configurations. For example in a

[†] Here our subsystem consists of only the height coordinate z of the atom. Since we know the energy associated with a given height without knowing the other configuration variables (such as the atom's momentum), we are allowed to use the height coordinate as a subsystem. Of course it is also correct to find the probability $f(z)$ starting from the full configuration probability $f(x, y, z, p_x, p_y, p_z)$. But as in the (c_1, c_2) note above, the effect of the other variables is simply to multiply $f(z)$ by an unimportant constant.

liquid containing two colloidal particles, we often wish to know the probability that these are separated by a displacement $\vec{r}$. In general the liquid has many configurations in which the two particles have this separation. And the number of configurations typically depends on the distance r. Solvent molecules are nestled and packed around each colloidal particle, altering the molecular arrangements near each particle. When two particles approach, the "spheres of influence" around them overlap. In the overlap region the solvent molecules must accommodate to two colloidal particles instead of one. Thus the number of arrangements or configurations of the liquid particles is altered. Figure 2.1 shows a schematic example of this effect. To find the probability that the particles have separation r, we must add the probabilities for all those configurations in the liquid with this separation. We denote this subset of configurations by c_r. Thus except for normalization this probability is $\sum_{c_r} f(c_r)$. When the two particles are very far away, the liquid near one is unaffected by the other, so that the probability is a constant. It is convenient to normalize our probability so that this constant is unity, the resulting relative probability is called the "pair distribution function" $g(r)$: $g(r) \equiv \sum_{c_r} f(c_r) / \sum_{c_\infty} f(c_\infty)$.

Depending on the nature of the colloidal particles and the solvent, the pair distribution function $g(r)$ may increase or may decrease from unity as the particles approach each other from infinity. To understand the meaning of these changes, we compare our fluid with the simplest possible one-particle system: an imaginary particle alone in an empty space and experiencing a potential $U(r)$. The particle is weakly coupled to some thermal reservoir at temperature T. We shall take a potential that vanishes at infinity. To find the $g(r)$ for this simple system is an easy task. This is because there is only one configuration at position $\vec{r}$. (We take the subsystem to be the coordinate of the particle; so its momentum need not be considered.) Thus the only r-dependent quantity in $f(c_r)$ is $e^{-U(r)/T}$. And thus $\sum_{c_r} f(c_r) = e^{-U/T}$ and $g(r) = e^{-U(r)/T}$.

One way to interpret the probabilities $g(r)$ for our colloidal system is to compare with the $g(r)$ of the simple system. In particular, one can find a potential $U(r)$ such that the $g(r)$ of the simple system matches that of the real system. Comparing the $g(r)$'s of the two systems, it is clear that they are equal if $U(r)$ is given by

$$U(r) \equiv -T \log(g(r)) = -T \log \left(\sum_{c_r} f(c_r) \right) + T \log \left(\sum_{c_\infty} f(c_\infty) \right)$$

of the real system. With this choice of U the simple system is equivalent to the real one as far as probabilities are concerned. Now we may give meaning to large or small values of $g(r)$. When $g(r)$ increases from one, the U of the system is negative: the pair acts as though there were an attractive interaction. Conversely, if $g(r)$ is smaller than 1, the two particles act as though they repel each other. In Chapter 5 to follow, we shall see how such effective interactions behave in practice.

Remarkably, this effective attraction or repulsion describes not only the probabilities of the two systems, but also their ability to do work. In the

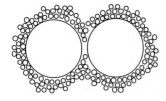

Fig. 2.1

Schematic picture of two colloidal spheres surrounded by small solvent molecules. Only the solvent molecules closest to the spheres are shown. In the upper picture, the colloidal spheres are far apart, and one sphere has little effect on the solvent molecules surrounding the other. In the lower picture, the colloidal spheres are close together and the arrangement of the solvent molecules is disrupted.

simple system the work required to bring the particle from infinity to r is evidently $U(r)^\dagger$. The analogous work in the real system is not obvious to deduce, since many parts of the system change when r changes. To evaluate this work, we resort to an indirect method. We apply an external force F in the x direction to pull the two particles apart. (This indirect method is easiest to imagine for a case where $U(r)$ is attractive and much larger than T, so that there is a large probability that the particles are close together.) We may do recoverable work on the system by increasing F slightly and thus increasing $\langle x \rangle$. We shall apply the same force to the real system and the simple system and show that the work done on the two systems is the same. Thus we can express the work done on the real system in terms of the U of the simple system.

When the force F changes, the resulting work δW is simply $F\delta\langle x \rangle$, as in the gravity example above. To know the work for a given shift of F it is sufficient to know the change in $\langle x \rangle$. The force affects the system through the microscopic energy $E(c)$. This $E(c)$ has the form $E_0(c) - Fx$, where $E_0(c)$ was the energy of configuration c before the force was applied.

To compute $\langle x \rangle$ we must average over all c. It is convenient to perform this average in two stages, first summing over all c's for a given displacement $\vec{r}$ and then summing (integrating) over r:

$$\langle x \rangle = \frac{\int d^3r \sum_{c_r} f(c_r)x}{\int d^3r \sum_{c_r} f(c_r)}. \tag{2.2}$$

We may write $f(c_r)$ as $\exp(-E_0(c_r)/T)\exp(Fx/T)$. Further,

$$\sum_{c_r} \exp(-E_0(c_r)/T) = g(r)\sum_{c_\infty}\exp(-E_0(\infty)/T).$$

Using these facts we can readily express the $\langle x \rangle$ in terms of the pair distribution function $g(r)$:

$$\langle x \rangle = \frac{\int d^3r\, x g(r)\exp(Fx/T)}{\int d^3r\, g(r)\exp(Fx/T)}.$$

Remarkably, the needed $\langle x \rangle$ can be found purely in terms of $g(r)$. Thus any two systems with the same $g(r)$ must have the same $\langle x \rangle$ and thence the same external work stored by a given force F.

Since our simple system has the same $g(r)$ as the real system, it must store the same energy under a given perturbation. In other words, the real system stores energy as though it were a simple system with potential $U(r) = -T\log(\sum_{c_r} f(c_r))$ and no other degrees of freedom. In particular the work required to bring the real system from a separation r_1 to a separation r_2 is the same as in the simple system, *viz.*

$$U(r_2) - U(r_1) = -T\log\left(\sum_{c_{r2}} f(c_{r2})\right) + T\log\left(\sum_{c_{r1}} f(c_{r1})\right)$$

This $U(r)$ is called the **potential of mean force**. It is a form of "free energy."

$\dagger$ It may not be so clear how this work is measured. One way is to subject our particle to an external potential $V_x(r)$ with a deep minimum that holds the particle at a specific $r = x$. Then we may move our particle from infinity to r_0 by manipulating V so that its minimum moves from infinity to r_0. We shall move slowly enough that the particle stays in equilibrium with the reservoir. The work required is the work done by the external force $F = -dV/dr$. At each point x *en route*, the particle is sitting arbitrarily near the minimum of $U + V$, so that $dU/dr = -dV/dr = F$. The work W done by F is $\int_\infty^{r_0} F(r)\,dr = \int_\infty^{r_0} dU/dr\,dr = U(r_0) - U(\infty)$.

In this example we have determined the work to change a degree of freedom from one definite value to another e.g. from r_1 to r_2. We have shown that this work can be determined using the relative probabilities of the variable e.g. $\sum_{c_r} f(c_r)$ when r is free to change in thermal equilibrium. The same reasoning can be carried out in general. We thus define the **free energy** $\mathcal{F}$ of a thermodynamic system as

$$\mathcal{F} \equiv -T \log \left(\sum_c f(c) \right). \tag{2.3}$$

The quantity in parentheses is called the **partition function**. The work required to change a system variable that is fixed for all c's in the sum is simply the change of $\mathcal{F}$.

Often one performs work on a system by perturbing its energy rather than by changing a variable from one definite value to another. In the above examples, the agent that changes the gravitational force g or the colloidal force F does work on the system, but a bit of care is needed in determining the work done. Let us take the colloidal system as an example. We shall imagine that the external force is exerted by a spring. We shall change the force F exerted on the colloidal particles by displacing the other end of the spring. Clearly, we are changing a system variable (the position of the spring) from one definite value to another. According to our reasoning above, the work done to displace the spring is the change in $\mathcal{F}$. However, only part of this work is done on the colloidal particles and the fluid surrounding them. The rest of it is done on the spring. In order to determine the work performed on the colloidal system by changing F, we must subtract the change in spring energy. Happily, there is a way to infer the work on the system in terms of $\mathcal{F}$ itself. We will discuss this in general terms.

We now consider a system with two contributions to the configuration energy $E(c)$. The first is the system's intrinsic energy, denoted $E_0(c)$. The second is the energy associated with the external influence that is to be altered to produce work. We shall write this external energy in the form $bB(c)$, where b is the parameter (analogous to F above) to be varied and $B(c)$ is some variable of the system, analogous to x above. Thus $E(c) = E_0(c) + bB(c)$. When we increase b by a small amount δb, we make configurations with large B more energetic and thus less probable. This means that $\langle B \rangle$ decreases. The work done on the system as a whole is $\delta\mathcal{F}$; the work done to change the external energy bB is (on average) $\delta\langle bB \rangle$. The remainder $\delta\mathcal{F} - \delta\langle bB \rangle$ is the work done on the intrinsic system. This is simply the change in the quantity $(\mathcal{F} - \langle bB \rangle)$.

We may verify that this expression for the work agrees with our directly determined value $F\delta\langle x \rangle$ from above. For this, we note that $\delta\mathcal{F}$ is related to $\langle B \rangle$:

$$\delta\mathcal{F} = \delta b \frac{d}{db} \mathcal{F}.$$

Using the definition of $\mathcal{F}$,

$$
\begin{aligned}
\frac{d}{db}\mathcal{F} &= \frac{d}{db}\left[-T\log\left(\sum_c e^{-(E_0+b\ B)/T}\right)\right] \\
&= -T\frac{-1/T\sum_c B(c)\ e^{-(E_0+b\ B)/T}}{\sum_c e^{-(E_0+b\ B)/T}}.
\end{aligned}
$$

The temperature factors cancel, and the remaining fraction is precisely $\langle B \rangle$. Thus

$$\delta\mathcal{F} = \delta b\langle B \rangle.$$

The change in our quantity $\delta(\mathcal{F} - b\langle B \rangle)$ can now be readily found:

$$\delta(\mathcal{F} - b\langle B \rangle) = \delta b\langle B \rangle - (\delta b\langle B \rangle + b\delta\langle B \rangle) = -b\delta\langle B \rangle.$$

In our colloidal example, the external variable was the coordinate x and the parameter b was $-F$. Thus $-b\delta\langle B \rangle \to F\delta\langle x \rangle = \delta W$, as anticipated.

The free energy $\mathcal{F}$ is much used to discuss systems in thermal equilibrium. It is worthwhile to note how our discussion is related to standard treatments.

1. One might worry that $\mathcal{F}$ is ill-defined, since it is based on the relative probabilities $f(c)$, whose normalization is arbitrary. If this normalization were changed by multiplying all $f(c)$ by a constant factor e^a, one readily sees from Eq. (2.3) that $\mathcal{F}$ is increased by an amount a. This ambiguity is harmless; as with any energy, only *changes* in $\mathcal{F}$ are significant. Still, it is conventional to fix the normalization by using $f(c) = e^{-E(c)/T}$ with no other prefactor. Then, for a system with a single configuration c, $\mathcal{F}$ is simply equal to $E(c)$.

2. In macroscopic systems one is often interested in that part of the work done on a system which does *not* increase its internal energy $\langle E \rangle \equiv \sum_c E_c f_c / \sum_c f_c$. In an isolated system the work done must be equal to the change in $\langle E \rangle$, in order to conserve energy. However a subsystem which can exchange energy with a reservoir may transfer energy to the reservoir when work is done on it. The energy transferred to an energy reservoir in this way is *heat*. The amount of heat transferred, denoted δQ, is evidently given by $\delta Q = \delta W - \delta\langle E \rangle$. It can be shown that this heat, like the free energy, can be expressed in terms of the normalized probabilities p_c. Specifically, $\delta Q = -T\delta S$, where the **entropy** S has the form

$$S = -\sum_c p_c \log p_c,$$

and p_c is the normalized probability $f_c/(\sum_c f_c)$. The entropy is a measure of the randomness of the system. Many of the concepts in this section can be expressed in terms of maximizing this entropy. However, in this book, issues of heat are not essential and thus the notion of entropy will play a minor role.

3. In thermodynamics textbooks, one usually distinguishes different types of free energy depending on what parts of the energy are considered as intrinsic (e.g. $E_0(c)$) and what parts (e.g. Fx) are considered external. The different free energies are denoted by different symbols. In this convention, our $\mathcal{F}$ is called the "Helmholtz free energy."

2.3. *Ideal gas free energy* A one-atom ideal gas of mass m is trapped in a cylinder of height h. The atom can exchange energy with the outside world, which is at temperature T. Gravity is negligible. The height can be reduced by moving a piston. The relative probability $f(c)$ of this system is evidently $f(x, y, z, p_x, p_y, p_z)$. (a) What is the function $f(x, y, z, p_x, p_y, p_z)$? Use the Boltzmann normalization: $f(c) = e^{-E(c)/T}$, since this is the proper one for computing a free energy, as done below. (b) How does f change if the height is expanded from $h/2$ to h? (Does the dependence on p_x, p_y, p_z change?) (c) The free energy $\mathcal{F}$ was defined as $-T \log(\sum_c f(c))$. Find the change in free energy upon changing the height from $h/2$ to h. This can be expressed as $-T \log[\sum_{c_h} f(c_h) / \sum_{c_{h/2}} f(c_{h/2})]$. (d) The change in $\mathcal{F}$ is supposed to be the work done in changing the height. Thus there must be a force on the piston given by $-d\mathcal{F}/dh$. Express this force in terms of T and h. (e) On the other hand, we can obtain this force using the conventional ideal gas law for pressure p: $pV = NT$ (here $N = 1$). Compare the force from this pressure p to that obtained in (d). You may assume the piston has cross-sectional area A. This problem shows the connection between the conventional temperature as defined by the ideal gas equation of state and the T appearing in Boltzmann's Principle. Now a gravitational energy mgz is added to the energy $E(c)$. The work to change h is still given by the pressure at the top times the area. But the density and thus the pressure at the top are now reduced by gravity. (f) Verify that the work δW to change h is still given by $\delta \mathcal{F}$.

2.1.3 Lattice gas

To illustrate the use of free energy, we discuss a model that will prove useful later: the "lattice gas." A **lattice gas** is a set of N particles that occupy the V discrete sites of a lattice. To specify a configuration, we list which lattice sites are occupied. No site may be occupied more than once. (Thus N must be no greater than V.) We characterize the density of particles by the "volume fraction" $\phi \equiv N/V$. The simplest lattice gas has no energy. Thus all the arrangements of the N particles on the V sites are equally likely. The free energy is thus $-T \log(\sum_c 1)$. To count the possible arrangements c is a tedious problem in combinatorics. We may avoid it by using the indirect method of the last subsection. We alter our system by applying an external energy that allows us to change ϕ. By keeping track of the work to change ϕ we can infer the desired free energy.

We focus on a single site of the lattice, which is filled with probability ϕ, and consider the work that would be required to empty the site. In order to empty it, we introduce an external energy: $E = \mu\phi$. (The field parameter μ is often called a **chemical potential**.) We may calculate the free energy of a single site in the presence of the μ energy. If the site is empty, this energy is zero. If the site is occupied, the energy is μ. Summing over both configurations, we have a free energy $\mathcal{F}_\mu = -T \log(1 + e^{-\mu/T})$. As we have seen previously, the derivative of $\mathcal{F}_\mu$ must give the average value of ϕ: $\langle \phi \rangle = d\mathcal{F}_\mu/d(\mu)$. The intrinsic work required to change $\langle \phi \rangle$ (not counting the unwanted contribution from the $\mu\langle\phi\rangle$ energy) must be given by the change of $(\mathcal{F}_\mu - (\mu)d\mathcal{F}_\mu/d\mu) = (\mathcal{F}_\mu - \mu\langle\phi\rangle)$.

We now calculate these quantities explicitly:

$$\langle\phi\rangle = \frac{d\mathcal{F}_\mu}{d\mu} \doteq \frac{e^{-\mu/T}}{1 + e^{-\mu/T}} = \frac{1}{1 + e^{\mu/T}}. \tag{2.4}$$

Thus the free energy is given by

$$\mathcal{F}_\mu - \mu\langle\phi\rangle = -T\log(1 + e^{-\mu/T}) + \mu/(1 + e^{\mu/T}).$$

We may express this free energy in terms of $\langle\phi\rangle$, by solving Eq. (2.4) for μ: $e^{\mu/T} = 1/\langle\phi\rangle - 1$ and $\mu = T\log(1/\langle\phi\rangle - 1) = T\log(1 - \langle\phi\rangle) - T\log(\langle\phi\rangle)$. Similarly $\mathcal{F}_\mu$ can be expressed in terms of $\langle\phi\rangle$:

$$\mathcal{F}_\mu = -T\log(1 - 1/[\langle\phi\rangle^{-1} - 1]) = T\log(1 - \langle\phi\rangle).$$

Thus,

$$\mathcal{F}_\mu - \mu\langle\phi\rangle = T\left[\log(1 - \langle\phi\rangle) - [\log(1 - \langle\phi\rangle) - \log(\langle\phi\rangle)]\,\langle\phi\rangle\right].$$

This simplifies to

$$\mathcal{F}_\mu - \mu\langle\phi\rangle = T[\langle\phi\rangle\log(\langle\phi\rangle) + (1 - \langle\phi\rangle)\log(1 - \langle\phi\rangle)].$$

This free energy gives the work to change $\langle\phi\rangle$ in the one-site system. In a system with V sites, the reasoning is equivalent. The sum on c for each site is independent, since the work is simply the sum of site free energies.

Combining these results, the work per site done in changing the average volume fraction from 0 to $\langle\phi\rangle$ is given by $T[\langle\phi\rangle\log(\langle\phi\rangle) + (1 - \langle\phi\rangle)\log(1 - \langle\phi\rangle)]$. If the number of sites V is large, the actual volume fraction, ϕ must be arbitrarily close to this average. Thus, in a large lattice gas the work W per site required to change to a *definite* ϕ from $\phi = 0$ is evidently $T[\phi\log(\phi) + (1 - \phi)\log(1 - \phi)]$. But, as we have seen, the work to change a coordinate of a system is simply the change in its free energy $\mathcal{F}$.

$$\mathcal{F}/V = T[\phi\log(\phi) + (1 - \phi)\log(1 - \phi)]. \tag{2.5}$$

Clearly the lattice gas differs significantly from an ideal gas. In an ideal gas each particle may occupy any position independent of the other particles. In a lattice gas the particles are obliged to avoid sites occupied by others. In the dilute limit of a single particle in a large volume V the volume fraction ϕ becomes $1/V$ and the lattice gas free energy of Eq. (2.5) has the limiting form

$$\mathcal{F} = -T\log(V),$$

like the ideal gas treated in the preceding Problem. In this limit the second term in Eq. (2.5) is negligible. But it becomes increasingly important as the volume fraction increases, making the lattice gas free energy larger than the ideal gas free energy. In the opposite limit, as the $\phi \to 1$, the lattice gas free energy becomes small again. A nearly full lattice is equivalent to a dilute gas of vacancies, since $\mathcal{F}(1 - \phi) = \mathcal{F}(\phi)$.

2.1.4 Approach to equilibrium

These relationships between work, probability and temperature presuppose that our many-particle system has attained a final equilibrium state. Whenever the system is altered by a change in the environment, it leaves equilibrium, and requires time to return. A fluid is never in equilibrium in all respects. For example, the organic molecules in the fluid are typically not in their most stable state, and given sufficient time they would decompose. On a given experimental time scale—say, a minute—many features of the system, like the profile of local composition, have ample time to reach an equilibrium state, i.e., a state that is independent of the most recent change in the environment. Other features, like the total number of molecules of a given type, retain their initial values, and make essentially no progress towards equilibrium. Degrees of freedom that reach their equilibrium statistical state are often called **annealed** variables; those that keep their initial values are called **quenched** variables. In studying a given fluid, one must have a sense of which features of the fluid are quenched and which are annealed. That is, one must have some way of estimating the timescale for a given feature to relax from an initial state to one of thermal equilibrium.

Relaxation can take place in a variety of ways. The cooperative modes of relaxation distinctive to structured fluids are an important subject to be discussed in later chapters. Here we deal with a more generic and widespread cause of slow relaxation called **activated hopping**†. Activated relaxation accounts for the gradual dissociation of a population of molecules or the condensation of a supercooled vapor to form droplets of liquid. In general, activated hopping occurs when two objects interact via a potential of mean force like that of Fig. 2.2.

As the graph suggests, it is quite improbable for the separation x to take on values near x^*, since such separations have high (free) energy, even in equilibrium. For example, if $U^* - U_0 = 10T \simeq \frac{1}{4}$ eV, the probability that the separation is x^* is $e^{10} \simeq 20,000$ times smaller than the probability that it is 0. Since visits to x^* are rare, it is natural that passage from x near zero to $x > x^*$ is slow. A quantitative understanding of this slow process was first developed by Eyring, Kramers [1], and many others [2]. If the two objects are initially at a separation x near 0, our problem is to find the average time required for it to cross the barrier at x^* as a result of random thermal fluctuations. The answer clearly depends on the energy profile $U(x)$. In addition we need information about the time scale for the fluctuations. The two molecules of our example move in response to collisions from neighboring molecules. Thus x changes as a result of many independent, small, random increments. To be sure, the potential U affects these random moves. Since lower values of U are more likely than higher ones, random steps that decrease U are more likely than those that increase it. Thus the separation x tends to be pushed away from the energy barrier and back towards the region around $x = 0$.

We may imagine the behavior of this pair of objects if an additional artificial force were applied which neutralized the effect of the potential U. For example, if our two molecules were part of a detailed molecular-dynamics simulation, we could determine the potential of mean force $U(x)$ explicitly and then impose an additional force $F(x)$ that would nullify $U(x)$,

† The term "activated" comes from the context of chemical reactions. Here the system at $x = x^*$ is thought of as a chemically active state, which is free to pass to more than one distinct final state.

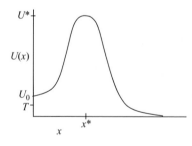

Fig. 2.2
Schematic plot of potential of mean force $U(x)$ versus separation x for two objects. If the two objects are initially together, with $x = 0$, with energy U_0, they cannot move apart without passing through a region of large potential U^* near the point x^*. The thermal energy T is marked on the vertical scale. Evidently a free energy much larger than T is needed if the objects are to move apart or together.

say, for all $x < x^*$. (Thus in the fictitious system, $U(x) = U(0)$.) Without a biasing potential the random moves of x can easily take it from its initial position to x^*. We denote the time to reach x^* with no potential as τ_0. Specifically, τ_0 is that time for which x has reached x^* with probability $\frac{1}{2}$. The inverse of this τ_0 is called the **attempt rate**. The Eyring–Kramers theory describes how the barrier increases the **escape time** from τ_0.

If the barrier is high, the objects must make many "attempts"—returns to $x = 0$—before reaching x^*. As time passes, the (un-normalized) probability $f(x)$ that the separation is x, approaches its equilibrium value for a greater and greater range of x around its starting point $x = 0$. After sufficient time, this range reaches x^*. Then $f(x^*)/f(0)$ approaches its equilibrium value of $\exp[-(U^* - U_0)/T]$. Whenever x does reach x^*, it may move to the right or to the left without bias. Thus with probability of roughly $\frac{1}{2}$, it crosses the barrier.

To see the effect of U, we first imagine the motion from the initial $x = 0$ state for a time τ_0 in the fictitious system with U nullified by the artificial force F. In a population of such molecules one encounters many possible motions. As noted above, this time τ_0 is sufficient for typical motions to reach x^*. One way to account for the effect of U is to form a new ensemble of motions starting from the motions of the fictitious system. One removes some of these motions from the fictitious ensemble to account for the time they spend in regions of high energy. These removals must assure that the overall probability for the separation to be x is the equilibrium probability $f(x)$, in the potential U. In the fictitious system all x's have the same equilibrium probability f_0. But in the real system, x occurs with equilibrium probability $f(x) \propto \exp(-U(x)/T)$. Evidently $f(x)/f_0 = \exp[-(U(x) - U(0))/T]$. In equilibrium one may thus recover the proper probabilities by simply throwing away all but a fraction $f(x)/f_0$ of all fictitious systems that end up at x. This same prescription is approximately correct even for our ensemble of motions in a finite time t. The probability that x has reached x^* during the motion is that of the fictitious system (*viz.* $\frac{1}{2}$) times the probability that the motion has survived the removal process. This probability is $f(x^*)/f(0) = \exp[-(U(x^*) - U(0))/T]$. Since the separation has reached x^* in time τ_0 with this probability, the average rate at which the separation crosses x^* is thus $\frac{1}{2}\tau_0^{-1} \exp[(-U(x^*) + U(0))/T]$. The pair of objects stay together for an average time that is the inverse of this rate. Over times much shorter than this escape time, the two objects may be considered as permanently bound together.

Clearly the most important influence on the escape time is the activation energy $U^* - U_0$. Escape over a barrier is exponentially slow if its height is much more than T. Increasing this height by an amount T increases the escape time by a factor e. The escape time also depends on the attempt time, but much more weakly. The Eyring–Kramers formula is applicable whenever the barrier height is much greater than T. Remarkably, the shape of the $U(x)$ function is immaterial in this limit. The formula has been derived under rather general conditions, as discussed in [2]. It has also been experimentally verified for a wide range of processes[†]. Though we have illustrated this behavior in the context of dissociation, it is applicable to any quantity x with a region of high free energy. Activated hopping is not the

[†] To check for activated kinetics one conventionally plots the log of the measured rate of e.g., dissociation against inverse temperature $1/T$. Activated kinetics gives a straight line on this "Arrhenius plot;" the slope is the activation energy $U^* - U(0)$.

only way a quantity like x can pass through a high-energy region. Another process is quantum tunneling. This requires no thermal fluctuations, but only the positional uncertainty all particles have by virtue of the wave nature of matter. Such tunneling phenomena are not generally important for the liquids we will consider, and we will ignore this and other quantum effects from here on.

Using the Eyring–Kramers theory, one may get a rough idea of what parts of the system can or cannot relax over the timescale of a given experiment. For example, the energy required to break a carbon–carbon bond is 2 eV, i.e., $80\,T$. The attempt time τ_0 is in the range of 10^{-12} s. or more, as discussed in the next section. From this one obtains an estimated escape time of some 10^{15} years. We shall need the notion of activated lifetimes when we discuss colloidal stability in Chapter 5.

2.4. *Metropolis dynamics: a concrete kinetic model*   It is useful to have a specific model to examine phenomena like dissociation and to gauge the validity of the Eyring–Kramers kinetic picture in specific cases. The **Metropolis algorithm** [3] used for **Monte-Carlo** simulations provides a simple example that is easy to implement on a programmable calculator or computer. Like the motion of real molecules, the Metropolis dynamics consists of a sequence of small steps. And after a long time, the Metropolis algorithm leads to the proper equilibrium Boltzmann distribution $f(x)$. Applied to our dissociation problem, the Metropolis algorithm works as follows. Time is divided into discrete steps of length Δt. The reaction coordinate changes by discrete, random increments of a fixed magnitude $\Delta x \ll x^*$. In the absence of a potential U, the motion is a sequence of random steps. At timestep t, the current $x(t)$ is changed by $\pm \Delta x$, with the $+$ or $-$ sign being chosen at random. In the presence of a potential $U(x)$, a given step results in a change of potential ΔU. The dynamics must be modified to account for the potential if it is to lead to the proper equilibrium distribution $f(x) \propto \exp(-U(x)/T)$. The modification is that after each random step the change of energy ΔU is computed. If the step is found to decrease U, the stepping process proceeds: the step is "accepted." But if the step is found to increase U, the step is "rejected" with probability $1 - \exp(-\Delta U/T)$. That is, the position x is reset to its value before the last step. To implement this on a computer, one chooses a random number between zero and one. If this number is larger than $\exp(-\Delta U/T)$, one resets the x coordinate; otherwise one retains the last increment of x and proceeds to the next random step.

If the range of x is limited, the entire range is eventually visited, and after a long time the probability $f(x)$ that the coordinate is at x reaches a constant distribution, independent of time t. One can show what this $f(x)$ must be, using the dynamical rule above. In general, one can express $f(x,t)$ in terms of f's at the previous timestep. In the absence of the potential U, if the coordinate is x at timestep t, it must have been $x - 1$ or $x + 1$ at the previous timestep. If it was at $x - 1$, then it moves to x with probability $\frac{1}{2}$. This $\frac{1}{2}$ is the conditional probability that the coordinate goes to x given that it was at $x - 1$. Thus $f(x,t)$ is the sum of two contributions for the two distinct states from the previous timestep that can reach x:

$$f(x,t) = \tfrac{1}{2}(f(x-1,t-\Delta t)) + \tfrac{1}{2}(f(x+1,t-\Delta t)).$$

If U is present, the conditional probabilities are modified. We suppose that at timestep $t - \Delta t$ the coordinate is x. Then the probabilities q_-, q_+, q_0 of going to $x - 1$, going to $x + 1$ or staying at x are:

$$q_- = \tfrac{1}{2}, \quad q_+ = \tfrac{1}{2}\exp(-\Delta U/T), \quad q_0 = 1 - q_- - q_+ = \tfrac{1}{2}[1 - \exp(-\Delta U/T)].$$

Here we define $\Delta U(x) \equiv U(x+1) - U(x)$. We have supposed for definiteness that U is increasing with x near the x in question. (If U is decreasing, the rules for q_+ and q_- are reversed, since we always accept a candidate move that goes downhill.) Combining the conditional probabilities from $x - 1$, x, and $x + 1$, we obtain the full probability $f(x,t)$.

$$f(x,t) = q_+(x-1)f(x-1,t-\Delta t) + q_0(x)f(x,t-\Delta t)$$
$$+ q_-(x+1)f(x+1,t-\Delta t). \tag{2.6}$$

This equation describes the "flow" of probability during one timestep. In an ensemble of many copies of the system, the (relative) net number of systems passing from x to $x+1$ in a timestep is called the probability current $j(x + \frac{1}{2})$. Evidently $j(x + \frac{1}{2}) = f(x,t-\Delta t)q_+(x) - f(x+1,t-\Delta t)q_-(x+1)$. (a) Express the change of f during a timestep, $f(x,t) - f(x,t-\Delta t)$ in terms of the currents $j(x - \frac{1}{2})$ and $j(x + \frac{1}{2})$ for any $q(x)$'s that have $q_- + q_0 + q_+ = 1$. (b) Find a relationship between $f(x)$, $f(x + 1)$ and the q's such that $j(x + \frac{1}{2}) = 0$. (c) When the system reaches equilibrium, $f(x)$ becomes independent of time. This can only happen if $j(x + \frac{1}{2}) = constant$ for all x. In the Metropolis process, this constant is zero. We may define $U(x = 0) = 0$ and $f(x = 0) = 1$. Find the distribution $f(x)$ for which $j = 0$ everywhere. (d) **(Harder)** Write a program to implement the Metropolis algorithm for the potential $U(x) \equiv T((x - 20)/3)^2$. Let x range from 1 to 40 in steps of 1. Start x someplace between 18 and 22. Define an array $N(1), \ldots, N(40)$, one element for each x. At each timestep (whether the move is retained or not) add one to the $N(x)$ for the current value of x. Thus $N(x)$ gives the number of timesteps when the system was at x. Plot this experimental distribution $N(x)$ after 10,000 timesteps and after 10^7 timesteps, and compare it with the equilibrium distribution $\exp[-(x - 20)^2/9]$. Find the average time for the system to reach $x = 26$ and compare with the Eyring–Kramers prediction. Begin by turning off U and determining τ_0 for some chosen starting value of x. Then restore U and measure the time using this same starting value.

2.5. *Equipartition theorem*   A thermodynamic system consists of a variable x and many other variables, labeled collectively c'. The energy $E(c)$ of a configuration has the form $E_0(c') + E_1(x)$. (Thus the x variable may be considered as a subsystem.) Further, the x energy is quadratic: $E_1 = bx^2$. (a) Show that $\langle E_1 \rangle \equiv (\sum_c E_1 f(c))/(\sum_c f(c))$ is simply $\frac{1}{2}T$. (Note that the sum on c can be immediately reduced to a mere sum on x.)

2.2 Magnitude of a liquid's response

In this section we discuss the quantitative behavior of liquids, including structured fluids, at the most primitive level. We account for the order of magnitude of the basic response properties in liquids. For example, we know that most simple liquids in everyday experience—water, gasoline, alcohol—have roughly similar viscosities, though they differ markedly in other respects. This impression is confirmed by a look at a property-table handbook. There one finds that most small-molecule liquids have viscosities of around 0.001 Pa s, i.e. 0.001 SI unit[†]. This size viscosity is a rough lower limit: those viscosities that differ greatly from this value are on the high side. Structured fluids like liquid polymers often have high viscosities. To understand the significance of this, we must understand how viscosity arises and what controls its magnitude.

[†] The cgs unit is called a poise; a poise is 10 Pa s.

The viscosity of a liquid may be defined by an experiment with the liquid between two parallel plates. If the top plate is slid over the bottom one with some (small) velocity v, the fluid moves parallel to the plates, with a velocity that increases linearly from zero at the bottom plate to v at the top. The velocity gradient dv/dz, or **shear rate** is constant throughout the liquid. Viscosities are typically measured at shear rates of 10–100/s. This is the order of shear rate encountered in pouring a cup of coffee. Shear rates larger than about 10^5/s are difficult to achieve in controlled, steady, laboratory flows.

It is useful to think of this velocity gradient in terms of the strain on a fluid element, a tiny cubical volume in the fluid. As the flow proceeds, the cube distorts into a parallelepiped. The amount of distortion of the parallelepiped—the lateral displacement divided by the height—is the **strain** γ which the fluid has undergone. This strain evidently increases at a constant strain rate $\dot{\gamma}$, which is identical to the velocity gradient.

A steady force on the top plate is needed to maintain the motion; this implies a stress in the fluid. Since the acceleration of every part of the fluid is zero, the forces on each fluid element of the fluid must sum to zero. If there is a force acting to the right at the top of a fluid element, there must be an equal force acting to the left at the bottom. The force on a small element must be proportional to its cross-sectional area. The force-per-unit-area or **stress** σ must be uniform from top to bottom since the fluid is not accelerating. Typically, the shear rate in each fluid element produces a stress σ proportional to the shear rate $\dot{\gamma}$. The constant of proportionality is the **viscosity** η: $\sigma = \eta\dot{\gamma}$.

The coefficient η also measures the dissipation rate. The work done per unit time to maintain the flow in a fluid element is the force times the velocity:

$$dW/dt = F \cdot \Delta v = (\sigma \Delta x \, \Delta y)(\dot{\gamma} \, \Delta z),$$

where $\Delta x, \Delta y, \Delta z$, are the dimensions of the element. Thus the dissipation rate per unit volume $\dot{w}$ is $\sigma\dot{\gamma}$ or

$$\dot{w} = \eta\dot{\gamma}^2. \tag{2.7}$$

Evidently, the dimensions of η are energy per unit volume times time.

Why should a fluid's dissipation be governed by a law like Eq. (2.7)? What determines the size of the parameter η? To discuss these questions, it is useful to imagine applying the flow between our plates in a different way. We imagine not a smooth motion at a fixed velocity, but a motion in sudden, tiny steps. We apply a small step strain, wait for a moment, and then apply another. These steps may be made so small that the dissipation they produce is the same as in the steady flow. But this stepwise picture gives us a way to examine the molecular basis of viscosity. For simplicity, we shall also imagine the simplest of fluids, a monatomic liquid like liquid Argon. (We should not need to treat too-specific features of the fluid, since viscosities depend but little on the molecular nature of a simple fluid.)

Figure 2.3 shows a fluid element before and immediately after a step strain. The sketch attempts to show an **affine** deformation of the centers—that which would occur if these were embedded in rubber and then distorted.

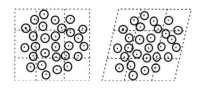

Fig. 2.3
The isotropic liquid at the left is suddenly subjected to a 20% shear on the right, thus pushing the molecules together from upper-left to lower-right.

Evidently the isotropic equilibrium arrangement of the atoms is distorted and made anisotropic by the strain. Producing this non-equilibrium state requires work; it costs a free energy proportional to the number of atoms in the fluid element. It is this stored energy that gives the energy scale in the viscosity, as we shall discuss below.

After the anisotropic state has been produced, it relaxes back to equilibrium. After some characteristic relaxation time the state of the liquid in the strained element is the same as it was before the strain was applied. The time required is the time for the distances between nearest neighbors to become the same in all directions. This is a few collision times, i.e., a few times the time required for an atom to travel between nearest neighbors.

With this scenario in mind we can estimate the magnitude of the viscosity. To find the order of magnitude of the transit time, τ we need to know the typical speed v of the atoms. We may focus on one component of the velocity, say v_x. The energy $E(c)$ of our system depends on v_x in a simple way: $E(c) = E(c') + E(v_x)$, where the variables c' are independent of v_x and the $E(v_x)$ is independent of the c'. In such cases, as we have seen above, the v_x variable may be considered as an independent subsystem. In our case the (kinetic) energy $E(v_x)$ is quadratic in the variable. Then one can readily calculate that $\langle E(v_x) = \frac{1}{2}T$, as shown in Problem 2.5. This **equipartition theorem** is true for any variable whose energy is quadratic. Using the equipartition theorem, the typical speeds of atoms in our simple liquid are evidently given by $mv^2 \simeq T$ or $v \simeq (T/m)^{1/2}$, where m is the atomic mass. For the light atoms found in air or common liquids, it is[†] a few hundred m/s. The distance between collisions is roughly an atomic diameter a—i.e., roughly 0.2 nm. We thus estimate that the relaxation time τ is of order $a/v \simeq 10^{-12}$ s [††].

For the tiny step strains in our experiment the stress σ must be proportional to the strain, as in any weakly deformed material. Accordingly, we define the "step-strain modulus" G_0 as the ratio of initial stress to the magnitude of the step strain: $\sigma \equiv G_0\gamma$.

The stored energy per unit volume w is the work done per unit volume by the stress force. We may compute it by integrating the force $\sigma \Delta x \Delta y$ on the top face with the displacement $dx = \gamma \Delta z$: $w\Delta x \Delta y \Delta z = \int_0^\gamma [\sigma(\gamma')\Delta x \Delta y]d[\gamma'\Delta z]$. Expressing σ in terms of the modulus G_0 we have $w = \frac{1}{2}G_0\gamma^2$. We see that G_0 has dimensions of energy per unit volume, like the dimensions of w, so that G_0 is roughly the free energy cost extrapolated to a strain of unity. This free energy is at least partly interaction potential energy between the atoms; the free energy per atom is roughly the cost of removing an atom. Such energies can be found from thermodynamic measurements on the liquid. But they can be crudely estimated on fundamental grounds. Our atoms condense into a liquid phase (in equilibrium with gas) because they attract each other. To form a phase, surface atoms moving away from the surface under thermal motion must be opposed by this attraction. The potential energy must thus be at least of order of the kinetic energy, which in turn is of the order of T. Were this is not true, the atoms could freely leave the liquid and the liquid would evaporate. But the interaction energy cannot be too much greater than T; otherwise the

[†] One shows in statistical mechanics that it is essentially the speed of sound in a gas made of these atoms.

[††] We shall use the symbols $\simeq$ and $\sim$ throughout this book. Our use of these symbols is exactly that of DeGennes' book on polymers, *op. cit.* These weak forms of equality are very useful for describing the scaling behavior we shall explore. The statement $A \simeq B$ means that (in the regime or limit under discussion) A/B is a numerical constant. This constant need not be close to 1. For example, a rod of length L and mass m has a moment of inertia $I \simeq mL^2$. The statement $A \sim B$ is weaker; it means that A/B remains finite in the regime or limit of interest. For example, rods of different length L and width a have moments of inertia $I \simeq ML^2 \sim L^3$ when $L \gg a$. Generally we use $\sim$ rather than the stronger $\propto$; $A \propto B$ indicates strict proportionality, with no implication of an asymptotic regime or limit.

kinetic energy of the atoms would be insignificant. They would take on an arrangement of minimum potential energy, and form a crystal. Thus the mere knowledge that our atoms are in a liquid phase tells us that their inter-action energy is not too different from T. The modulus G_0 is thus of order T/a^3. In this estimate we neglected the non-potential part of the free-energy of distortion. This "entropic" part arises because fewer configurations are available to the atoms after the step strain, thus reducing $\sum_c f(c)$. A unit of strain reduces the number of configurations by a finite factor for each atom, so that the free energy $-T \log(\sum_c f(c))$ changes by an amount of order T per atom. Its contribution to the modulus G_0 is comparable to that of the potential energy.

With these numbers in hand we may now estimate the rate of dissip-ation $\dot{w}$. Each time a step strain is made, the energy density w stored is $\frac{1}{2}G_0\gamma^2 \simeq T/a^3\gamma^2$. This energy is dissipated irrecoverably as the liquid relaxes to its equilibrium state in time τ. The dissipation rate $\dot{w}$ over this time is of order $T/a^3\gamma^2/\tau$. If we repeat the step strain at intervals τ, we obtain a steady dissipation of this order. The shear rate $\dot{\gamma}$ is then γ/τ, and $\dot{w} \simeq T/a^3\tau\dot{\gamma}^2$. This has the proper form for viscous dissipation; by com-paring with Eq. (2.6), the viscosity η may be read off. Using $(1/40)\,eV$ $(= 1.6 \times 10^{-19}/40\,J)$ for T, $10^{-12}\,s$ for τ and $0.2\,nm$ for a, this yields an estimate of $10^{-3}\,Js/m^3$ for η. That is, $\eta = 0.001$ in SI units, in agree-ment with the measured values quoted at the beginning of this section. (In view of the crudeness of our estimates, this agreement is somewhat fortuitous!)

The mechanism for viscosity sketched above is somewhat altered for molecular liquids. Let us imagine what happens to the viscosity as we pass from the simplest hydrocarbon, methane, to the two-carbon chain ethane to the three-carbon chain propane, and so forth. As we do this, we may have to change the temperature so as to maintain our sample in the liquid state. For definiteness we shall maintain a temperature, say, 1% above the melting temperature. This temperature rises as the molecules get larger, but only to a limited extent: it rises from roughly 200–400° K. The liquid density is also roughly constant; the atoms stay roughly at the separation that minimizes their interaction potential. This potential is virtually independent of temperature. Thus, it remains roughly true that the interaction energy per atom is roughly T. Accordingly, the energy scale of the viscosity should not be greatly altered as the molecules become larger. The timescale for a distortion to relax is also not greatly altered. Much of the distortion of the hydrocarbon fluid can be relaxed by few-atom local motions of the flexible hydrocarbon chains. (This flexibility is necessary to maintain a liquid phase; if the molecules are too rigid, they pass directly from a solid to a gaseous state.) Since each atom moves at thermal velocities as calculated above, the time τ to move an interatomic distance is not qualitatively changed as the molecules grow. We conclude that the order of magnitude of the viscosity of a molecular liquid should not be greatly different from that of an atomic liquid, as is observed. For very long hydrocarbon chains, the assumptions made in this paragraph begin to break down and large changes in the viscosity occur. The origin of these changes is an important subject in the chapter on polymers.

2.6. *Rubber-band modulus* The postal service rubber bands you find on the sidewalk have a cross-sectional area of 5 mm^2. They consist of hydrocarbon polymers with a molecular weight per atom of about 5. The density is about 1000 kg/m^3 (1 g/cm^3). Under a 0.200 kg load, such a rubber band elongates by a factor of 2. (a) What is the modulus of this rubber band? (b) What is the modulus expressed in atmospheres of pressure? (c) Under a factor-of-two elongation (unit strain) using the density, Avogadro's number and the average mass per atom, how many atoms store energy T? Order-of magnitude estimates are fine.

There is a major effect that influences the viscosity of small-molecule liquids, which we have not discussed. This is the so-called glass transition, and is most important in molecules of irregular shape. As these liquids are cooled, the viscosity grows too large to be measurable as the liquid cools towards a characteristic temperature T_g. One can certainly not account for this divergent viscosity by the arguments discussed above. However, one may still ask whether the large viscosity is due to a large modulus or a long relaxation time. The answer is the latter. A liquid near its glass transition cannot equilibrate from an imposed distortion in a few atomic collision times. The molecules have so little room to move and are so interlocked that stress takes exceptionally long to relax. This phenomenon is important for us in two ways. First, many of the background solvents of our structured fluids have experimentally important glass transitions. This often provides a useful means of controlling their viscosity. Second, large fluid structures often themselves lead to high viscosities. There are some parallels between this type of high viscosity and the glass transition, as we will discuss in due course.

In accounting for the viscosity we have defined the fundamental response of liquids. We've also identified the lengths, times, and energies that characterize liquid structure and dynamics on the atomic scale. These conditions set the backdrop against which the distinctive behavior of structured fluids occurs.

2.3 Experimental probes of structured fluids

Physicists are increasingly interested in structured fluids in large part because the collective structures in them have become increasingly accessible to experimental probes. Before discussing specific phenomena in particular fluids, it seems good to have in mind how the predicted behavior might be observed. We shall see that though some of these experimental methods are common to condensed matter studies, many are distinctive to structured fluids.

2.3.1 Macroscopic responses

As in all condensed matter, the most apparent properties of structured-fluid systems are their macroscopic responses: we perturb a macroscopic sample in some global way and measure something that happens throughout the sample. The primary case-in-point for structured fluids is the so-called dynamic modulus $G(\omega)$. As discussed in the previous section, the modulus is the ratio of stress to strain. The **dynamic modulus** is the ratio of stress to

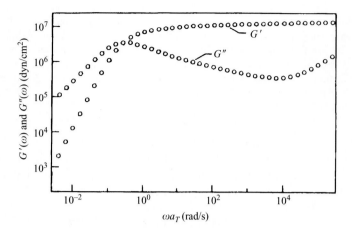

Fig. 2.4
Complex dynamic moduli of a
high-molecular-weight polymer melt,
from [4]. Reprinted with permission
purchased from *Rubber Chemistry and
Technology* © 1987 Rubber Division,
American Chemical Society, Inc. G' labels
the in-phase (storage) modulus; G'' labels
out-of-phase (loss) modulus. The angular
frequency ω is multiplied by a
temperature-dependent factor a_T to show
the behavior that would occur at 27 °C.

strain when an *oscillating* strain at frequency $\omega/(2\pi)$ is applied. In general the stress is not in phase with the strain, so that the dynamic modulus is a complex number. In a simple liquid with a characteristic step-strain modulus G_0 and relaxation time τ, both G_0 and τ may be inferred directly from the dynamic modulus function. Figure 2.4 shows a classic example: the dynamic modulus of a polymer melt. Many structured fluids we shall meet have several important frequency regimes. The dynamic modulus gives complete information about the response to a (small) arbitrary time-dependent strain. Another equivalent response function is often discussed: the step-strain modulus. This is the stress remaining at time t after a small step strain has been applied. Dynamic moduli are routinely measured in the range from 0.01 to tens of hertz. In many structured fluids the characteristic relaxation rates may be brought into this range e.g. by varying the temperature. Alongside this distinctive dynamic modulus is the more generic dielectric relaxation—i.e., the complex dielectric function $\epsilon(\omega)$. It is used to probe the characteristic relaxation timescales in a structured fluid, with great dynamic range but without great specificity.

2.7. *Jello modulus* Straight, unflavored gelatin is mostly a flexible biopolymer called collagen. It has an average molecular weight per atom of 10 or so. The point of this problem is to see how effective these molecules are at giving elasticity to water. We want to know how much gelatin molecule (what molecular weight) stores about T of energy under a unit of strain. The idea is to see if there is about T per atom, T per 10,000 atoms or what. To estimate this, one may use the following information. Five grams of dry gelatin in 0.5 kg of water makes a gel of the consistency of normal Jello. A 5 cm cube of such a gel on a plate oscillates at about 1 cycle/s. For simplicity, we can assume that the amplitude of the strain is about unity—i.e., the top of the cube goes back and forth 5 cm from its resting position. From this frequency we may estimate the stored energy in the gel. Since the piece of Jello is a harmonic oscillator, the stored energy is the same as the kinetic energy $\frac{1}{2}mv^2$. The typical velocity is 5 cm/s, grossly speaking. (a) From this data estimate the number of atoms of gelatin responsible for storing each T of elastic energy. (b) *(Harder)* The oscillations of this Jello stop in a few cycles. That means a substantial fraction of the initial kinetic or potential energy is dissipated in one second. This dissipation should somehow arise from viscous drag in the water. Using the viscosity of water (10^{-3} SI units), calculate the dissipation rate in a 5-cm cube of pure water at a shear rate of 1/s. Comment on whether this estimate accounts for the damping of the Jello.

In solutions, the modulus generalizes in an interesting way. A solution containing dispersed objects resists compression of these objects into a smaller volume of solvent. This **osmotic pressure** is evident if one encloses the concentrated solute in a membrane which can pass the solvent, but not the solute: solvent flows into the enclosure increasing the pressure there. Present day osmometers use more sensitive methods, such as e.g. the reduction in vapor pressure of the solvent or the depression of the freezing point due to the presence of the dispersed objects [5]. This measurement provides a way of inferring how many solute particles are present, and of verifying that they are well dispersed. Other systems, like polymer solutions, have distinctive predicted behavior of their osmotic pressure, which can be tested by this means.

2.8. *Osmotic sensitivity*   Many commercial osmometers for water solutions work by lowering the temperature until ice crystals begin to form. The solute has the effect of depressing the freezing point, and this depression allows one to measure the osmotic pressure. The solute particles are virtually all excluded from the ice crystals. Thus when ice forms, the fluid volume accessible to the solute is reduced. Each molecule in the water, displaces a volume v_w. When it transforms into ice, it reduces the water volume by v_w. The work per molecule done against the osmotic pressure Πv_w. In order for the molecule to be in equilibrium with ice, its free energy in the ice must increase by this amount. For this purpose we may estimate that the free energy is of order T, so that the needed shift in temperature ΔT required to provide the needed increase in temperature is given roughly by $\Delta T \simeq \Pi v_w$. (a) Find v_w in nm^3 given its mass density and molecular weight. (A useful benchmark is that $1\,\mathrm{g/cm}^3 = 0.6023\,\mathrm{amu/Å}^3$.) (b) If an instrument is capable of measuring $\Delta T = 0.001°$, what is the largest volume per solute particle V that it can detect?

One of the most important macroscopic properties in a structured fluid is almost trivial to measure. One simply determines whether or not the system is stable as a single phase. Phase stability is by no means a foregone conclusion in a solvent with polyatomic solutes. In some microemulsion and colloid systems the conditions for single-phase stability and the nature of the coexisting phases when a single phase is not stable are central questions for study. We shall see that phase stability reflects distinctive conditions in the large-scale spatial structure.

Recently the measurement of mechanical response of a fluid has been extended in space as well as time. Forces can be measured across just a tenths of a nanometer of a fluid using the Tabor/Israelachvili **surface forces apparatus** (Fig. 2.5). This bold and simple device relies on the flatness of cleaved crystals of mica. Such a surface can be atomically flat over square millimeters of surface. Thus two such surfaces may be brought to e.g. a nanometer spacing over many square microns. The surface forces apparatus is simply a pair of spring-mounted surfaces immersed in a fluid with optical means for measuring the distance between them. With this device it has proven possible to measure normal and shear forces at distances from 1 to 100 nm separations. The technique is sensitive: it routinely measures surface interaction energies amounting to less than T per $(30\,\mathrm{nm})^2$ area.

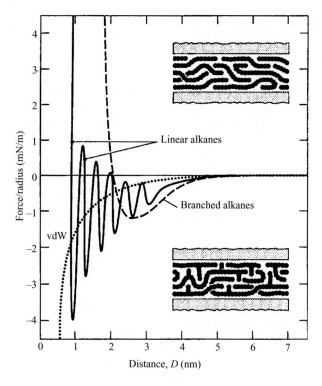

Fig. 2.5

Data from Jacob Israelachvili *et al.* from the surface forces apparatus [6]. The liquid between the mica plates is tetradecane, an unbranched hydrocarbon chain 14 carbons long. Straight, rising segments of the curve are unstable regions of the spring balance, and contain no measured data.

The apparatus was used to exhibit molecular-layer oscillations in forces in simple fluids. It has also shown forces due to polymers and surfactants adsorbed on the mica plates. Many of the triumphs of this technique are reported in the recommended text by Israelachvili.

2.3.2 Probes of spatial structure

Several physical probes explore specifically the arrangement of parts of a fluid in space. They all involve sending particles or waves through the fluid and monitoring how they are diffracted, reflected, or absorbed. There are three basic types: microscopes, scattering experiments, and what one may call depth probes. Much of the available qualitative information about structure comes from electron microscopy. As we have seen in the previous chapter, these can yield direct evidence of structure down to 10 nm and less. The principle is the same as an ordinary transmission optical microscope. A beam of particles from a point source is passed through a thin, partially absorbing sample, and onto a screen or detector. The more absorbing parts of the sample thus cast shadows on the screen. But both obtaining electron microscope images and interpreting them is a tricky art. This trickiness is well epitomized by two ironic laws of electron microscopy from the microscopists' folk tradition: (i) It is impossible not to get an image, and (ii) Not everything you see is an artifact.

The requirements of electron microscopy are essentially incompatible with those of the liquid state. Structured fluid samples are typically prepared by rapidly cooling them to cryogenic temperatures. The samples thus frozen are sliced in an ultramicrotome for transmission studies. In order to enhance the absorbing contrast, it is usually necessary to "stain" the sample with chemicals that absorb strongly and that are attracted to certain molecules in the sample. Or they may be fractured and the resulting fracture surface treated to reveal its topography. Often (as in the microemulsion pictures in the Overview chapter) feats of ingenuity are needed to cool the liquid fast enough to freeze in the undisturbed liquid structure[†]. Another problem often encountered is damage by the electron beam, especially when high magnification is needed.

The discovery of **scanning probe microscopy** has enabled new ways to examine structured fluids on surfaces. The classic example of this technique is **scanning tunneling microscopy (STM)** [7]. In STM one breaks a piece of metal to produce a sharp asperity and then scans this "tip" over the surface to be analyzed. One applies a voltage to the tip; thus a "tunneling" current flows when the tip comes within a few tenths of a nanometer of the substrate. The current is a very sensitive function of the distance of closest approach. By adjusting the height of the probe above the surface as it moves across the surface, one may keep this current constant, so that the distance above the surface is constant. Thus the probe traces out the height of the surface as a function of position. The vertical and lateral resolution may be as small as a tenths of a nanometer. One may use the principle of the STM even when the substrate is not conducting, as in most surfaces involving hydrocarbon liquids. In the **atomic force microscope (AFM)**, the tip is made to respond to the mechanical pressure of contacting the substrate rather than to a current. The ability to see structured fluid surfaces with such scanning tip probes is rapidly being developed (see, e.g., Fig. 2.6).

A significant limitation of scanning tip microscopy is its lack of temporal resolution. An image containing the information of a single television frame requires many seconds to scan. This is very slow relative to the

[†] Slow freezing invites crystallization and associated disruption of the original structure. A slowly frozen sample may resemble the original liquid no more than slowly refrozen ice cream resembles fresh ice cream.

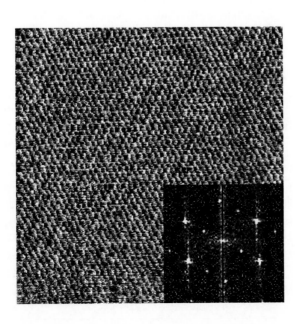

Fig. 2.6
AFM image (20 nm by 20 nm) of a three-monolayer-thick film of the surfactant cadmium aracidate (CdA_2) on mica, courtesy of J. Zasadzinski. Lighter colors correspond to higher areas, and the peak-to-valley height modulations about 0.2 nm. The two-dimensional fourier transform is inset. See [8].

relaxation times of liquids, as we have seen. Since the measurement requires mechanical movement of the macroscopic tip, even local measurements without scanning are limited to acoustic frequencies. Another problem is that these microscopes are best suited to measuring a topography—a height as a function of lateral position. Adsorbates like polymers and surfactants encountered in structured fluids have a richer structure, not well expressed as a topography. Using scanning tip microscopy for such a system is rather like trying to infer the structure of a forest by landing blimps on the treetops.

A complement to microscopy is the statistical information about spatial structure provided by scattering. A plane wave (electromagnetic or Shrödinger) is sent through the sample. Any inhomogeneities in the interaction between the matter and this wave result in spherical scattered waves propagating outward from the inhomogeneities. The spherical waves add together to give a net wave intensity in directions away from the incident beam direction. Inhomogeneities as large as the sample itself do not appear to the beam as inhomogeneities at all, and the incident beam propagates through it without scattering. Conversely, isolated pointlike inhomogeneities produce an isotropic spherical scattered wave. Scatterers of intermediate size produce scattered waves of finite angular spread.

The angular width θ of the scattered waves is related to the size R of the scatterers by the well-known diffraction law studied in freshman physics: $R \simeq \lambda / \sin(\frac{1}{2}\theta)$, where λ is the wavelength of the waves. This simple relationship between size of objects in the sample and the characteristic angles of scattered waves can be refined into a powerful probe. We will have much more to say about the interpretation of scattering in coming chapters. To illustrate this power, Fig. 2.7 shows a diffraction peak obtained by scattering x-rays from a surfactant–oil solution. The spacing of the diffraction peaks and the dependence of their position on concentration shows that the surfactants are arranged in a stack of regularly-spaced sheets about 6 nm apart. The shape of the diffraction peak shows a more subtle thing: it shows how the sheets fluctuate away from perfect order. The falloff of intensity with distance from the peak is a power law. The power varies continuously and predictably with concentration.

Fig. 2.7

Scattering data obtained at National Synchrotron Light source by Didier Roux and Cyrus Safinya [9] on a lamellar phase of 20% surfactant, oil, and water. The ordinate $I(q)$ is the intensity of scattered radiation, in arbitrary units. The quantity q on the abcissa means $2\sin(\theta/2)/\lambda$, where θ is the deflection angle of the x-rays. Three sets of data are for three different amounts of water. The solid curves show the predicted power-law dependence on q.

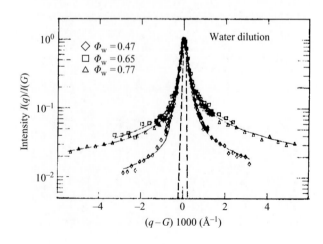

2.9. *Scattering assay*   A HeNe laser beam (wavelength about 500 nm) passes
    through a glass container of unknown solution and onto the wall. We observe
    that there is a faint glow of laser light around the central spot. The angular width
    of this glowing region is about 5°. From this we infer that there are objects in
    the solution which are much larger than atoms. (a) About how large are they?
    (b) A second solution, similar to the first, shows the same type of scattering,
    except that the angular width is now only $2\frac{1}{2}°$. By what factor are these objects
    larger or smaller than the ones in (a)?

If the scatterers move, the scattered intensity fluctuates in time. By
analyzing these intensity fluctuations, one may infer something about how
the scatterers move. The principle is similar to that behind diffraction itself.
If in a given time the scatterers move much less than one wavelength in the
medium, the phase of its scattered waves on the detector is changed by a
only a small amount, and the net intensity is little affected. The intensity is
virtually constant over such a short time. Given longer time intervals, the
scatterers may move as much as a wavelength. Then the resultant intensity is not constant but fluctuates. Thus the **correlation time** over which the
intensity stays constant measures the time for the scatterers to move about a
wavelength. Thus **dynamic scattering** is a powerful probe of motion as well
as structure [9].

The waves most commonly used in scattering studies on structured fluids
are visible light, x-rays, and neutrons. Visible light is of limited usefulness,
because most structures of interest are smaller than optical wavelengths.
Still it can give limited information about overall sizes and long-distance
diffusive motions. Dynamic visible-light scattering can probe correlation
times from a microsecond out to (at least) several seconds. X-rays have
typical wavelengths of roughly a tenth of a nanometer—smaller than the
structures of interest. But the high quality of the best x-ray beams permits
scattering measurements at very small angles, corresponding to distances
$\lambda/\sin(\frac{1}{2}\theta)$ of 10^3 nm. (It is not trivial to use such beams; one must go to a
national synchrotron source.) X-rays are not routinely used to infer motion.
The reason lies in the quantization of the waves. For a quantized wave to
sense motion by the means discussed above, more than one quantum should
scatter off each scatterer during the correlation time. For x-rays this would
ordinarily entail the deposition of too much energy and momentum for the
sample to sustain[†].

Modern neutron sources have wavelengths of a few tenths of a nanometer.
The neutrons are produced by a nuclear chain reaction (in a nuclear reactor)
or a reaction with a charged particle beam from an accelerator[††]. The neutron
radiation thus produced is incoherent thermal radiation, like the light from
a light bulb. It is far from the angular or energetic purity of a laser beam or
a synchrotron source. Thus the neutron beam must be collimated and made
mono-energetic by great sacrifices in intensity. Only a few facilities in the
world have enough intensity to do adequate scattering studies. Despite these
obstacles, neutrons offer some unique advantages. The main one is that they
permit **isotopic labeling**. One may significantly alter the scattering power
of an atom by neutrons if one substitutes a different isotope of the atom.
Such a change has virtually no effect on its behavior in a fluid. Thus, by
judicious attachment of different isotopes to different parts of a structured

[†] Recently it has proven possible to perform
dynamic scattering with x-rays in favorable
cases, despite the disadvantages noted in the
text [10].

[††] The apparatus is called a **spallation
source**.

fluid, e.g. a colloidal core, the surfactants around it, and the solvent, one may get much information about the spatial correlations of each component and about cross-correlations between components.

Neutrons also give dynamic information. If a neutron scatters off a moving object, its kinetic energy and speed are altered. The distribution of speeds in the scattered beam can be elegantly and sensitively measured by sending this beam through a magnetic field. The magnetic field causes the magnetic moment of the neutrons to oscillate with a specific frequency. The **neutron spin echo** technique uses these oscillations to infer motions of the scatterers in this frequency range. Frequencies between 10^6 and 10^9/s or so can be measured this way.

Waves and particles are used in another way to probe spatial structure near surfaces. Specifically, they can sense the profile of some atomic species with depth. The simplest of such method is to sense the reflectivity of a wave incident on the surface. The profile of scatterers results in a profile of index of refraction seen by the wave. This in turn shapes the reflectivity as a function of angle, especially near the total-internal-reflection angle. The method senses depths as small as a tenth of a wavelength and depths as large as 100 wavelengths. It is not trivial to infer the index-of-refraction profile from the reflectivity spectrum. One must assume various profiles and calculate their spectra until a spectrum matching the measured one is obtained. Sometimes two rather different profiles can give similar spectra so that interpretation of the data is ambiguous. A less ambiguous but more brute-force probe of surfaces is called **Rutherford backscattering** [11]. There the probe is a beam of massive particles like alpha particles, whose wavelength is negligible. When a mono-energetic beam is directed onto a surface, the particles backscatter elastically from the nuclei in the substrate, losing a specific amount of energy that depends on the mass of the nucleus hit. One may infer the depth of a backscattering event by the further energy loss of the emerging particle on its way in and out. This method requires a small room-sized particle accelerator. The method is routinely used to sense the profile of composition in solid samples to depths of the order of 10 nm.

2.3.3 Probes of atomic environment

There is more to be learned about a fluid than simply where the atomic species are and how fast they are moving. For many purposes it is valuable to know how parts of a molecule are oriented, how fast this orientation changes, what atomic species are near a given atom and how near. A variety of nuclear and electronic resonance methods are available to sense these things. **Magnetic resonance** senses the frequencies of rotation of atomic-scale magnetic moments in a large static magnetic field. When a small oscillating electric or magnetic field of the same frequency is applied, resonant absorption can be detected. **Nuclear magnetic resonance (NMR)** [12] senses the electric and magnetic field at a specific type of nucleus. The resonant frequencies are shifted or split by static fields. Time-varying internal fields can also limit the lifetime of the resonant state. An important example of NMR used in practice is to replace hydrogen by deuterium at specific points on chain molecules. Then by measuring the effect on the resonance line, one may sense the average orientation of that part of the chain in space. As Fig. 2.8 shows, the changes in the resonance can reveal subtle changes in the atomic orientations due to overall deformation of the fluid. One can sense the effect of a 10% stretch of a rubber in this way; likewise, one can sense the alignment of specific carbon–deuterium bonds in a surfactant monolayer.

Another class of local probes involves optical fluorescence of small dye molecules [14]. The dye molecule is chemically attached to the macromolecule to be probed. For example, fluorescence may be excited with a pulse of light and its intensity or polarization are monitored with time. The change of polarization senses the dye molecule's rotation from its initial orientation. The time dependence of intensity is of interest because the fluorescent decay can be strongly affected by the presence of quenching molecules. From the intensity decay one may infer the rate at which excited molecules encounter quenching molecules. By attaching the dye and quencher at suitable places, one may use this as a probe of mobility, and of where the dye molecule likes to sit. A variant of the quencher molecule is an excimer-former—usually a second dye molecule. **Excimers** are bound states of two dye molecules. The proximity of the second molecule shifts the fluorescence in a specific way. Fluorescence probes of this kind are very sensitive and flexible in their time scale. Temporal responses have been measured on scales ranging from fractions of nanoseconds to many seconds.

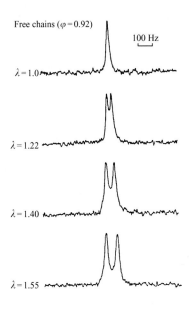

Free chains ($\varphi = 0.92$)

100 Hz

$\lambda = 1.0$

$\lambda = 1.22$

$\lambda = 1.40$

$\lambda = 1.55$

Fig. 2.8
Nuclear magnetic resonance spectrum of a rubber after [13]. The rubber is a polymer made of hydrogen, carbon, oxygen, and silicon (poly dimethyl siloxane) in which the hydrogen atoms have been replaced by deuterium atoms. The plot shows absorption versus frequency of oscillation. A strip of the rubber was stretched in the direction of the static field by various elongation factors λ shown at the left. The splitting of the resonant peak indicates a resulting slight anisotropy in the neighborhood of the deuters. The splitting is proportional to the amount of stretching. The frequency of the unsplit peak is 13 MHz. The bottom curve shows a splitting of about 100 Hz.

Solution to Problem 2.1

The outer part of the system (the "reservoir") has $N - k \equiv m$ balls in $U - V \equiv v$ sites. The number of distinct configurations with m balls $\sum_{\{d_m\}} 1$ will be called the "partition function" z_m. We wish to infer z_{m+1} given z_m. Any $m + 1$-ball configuration may be made from some m-ball configuration by adding one ball. Thus to count all $m + 1$ configurations we may consider each m-ball configuration and ask how many $m + 1$-ball configurations can be made from it.

(a) Whatever m-ball configuration we start from, there are exactly v $m + 1$-ball configurations that can be made from it, since every possible site to add the new ball results in a distinct configuration.

(b) Each of these $m + 1$-ball configurations could have been made from $m + 1$ different m-ball configurations. If we remove any one of the $m + 1$ balls, we get one m-ball configuration. If $v \gg m$, the chance of two balls at a site is negligible and so essentially all these m-ball configurations are distinct.

(c) Not all the configurations found in (a) are distinct: one can arrive at the same $m + 1$-configuration starting from many different m configurations. In fact, if we generate all the $m + 1$-configurations separately from each m configuration, we will end up generating each $m + 1$ configuration $m + 1$ times—once for every m configuration that could have given the $m+1$-configuration in question. Thus of vz_m configurations generated, a factor $m + 1$ are redundant: they are duplicates of others. The number of distinct $m + 1$ configurations z_{m+1} is thus $(v/(m + 1))z_m$. We recall that the desired probability $p(c_k) = (const)z_{N-k}$. Thus we have found that $p(c_{k+1})/p(c_k) = z_{N-k}/z_{N-k-1} = (N - k)/v$. Since $N \gg k$, this ratio becomes N/v and is independent of k, as we wished to show. (If the reservoir is not dilute, the ratio $p(c_{k+1})/p(c_k)$ changes. It becomes a function of the volume fraction N/U of the reservoir, because the number of

m-ball configurations obtained from an $m + 1$-ball configuration is no longer $m + 1$. Still, the $p(c_{k+1})/p(c_k)$ remains independent of k.)

(d)
$$p(c_k) = p(c_{k-1})(N/v) = p(c_{k-2})(N/v)^2 = \cdots$$
$$= p(0)(N/v)^k = p(0)e^{k \ln(N/v)}.$$

This can be written $p(c_k) = p(0)e^{-k/K}$ where $K = 1/\ln(v/N)$. Note that $p(c_k)$ is the probability of a *given* configuration of the subsystem. To find the overall probability that the subsystem has k particles, one must add up the probabilities for all the k-particle configurations of the subsystem.

References

1. See e.g. N. G. van Kampen, *Stochastic Processes in Physics and Chemistry* (Amsterdam, New York: North-Holland, 1992).
2. S. Glasstone, K. J. Laidler, and H. Eyring, *The Theory of Rate Processes*, (New York: McGraw Hill, 1941); H. A. Kramers, *Physica* **7** 284 (1940); S. Chandrasakhar, *Rev. Mod. Phys.* **15** 1 (1943).
3. See e.g. H. Gould and J. Tobochnik, *An Introduction to Computer Simulation Methods* (New York: Addison Wesley, 1988).
4. D. S. Pearson, *Rubber Chem. Technol.* **60** 439 (1987).
5. C. Morris and H. Coll, *Determination of Molecular Weight* (New York: Wiley and Sons, 1989).
6. Reprinted from Jacob N. Israelachivili, *Intermolecular and surface forces,* 2nd ed. (London; San Diego, CA: Academic Press, 1991). p272. © 1991 with permission from Elsevier.
7. R. Weisendanger and H. J. Guntherodt, *Theory of STM and Related Scanning Probe Methods*, Springer Series in Surface Sciences, V 29 (1993).
8. J. A. Zasadzinski, R. Viswanathan, L. Madsen, J. Garnaes, and D. K. Schwartz, *Science* **263** 1726 (1994).
9. B. J. Berne and R. Pecora, *Dynamic Light Scattering: With Applications to Chemistry, Biology, and Physics* (New York: Wiley, 1976); *Dynamic Light Scattering: Applications of Photon Correlation Spectroscopy*, ed. R. Pecora (New York: Plenum Press, 1985).
10. A. Q. R. Baron, H. Franz, A. Meyer, R. Ruffer, A. I. Chumakov, E. Burkel, and W. Petry, *Phys. Rev. Lett.* **79** 2823 (1997).
11. For this and related methods, see T. P. Russell, *Ann. Rev. Materials Sci.* **21** 249 (1991).
12. See e.g. V. D. Fedotov and H. Schneider, *Structure and Dynamics of Bulk Polymers by NMR-Methods*, NMR Basic Principles and Progress series, Vol. 21 (Heidelberg: Springer-Verlag, 1989).
13. B. Deloche and A. Dubault, *Europhys. Lett.* **1** 629 (1986).
14. See e.g. C. E. Hoyle and J. M. Torkelson, eds., *Photophysics of Polymers*, ACS Symposium Series Vol. 358 (Columbus, Ohio: American Chemical Society, 1987).

Polymer molecules

3

In this chapter we have our first in-depth encounter with a specific fluid structure: the flexible chain molecules known as polymers. The basis for much of their distinctive behavior lies in their flexibility and randomness. This chapter focus on the properties of individual molecules that embody this flexibility and randomness. The next chapter focuses on the further properties that emerge when large numbers of polymers interact in a liquid. There we will also treat the basic *motions* of a polymer solution: diffusion and flow. The Preface notes a number of other texts covering these polymer phenomena.

We begin this chapter with an informal introduction to some common monomers and the ways they can be joined to make a polymer chain. Next we use the idealized example of a random-walk polymer to demonstrate scaling and renormalization. The following section describes the interior structure of polymers, emphasizing their fractal properties and dilation symmetry, and showing how these features can be revealed by scattering experiments. Next, we study the impact of self-avoidance on the structure of a polymer. We find that self-avoiding polymers have fractal structure but the scaling exponents are altered by the self-avoidance constraint. We discuss another form of self-interaction that occurs when the polymer is electrically charged; this leads to dramatic changes in its structure.

3.1 Types of polymers

The simplest polymer one can imagine is a chain of carbon atoms each holding two hydrogen atoms: $CH_3-[CH_2-]_n CH_3$. This polymer, **polyethylene**, is also one of the most ubiquitous. The degree of polymerization n is typically 10^3–10^5. This is the material of the cheapest plastic bags. Polyethylene, more than other types of plastic, feels waxy. This is no co-incidence: polyethylene differs from paraffin wax only by having a larger value of n.

This polymer is the simplest in a rich variety of polymer types. In this section we survey the various types of polymers that exist, to give a sense of the range of molecules that can be made. First, I will discuss several common monomers and the properties associated with them. Then I will describe the main architectural variations in how monomers may be put together. Finally I will discuss the two main types of chemical reactions that produce polymers.

3.1.1 Monomers

[†] For reference, carbon in organic molecules has a valence of four; i.e., it makes four bonds to adjacent atoms. Two of these may go to the same atom, forming a double bond. Silicon also has a valence of four. Nitrogen has a valence of three. Oxygen and sulfur have a valence of two. Hydrogen has a valence of one.

[††] We return in Chapter 7 to a more quantitative discussion of molecular mixing.

[†††] A polymer is generally named after the monomer from which it is made. For example the ethylene monomer has four hydrogens surrounding two double-bonded carbons. In polymerization, the second carbon–carbon bond of ethylene becomes attached to the end of the polyethylene chain. Sometimes a polymer has two different names, because either of two monomers can be used to make it.

The simplest and most common polymers are **hydrocarbons**, i.e., molecules consisting entirely of carbon and hydrogen. It is useful to classify the different types according to their **polarity**, i.e., the degree of electric polarizability. The least polarizable hydrocarbons are the saturated hydrocarbons, where each carbon is bonded to four different atoms (i.e., there are no double bonds)[†]. The bonds in a saturated hydrocarbon are very stable, symmetrical, and difficult to deform. Small saturated hydrocarbons are oils and waxes. All these species mix well with each other but less well with the more polar species to be described below. They mix poorly with water[††]. **Polypropylene** is the main example besides polyethylene. It has a CH_3 group replacing one of the hydrogens on every second CH_2 group of polyethylene[†††]. Polyethylene and polypropylene are structurally simple enough to crystallize readily. This contrasts with another very common polymer, **polystyrene**, the material of cheap plastic spoons and styrofoam cups. Figure 1.8 shows a typical configuration of polystyrene. It is related to polypropylene by replacing each CH_3 group of polypropylene by a benzene C_6H_5 ring. The bulky rings make crystallization difficult; thus polystyrene solidifies in a glassy rather than a crystalline state.

Polystyrene is an **unsaturated** hydrocarbon. That is, it contains double bonds (between some carbon atoms and their neighbors). Some of these redundant bonds could be used to attach additional hydrogen atoms, thus "saturating" the molecule with hydrogen. The redundant bonding in polystyrene resides in the delocalized electrons in the benzene rings. These, like other double bonds, are more polarizable than simple bonds. This means that polystyrene (at high molecular weight) is not miscible in saturated polymers. Two other important unsaturated polymers are natural rubber or **polyisoprene**, and the synthetic rubber **polybutadiene**. Several common hydrocarbon polymers are sketched in Fig. 3.1.

The inclusion of elements beyond carbon and hydrogen makes for asymmetric bonds with carbon, and often increases polarity. For example, one may make a carbon chain with oxygen at every third position. The simplest case is $[-CH_2-CH_2-O-]_n$—**polyethylene oxide**[§]. Polyethylene oxide is sufficiently polar to be soluble in water. Polymer backbones can be made with no carbon at all. The most common case is **polydimethyl siloxane**. This is the material of silicone rubber, sealants, and lubricants. It has a backbone of alternating silicon and oxygen atoms. Each silicon has in addition two methyl (CH_3) groups. **Polyesters** such as Dacron[TM] have the generic structure $[-R-COO-]_n$, where the R is an arbitrary molecular substructure called a residue and COO is a carbon with a double-bonded oxygen and a single-bonded oxygen taking three of its four available bonds. The last bond is with the R group. The double-bonded oxygen may form weak **hydrogen bonds** with any covalently-bonded hydrogen present, e.g., in the R group. Another example is the **polypeptides**; all proteins are polypeptides. Their backbone has the form $[-N-C-C-]_n$, with an H taking the third bond of the nitrogen, an H and a variable R taking the remaining two bonds of the middle carbon, and a double-bonded O taking the two bonds of the last carbon. In many polypeptides (with favorable

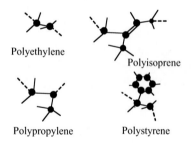

Polyethylene

Polyisoprene

Polypropylene

Polystyrene

Fig. 3.1
Four common hydrocarbon polymers. Black dots represent carbon atoms. Solid lines represent chemical bonds. A hydrogen atom is at the end of each bond where no carbon atom is shown. Dashed lines represent bonds from the preceding monomer in the chain. Variants of polybutadiene are shown in Fig. 3.3.

[§] also known as poly-oxy-ethylene, or poly-ethylene glycol (PEG).

R groups) this oxygen forms a hydrogen bond with the hydrogen several monomers down the chain, thus forming a helical structure called the **alpha helix**. The alpha helix is one example of weak, cooperative bonding within a chain which favors a certain configuration or crosslinking topology. This type of bonding is called **secondary structure** (in proteins) or **association** (in synthetic polymers). Secondary structure can often be disrupted reversibly without disrupting the primary, covalent bonds of the polymer.

Other types of monomers respond more strongly to electric fields than can be described by a polarizability. **Conjugated** polymers like **polyacetylene** have alternating single and double bonds along the backbone. Moreover, translating all the double bonds forward by one unit along the structure results in an identical structure. The position of the double bonds represents a broken symmetry. By doing this translation, one has transported an electron in each double bond one step along the chain. This means that such polymers may readily be made into electrical conductors. This requires doping, the removal or addition of a small fraction of the electrons. Conjugated polymers can have electron mobilities comparable to those found in metals. This conductivity can be made to occur in the liquid state and in crosslinked solutions [1].

Polymers can be synthesized with ionic side groups. A common example is **polystyrene sulfonate**. This derivative of polystyrene has an SO_4^- ion attached to each benzene ring. The neutralizing counterion is typically a metal like Na^+. In a polar solvent like water, the electrostatic binding energy of the counterions is smaller than the thermal energy T, and accordingly most of the ions dissociate and move freely in the solvent. The polymer is left with a large net charge. Such polymers are called **polyelectrolytes**. We will see later in this chapter that the charge has a dramatic effect on their shape and properties.

3.1.2 Architecture

In polymerization monomers like those catalogued above are joined end to end. The result is generally a flexible chain molecule, in which each single bond can rotate. That is, the bond acts like a rigid rod, with the atoms at the two ends attached as if on freely rotating bearings. The bonds of a carbon atom point symmetrically outward at fixed angles of about $109°$. A double bond may be regarded as a pair of rods; it does not allow rotation. Even in a flexible polymer chain certain aspects of the structure are locked in and unchangeable, over and above the fixed structure of the constituent monomers. This structure is determined at the moment each monomer joins the chain. Two types of locked-in structure are important: **isomerism** and **tacticity**. Isomerism concerns the orientation of the two ends of the monomer along the chain. Each monomeric link is attached to the chain at its two ends. In general the two are not equivalent, so that monomers may be attached "head to tail" in the same orientation, or "head to head" in opposite orientations. How much of this isomeric irregularity occurs depends on the details of the polymerization reaction. Of the polymers shown in Fig. 3.1, only polyisoprene has isomeric variation of this type.

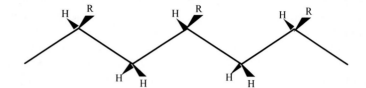

Fig. 3.2
A generic polymer chain with a C_2 repeat
unit along the backbone. Bonds directed
out of or into the backbone plane are shown
as triangles. The residues R are at the
isotactic positions; they are all on the same
side (the front) of the backbone plane.

Cis-trans isomerization refers to the atomic configuration around a pair of double-bonded carbons. An example is shown in Fig. 3.3.

Tacticity is a second way in which the structure of a polymer chain may be irregular. To demonstrate what tacticity means, we consider a polymer of the form $[-CH_2-CHR-]_n$, like the polypropylene of Fig. 3.1. We rotate the bonds so that the carbon backbone lies in the plane of the paper from left to right. Since the successive bonds are at 109°, tetrahedral angles, the backbone makes a zigzag line. We suppose the isomerism is head-to-tail, so that every other carbon—say the upper ones—have an R group. As shown in Fig. 3.2 the $C-R$ and $C-H$ bonds must stick obliquely out from and back from the page. An **isotactic** structure is one in which all the R groups lie on the same side of the backbone—i.e., all out of or all into the page. In a **syndiotactic** structure the R groups alternate along the backbone: every other one points out of the page. **Atactic** structures have a random positioning of the R's. Evidently tacticity is a permanent property of the chain structure; it cannot be changed by rotating bonds. Tacticity has a noticeable effect on the miscibility of polymers and on their ability to crystallize. It can be controlled to some degree by the polymerization conditions like temperature, choice of catalyst, etc.

More profound variations along the chain are those in which the monomeric precursors are chemically different. A common example is **polyethylene–propylene**, made by **copolymerizing** ethylene and propylene monomers. The sequence of monomers obtained is quite disordered, though not completely random. The disorder is sufficient to prevent the crystallization that occurs in ethylene or propylene homopolymer. Thus these copolymers remain liquid well below room temperature. Liquid polymers when crosslinked have the springy deformability of rubber. Polyethylene–propylene is one of the cheapest polymers and is used in great quantities as a rubber[†]. Nowadays it is possible to control the sequence of monomers in a copolymer in some ways. An important case is the **diblock** or **triblock copolymer**—two or three homopolymer chains joined end to end. When the two copolymerized species are mutually immiscible, dramatic self-organization like that shown in Chapter 1 can occur. We postpone our discussion of these phenomena to the "Amphiphilic polymers" section in Chapter 7.

Many variants of the linear-chain topology occur. Liquid polymers are converted to rubber by crosslinking: adding chemical groups (like sulfur) that join two chains together. Often chains branch as they polymerize. Nowadays, using favorable monomers, it is possible to build polymers with virtually any desired topology. This includes rings, stars with up to 60 arms, and H-shaped molecules [2].

[†] The commonest and cheapest rubber is styrene butadiene copolymer.

3.1.3 Polymerization

To create a polymer, one must find a chemical reaction mechanism that joins the monomers together. The organic chemistry of these reactions [3] is similar to that of small molecules, and we shall not discuss it here. But some important features of polymerization can be appreciated without organic chemistry. For example, statistical fluctuations are important in the many-body process of polymerization. These produce variations in isomerization and tacticity along the chain. More centrally, they produce variations of molecular weight from chain to chain. We discuss three types of polymerization to illustrate the possibilities.

The least variability occurs in **biosynthesis**, the natural process by which proteins are synthesized in living organisms. By recombinant DNA technology the natural processes can now be commandeered to produce peptide copolymers with the precise sequence desired [4]. These are virtually identical in structure and molecular weight from chain to chain. This kind of synthesis is still experimental as a means of producing synthetic polymers and is restricted to polypeptides and DNA-like **polysaccharides**.

Some of this molecular uniformity occurs in the important type of polymerization called **addition polymerization**. In this method a catalyst **initiates** polymerization in a solution of monomers. This turns a small fraction of the monomers into a chemically active form such as a free radical. An example is the polymerization of styrene monomers $CH_2 = CHR$. In the active form the double bond is broken and made reactive. This active form is able to combine with a second monomer. The activity of the first monomer propagates to the second. Thus it is able to react with a third, and so forth. Ideally many monomers are joined in sequence before the activity of the chain end is lost. Finally the reaction is terminated by deliberately adding a chemical that inactivates the active ends. Another example of addition polymerization is shown in more detail in Fig. 3.3.

The successive addition of monomers to the chains is a statistical process; still, it results in relatively uniform chains when these chains are long. The distribution of lengths $P(n)$ becomes relatively narrower as the mean chain length $\langle n \rangle$ gets larger. This results from the law of averages. We may see how this works by considering the time T required to make an n-mer. Evidently T is the time t_2 required to attach the second monomer to the first, plus the time t_3 for the third monomer, and so on up to t_n. Of course each of these t_i varies widely from chain to chain, owing to the statistical fluctuations in the system when the ith monomer was being added. These variations are reflected in the total time T, by the familiar law of propagation of errors: $\Delta T^2 = \sum \Delta t_i^2$. The variations of the t_i are roughly the same for all i, so that $\Delta T^2 \simeq n \Delta t_n^2$. Thus $\Delta T \simeq \langle T \rangle n^{-1/2}$; the relative variation in T becomes progressively smaller as n increases.

This increasing uniformity contrasts with the results of **condensation polymerization**. Here instead of each chain having a single active end that reacts only with monomers, many chain ends may react with one another. An example is the **polyamide** nylon, in which any two chain ends may join, producing a water molecule as a by-product (Fig. 3.4).

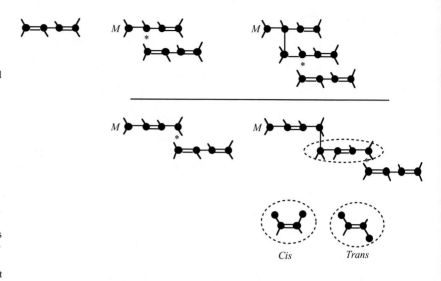

Cis   *Trans*

The kinetics of this reaction involves many processes contributing to the change of the n-mer concentration $[n]$. For simplicity we assume that all n-mers are equally reactive with one another. Each reaction of i-mer with j-mer to produce n-mer is a second-order reaction, whose rate is proportional to the amount of each reactant present, i.e., to $[i][j]$. Considering all these reactions that affect the number of n-mers, we have:

$$\frac{d[n]}{dt} = K \left(\frac{1}{2} \sum_{k=1}^{n-1} [k][n-k] - [n] \sum [k] \right). \tag{3.1}$$

$[n]$ increases owing to the reaction of $(n-1)$-mers with 1-mers, $(n-2)$-mers with 2-mers, etc. $[n]$ also *decreases* owing to the reaction of existing n-mers with anything to produce longer chains. We have supposed that all these reactions occur at some uniform rate governed by the single constant K. This assumption is justified under certain conditions.

It happens that this formidable-looking set of coupled rate equations has a simple solution, as one may readily verify. It is exponential for long times and large $\langle n \rangle$: $[n] \rightarrow (constant) \exp(-n/\langle n \rangle)$. Thus the relative width of the molecular weight distribution does not become narrower with time, in contrast to addition polymerization.

A third category of polymerization is called **living polymerization**. Here the polymerization reaction is reversible, so that the various chains are free to exchange monomers amongst themselves. This freedom to exchange results in a broad distribution of chain lengths, much like that of condensation polymers. We shall return to living polymers in Chapter 7, in our treatment of wormlike micelles. In the sequel we shall encounter many reasons why the molecular weight distribution is important.

3.1. *Polycondensation*   In a polycondensation reaction, the reaction rate of k-mers with l-mers was proportional to the product of their concentrations $[k][l]$. Thus,

H—O—C—C—C—C—C—O—H

(chemical structure with H atoms and O)

Fig. 3.4
A polycondensation reaction. An amine NH_2 end reacts with an alcohol COH end by eliminating the circled atoms to form a water molecule, thus joining the two chains together. The polymers thus produced are called nylon 6-6.

considering all the reactions which increase or decrease the number of n-mers, the rate of increase of the concentration $[n]$ obeys Eq. (3.1) in the text, where K is an overall rate constant. In analyzing this equation, you may assume that n, k, and l are everywhere large enough that they may be considered as continuous variables running from 0 to ∞ instead of discrete variables running from 1 to ∞. (a) Show that the total mass $\sum n[n]$ is independent of time, as it must be. (b) Show that there is a solution of the form $[n](t) = [0](t)\exp(-nc(t))$, where c and $[0]$ are some function of t. (Since $\sum n[n] = [0]/c^2$, $[0]$ must be $(const)c^2$.) How does c vary with t? (c) How is c related to $\langle n \rangle \equiv (\sum n[n])/(\sum[n])$?

3.2 Random-walk polymer

All of the chain polymers described in the last section have some randomness in their spatial arrangements. This randomness is least in a self-associating species like a protein. Proteins are often frozen into a specific configuration with a specific size and shape. The randomness in each atom's position then consists of small excursions from its average position, like the vibrational fluctuations in a solid. Much greater configurational randomness occurs when a non-self-associating molecule like polystyrene is dispersed in a good solvent like toluene. In this section we explore the results of this randomness. Here there is no specific size and shape; instead, a statistical description is necessary. We shall also explore the characteristic energy scales of these polymers—i.e., the work required to deform them from their natural random state.

We may begin to explore the spatial randomness of a polymer by looking at the smallest polymer: a four-carbon segment of polyethylene. A fundamental geometric property to examine is the distance between the two terminal carbons. This distance is not definite because the carbon–carbon bonds are free to rotate;[†] the only restriction is that they maintain the 109° angle between successive bonds. We may assume that the first two bonds lie in the plane of the paper and that the first bond is horizontal. Then the third bond does not in general lie in this plane. As this bond swings around into its various possible positions, the end-to-end distance changes. The maximum end-to-end distance occurs when the third bond is parallel to the first in a *trans* configuration; the minimum occurs when this bond is rotated half-way around, into a *cis* configuration Fig. 3.5. Thus there is a probability distribution in the end-to-end distance r. There is also a distribution of angles between the first and third bonds, ranging from 0 degrees to $2 \times 109°$.

A fourth bond adds to this randomness. For each configuration of the three-bond chain, the fourth bond may swing around a complete circle of

[†] In fact not all rotations are equally likely because of subtle chemical interactions between adjacent bonds. But such subtleties may be ignored for our purposes.

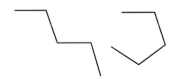

Fig. 3.5
Two configurations of a four-bond hydrocarbon polymer with maximal and minimal end-to-end distances. The left-hand chain is all *trans*; the right-hand chain is all *cis*.

angles as the third one did. Here the last bond may point in any direction relative to the first, and these directions are nearly equally likely. With the addition of another bond or two, the directional randomness of the last bond is nearly complete. To discuss the randomness in a long polymer, it is convenient to imagine that its building blocks are segments which are several atoms long, so that the directions of the beginning and the end of each segment may be assumed statistically independent[†].

Another feature appears when our hydrocarbon chain exceeds five bonds in length: it may hit itself. There are some rotational configurations of the bond angles that would place one carbon atom on top of another. Clearly these configurations are absent in real polymers: the polymer must avoid itself. This requirement is clearly separate from the requirements on the successive bond angles or **local structure**. Local structure can be expressed in terms of random variables (e.g. our bond angles) that are statistically independent for each bond or small segment. The **self-avoidance** constraint cannot be expressed in this way. A major topic of this chapter is to analyze the effects of self-avoidance. For the moment, we shall ignore it. We shall depart from the real world and imagine **phantom** polymers which are free to intersect themselves. After their behavior is understood, we will be in a position to attack the constraint of self-avoidance.

3.2. *Law of Averages for* $\langle r^2 \rangle$ A polymer in one dimension consists of n vectors $\vec{a}_1, \vec{a}_2, \ldots, \vec{a}_n$ of length 1 that point randomly left or right. Each vector is independent of the others. (a) Find $\langle \vec{a}_5 \rangle$, $\langle \vec{a}_5 \cdot \vec{a}_5 \rangle$, and $\langle \vec{a}_4 \cdot \vec{a}_5 \rangle$. (b) Using averages such as these, and noting that $\vec{r} = \sum_i \vec{a}_i$, find $\langle \vec{r} \rangle$ and $\langle (\vec{r})^2 \rangle$ as a function of n.

3.2.1   End-to-end probability

Although the end-to-end distance of our chain varies over a wide range, there must be some way to characterize it statistically, even for very long chains. We are thus led to look at the limiting behavior of the end-to-end probability distribution $p(n, r)$, defined as the probability that the end-to-end vector has some particular value $\vec{r}$ in a chain of n segments. We normalize $p(n, r)$ according to $\int d^3r\, p(n, r) = 1$, so that p is a probability-per-unit-volume.

We may define how this $p(n + 1, r)$ must behave for all r if we know $p(n, r)$ for all r. This is because any $n + 1$-segment chain can be thought of as an n-segment chain with a single segment joined to its end. Here as indicated above, we shall think of each segment as a sequence of several atoms. We suppose that the sequence is long enough so that the directions of its two ends are independent. That means that the segment itself has an end-to-end probability $p_0(r_1)$ that depends only on the magnitude of $\vec{r}_1$. It also means that this $p_0(r_1)$ is unaffected by the arrangement of the n-segment chain attached to it. In that case, we may regard the occurrence of an $n + 1$ chain with vector $\vec{r}$ as the joint occurrence of two independent events: (1) last segment has some vector $\vec{r}_1$, and (2) the first n-segment part of the chain has the vector $\vec{r} - \vec{r}_1$. The probability of this joint occurrence is $p(n, \vec{r} - \vec{r}_1) p_0(\vec{r}_1)$. To obtain the full $p(n + 1, r)$ we must add the probabilities for

[†] In some cases this directional randomness cannot be assumed. An important one is the case when the backbone is rigid rather than flexible. Important examples are DNA and RNA helices and straight-chain hydrocarbons like polyacetylene. Here each segment has nearly the same direction as its predecessor. The direction can only become random by the addition of many of these random small increments over a long length of chain. Chains shorter than this **persistence length** but much longer than a monomer have a universal statistical behavior more complicated than that of the random-walk polymers treated in the text. Rigid chains in this regime are called **wormlike chains**. [5].

all occurrences that give the vector $\vec{r}$:

$$p(n+1,r) = \int d^3r_1 \, p_0(r_1) p(n, \vec{r} - \vec{r}_1).$$ (3.2)

This is a clear-cut prescription for determining p for arbitrary n given p of a single segment. Without some specific input regarding $p_0(r)$ this seems to be as far as we can go. On the other hand, intuition tells us that we should be able to say *something* further about $p(n,r)$ by asking for only part of the full behavior implicit in Eq. (3.2): namely, the limiting behavior for arbitrarily large chain-length n. This problem is archetypical of the quantitative issues in this book. We will be repeatedly interested in how some quantity behaves when some parameter measuring the system size goes to infinity.

Accordingly, we try to simplify Eq. (3.2) in the limit of large n. The equation tells the effect of incrementing n by 1. When n becomes large, the effect on p must become small. We may examine the small change in p directly by subtracting $p(n,r)$ from each side of Eq. (3.2).

$$p(n+1,r) - p(n,r) = \int d^3r_1 \, p_0(r_1)[p(n, \vec{r} - \vec{r}_1) - p(n,r)].$$ (3.3)

The small difference on the left can evidently be treated as a derivative. The same should be true for the spatial difference of p's in brackets on the right. For large n we expect that the typical distances r for which p is significant are much greater than the distances r_1 for which the integrand has a significant contribution. Thus the difference in [...] involves relatively small spatial displacements r_1. Over such small distances, we expect $p(n,r)$ to be gently varying. Thus we use a Taylor expansion to represent this variation:

$$p(n, \vec{r} - \vec{r}_1) = p(n,r) - \vec{r}_1 \cdot \nabla p(n,r) + \tfrac{1}{6} r_1^2 \nabla^2 p(n,r) + \cdots$$ (3.4)

Using these derivatives, Eq. (3.3) becomes

$$\frac{dp}{dn} = -\nabla p \cdot \int d^3r_1 \vec{r}_1 \, p_0(r_1) + \frac{1}{6}\nabla^2 p(n,r) \int d^3r_1 r_1^2 \, p_0(r_1) + \cdots .$$ (3.5)

We note that the first integral is the average of r for a one-segment chain: $\langle \vec{r} \rangle_1$ and the second is $\langle r^2 \rangle_1$.

We have taken a case where the step distribution $p_0(r)$ is isotropic; thus the first integral must vanish. The same is true with all odd powers and their associated odd-order derivatives. Thus our equation for p reduces to

$$dp/dn = \tfrac{1}{6}\langle r^2 \rangle_1 \nabla^2 p(n,r) + const. \ \text{``}\nabla^4\text{''}p + \cdots .$$ (3.6)

Here we have indicated the form of the first nonvanishing term beyond the ∇^2 term; the "∇^4" denotes various fourth-order derivatives.

We can deduce much about this equation without solving it, by investigating its asymptotic behavior as $n \to \infty$. But it will not do simply to evaluate $p(\infty, r)$. We would find that this is zero for all r. We are in the paradoxical situation of taking n arbitrarily large, but not infinite. Since it is arbitrarily large, the specific value of n cannot be important. In the regime of interest the behavior of p is the same whether we consider n, $n/1000$, or $1000n$. That is, p should show the same behavior when n is multiplied by an arbitrary factor λ. By assuming such a limiting behavior, we are implicitly postulating some kind of *symmetry*; namely, invariance under multiplicative scaling of n.

We anticipate that a change of n by some factor λ preserves the *shape* of $p(n, r)$, but not its specific value. That is, $p(\lambda n, r)$ should differ from $p(n, r)$ by a mere change of scale on the r axis and the p axis. We express the r scale factor as $\mu(\lambda)$ and the p scale factor as $\eta(\lambda)$. Then our anticipated λ variation takes the form $p(n, r) = \eta(\lambda) p(\lambda n, \mu(\lambda) r)$.

We may use Eq. (3.6) to decide whether $p(n, r)$ indeed has an asymptotic behavior of this form. Substituting the proposed form,

$$\frac{d\eta p(\lambda n, \mu r)}{dn} = \frac{1}{6} \langle r^2 \rangle_1 \nabla^2 \eta p(\lambda n, \mu r) + const. \text{``}\nabla^4\text{''} \eta p(\lambda n, \mu r) + \cdots .$$

$$(3.7)$$

To see the effect of λ and μ on the equations, we define the rescaled λn as $\tilde{n}$, and the rescaled μr as $\tilde{r}$. Similarly, $\eta p \equiv \tilde{p}$. We may readily express Eq. (3.7) in terms of these **renormalized** variables. Evidently, $d/dn = \lambda d/d\tilde{n}$ and $\nabla_r^2 = \mu^2 \nabla_{\tilde{r}}^2$. Likewise, any fourth-order derivative satisfies $\nabla_r^4 = \mu^4 \nabla_{\tilde{r}}^4$. In these new co-ordinates,

$$\frac{d\tilde{p}}{d\tilde{n}} = \frac{1}{6} \langle r^2 \rangle_1 \frac{\mu^2}{\lambda} \nabla_{\tilde{r}}^2 \tilde{p}(\tilde{n}, \tilde{r}) + const. \frac{\mu^4}{\lambda} \text{``}\nabla_{\tilde{r}}^4\text{''} \tilde{p} + \cdots . \qquad (3.8)$$

We may now freely take n to be arbitrarily large; we simply select a compensating scale factor λ so that $\tilde{n}$ remains finite. Evidently, $\lambda \to 0$ in this limit. In order for Eq. (3.8) to remain finite in this limit, μ must evidently change with λ. If the ∇^2 term is to remain finite, we must have $\mu^2/\lambda = constant$. With this choice, the ∇^4 term has a factor $\mu^4/\lambda = \lambda$. The ∇^4 term is arbitrarily small, so that it can be neglected. Such quantities that become arbitrarily small when an asymptotic limit is taken are termed **irrelevant variables**. We may clearly extend this reasoning to show that all higher orders in ∇ are irrelevant in this same way[†]. Likewise, just as the ∇^2 term involved a moment $\langle r^2 \rangle_1$, one can readily show that the ∇^4 terms have factors of the form $\langle r^4 \rangle_1$. By our reasoning these moments of $p_0(r)$ are also irrelevant for the asymptotic $p(n, r)$. Indeed, the only feature of $p_0(r)$ that *is* relevant is $\langle r^2 \rangle_1$[††]. By similar reasoning we may determine the scale factor η for p. A simple method is to note that $\int d^3 r p(n, r) = 1$: the probability that the end monomer is *somewhere* is a certainty. Inserting the scaling form and noting that $\int d^3 r = \mu^{-3} \int d^3 \tilde{r}$, we infer $\eta = \mu^3 = \lambda^{3/2}$.

The requirement that n and r must scale together means that the asymptotic $p(n, r)$ is really only a function of one variable. For any given n and r, we are free to set $\lambda = 1/n$. Then recalling that $\mu \propto \lambda^{1/2}$,

[†] Technically, the n derivative on the left side of our equation is also an approximation. Like the spatial derivatives on the right side, it should be replaced by a full Taylor expansion. But the higher terms not shown here may be readily shown to be irrelevant like the higher spatial derivatives.

[††] One may ask why we chose $\mu(\lambda)$ to make the *first* term in our equation finite. By choosing $\mu \propto \lambda^{1/4}$ we can make the second term finite instead. But this choice does not lead to a consistent equation. Now the first term has a factor $\lambda^{-1/2}$. Instead of being irrelevant, the first term diverges, dominating the one we have rendered finite. We must thus reject this choice of μ.

$p(n,r) = n^{-3/2} p(\tilde{n}, \tilde{r})|_{\tilde{n}=1, \tilde{r}=rn^{-1/2}}$ [†]. By use of the scaling symmetry, we have reduced our unknown function of two variables, $p(n,r)$, to an unknown function f of a single variable:

$$p(n,r) = n^{-3/2} f(rn^{-1/2}).$$

For any $\vec{r}$ in the n step polymer, the corresponding $\vec{r}$ in the λn step polymer is $\lambda^{1/2}$ times as big. From this scaling law, we can readily determine how any average r characterizing our polymer must scale with n. The most common average to consider is the second moment $\langle r^2 \rangle \equiv \int d^3 r\, r^2 p(n,r)$. Using the scaling law we can treat the arbitrary p'th moment:

$$\langle r^p \rangle_n = \int d^3 r\, n^{-3/2} r^p f(rn^{-1/2}),$$

$$= \int d^3 (rn^{-1/2})[(rn^{-1/2})^p n^{p/2}] f(rn^{-1/2}),$$

$$= n^{p/2} \int d^3 \hat{r}\, \hat{r}^p\, f(\hat{r}), \tag{3.9}$$

where the dummy variable $\hat{r}$ stands for $rn^{-1/2}$. Since the integral is independent of n, we infer that any measure of the polymer size of the form $[\langle r^p \rangle]^{1/p}$ scales the same with n:

$$[\langle r^p \rangle_n]^{1/p} = K_p n^{1/2}. \tag{3.10}$$

The constants K_p are numerical factors related to the $\hat{r}$ integral in Eq. (3.9).

This same type of scaling appears whenever we do mathematics with dimensioned quantities like mass, length, and time. Here we know that the number representing, say, the time in our formulas has no significance in itself. The formulas must give the same physical result whether we use one unit of time or some other unit λ times smaller. Thus the number representing time can be multiplied by an arbitrary factor λ without changing the physical result of the formula. This is the familiar requirement of dimensional consistency: the answer must be unchanged when all quantities are scaled in accordance with their time dimensions by an arbitrary factor λ. Mathematically, this is a requirement of **homogeneity**. If we express our formula in the form $f(t, a, b, c, \ldots) = 0$, where t is the time variable, then dimensional analysis amounts to the following invariance property: for each physical quantity a, b, c, etc. in our formula there are exponents α, β, γ, etc. such that

$$f(\lambda t, \lambda^\alpha a, \lambda^\beta b, \lambda^\gamma c, \ldots) = 0. \tag{3.11}$$

The exponents, usually integers or simple fractions, are the time dimensions of the quantities a, b, c, etc. Though n in our polymer has no physical dimensions, it has something in common with a dimensioned variable: namely, the invariance of the results under changing the scale factor λ. The resulting symmetry is the same homogeneity familiar in dimensional analysis: thus asymptotic quantities like n behave as though they had dimensions. This will prove true in the more subtle forms of scaling to be treated later.

[†] Here $p(1, \tilde{r})$ means the asymptotic function p evaluated for $\tilde{n} = 1$. It does not mean the monomeric $p_0(r)$ from which $p(n,r)$ was constructed.

Returning to our polymer, the asymptotic equation now has a simple and familiar form. It is the heat equation, the equation that describes how a diffusing substance spreads out with time[†].

$$dp/dn = 1/6\langle r^2\rangle_1 \nabla^2 p(n,r),\qquad(3.12)$$

Here we have set $\lambda = \mu = 1$ now that they have served their purpose. One may readily verify that the solution is a Gaussian:

$$p(n,r) = [2\pi n\langle r^2\rangle_1/3]^{-3/2}\exp\left(-\frac{3}{2}r^2/(n\langle r^2\rangle_1)\right).\qquad(3.13)$$

[†] The polymer scaling is familiar in the context of diffusion. It is the counterpart of the fact that the distance a diffusing substance spreads is proportional to the square root of the spreading time.

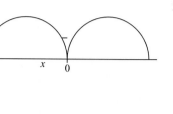

3.3. *Toy renormalization* In the text we deduced a way of inferring polymer behavior by analyzing the scaling properties of the equation without solving it. This problem is meant to show the same procedures in the familiar context of classical mechanics. Here we wish to find the limiting properties of a system's motion as the energy of a particle goes to zero. The potential energy function $U(x)$ of a point particle has the form of two semicircles side by side: $U(x) = U_0\sqrt{1 - (1 - |x|)^2}$. A particle with energy $E \ll U_0$ is trapped in the region near $x = 0$. It is obliged to oscillate with some period τ. The object is to find how the period τ varies as the energy $E \to 0$. (a) Write conservation of energy for the particle in the region $x \ll 1$. Notice that in this region $U(x) \simeq U_0\sqrt{2|x|}(1 - \frac{1}{4}|x|)$. (b) In order to find the limiting behavior as $E \to 0$, define $\tilde{E} = \lambda E$ so that $\tilde{E}$ will stay finite when $E \to 0$ when expressed in terms of the ˜variables. We wish to find a corresponding time scale $\tilde{t} = \mu t$ and distance scale $\tilde{x} = \eta x$ such that the equation of motion will stay finite when $\lambda \to \infty$. (c) Restate the conservation of energy from (a) in terms of the ˜variables. (d) How must μ and η vary with λ in order for the terms in the equation to stay finite? (e) How do the maximum displacement in x and the temporal period τ vary with E as $E \to 0$? (f) If μ and η vary with λ as found in (b), does the correction in $\frac{1}{4}|x|$ become small or become large?

This function describes the end-to-end distance of long polymers. Remarkably, the microscopic structure of the chain, given by $p_0(r)$, has little impact on $p(n,r)$; the only relevant feature of this $p_0(r)$ was its mean-square average $\langle r^2\rangle_1$. Since virtually any $p_0(r)$ leads to the same asymptotic $p(n,r)$, we say that the Gaussian $p(n,r)$ is **universal**.

It is worth noting that this Gaussian is not the asymptotic $p(n,r)$ in every sense. To see this, we might consider our hydrocarbon polymer for very large extensions r. The polymer has a maximal extension where the backbone bonds form a coplanar, zigzag line. Beyond this distance $p(n,r)$ is strictly zero, so that the Gaussian form is qualitatively wrong for such values of r. This r_{max} is evidently proportional to n, so that it scales entirely differently from $\sqrt{\langle r^2\rangle_n}$. This illustrates the point that asymptotic scaling behavior depends on the question asked. In determining the Gaussian scaling, we asked for the behavior of $p(n,r)$ for "typical" r values with appreciable probability. The r_{max} is extremely atypical; the probability that r is any finite fraction of r_{max} is exponentially small in n, as one may verify using Eq. (3.13). For our purposes in this book, we shall be interested in typical polymer configurations, so that the Gaussian scaling is the right regime to consider. But there are some questions of physical importance where the Gaussian treatment is not sufficient. The exercise on the **coil-stretch transition** gives an example.

3.4. *Short lattice chains*   A four-step polymer in one dimension consists of left-
and right- directed bonds of unit length. (a) How many configurations of the
chain have the end four units to the right of the beginning? How many have
the end three steps to the right? Two? One? Zero? You may find it easiest to
draw a picture of all 2^4 configurations. (b) Find $\langle r^4 \rangle / \langle r^2 \rangle^2$ for this polymer
and compare with 3, the corresponding ratio for a one-dimensional Gaussian
chain.

Knowing the probability $p(n, r)$ gives us information about the energy
needed to distort the polymer. We recall that any system with a given relative
probability $f(r)$ had a free energy given by $-T \log f$. We saw that this free
energy is equivalent to a potential energy: if r is changed from one definite
value to another the work required is the change in this free energy. For our
polymer, the free energy

$$U(r) = -T \log(p(n, r)) = \frac{3}{2} \frac{T r^2}{\langle r^2 \rangle}. \tag{3.14}$$

The polymer stores energy like an ideal spring with zero unstretched length.
As with the spring, the force required to elongate is proportional to the
distance. In magnitude, the energy U is of order T for typical extensions
$r \simeq \langle r^2 \rangle^{1/2}$. This spring energy appears in many forms in polymer liquids. It
is responsible for the rod-climbing behavior of Fig. 1.6. It is also responsible
for the microphase separation patterns seen in block copolymers, like that
of Fig. 1.1.

3.5. *Coil-Stretch Transition*   An **elongational flow** field may be made by aiming
two round nozzles at each other underwater, then sucking water out through
both nozzles at the same rate. Any polymers at the center of this contraption
are pulled towards both nozzles at once. The effect is roughly equivalent to a
force pulling on each end, of magnitude bz^2, where z is the distance from the
center along the nozzle axis. The coefficient b is proportional to the elongation
rate $\dot{\gamma}$. One can detect the amount of elongation z induced in the polymers by
the flow. (a) What energy $W(z)$ would a free particle starting at the origin gain
if it moved to z under the action of this force? This energy is *part* of the energy
of any polymer ending at z in the presence of the flow. (b) Denote the $\langle z^2 \rangle$ of
the polymer without flow as Z^2. Treating the polymer as a spring, as explained
in this section, sketch the work $U(z)$ required to elongate the chain to length
z as a function of z for small b, counting both the elastic free energy of the
polymer and energy in (a), and write its functional form. (c) This energy $U(z)$
attains a maximum for some elongation z^* and energy U^* (relative to its $z = 0$
value). How does the height of this maximum vary with b, when b is small? If
the height U^* is adjusted to be the thermal energy T, what is the corresponding
position z^* of this maximum in terms of Z? (We have seen that this maximum
amounts to an *activation barrier*. When it is higher than about T, passage over
the barrier becomes very slow.) (d) If this polymer were in thermal equilibrium
at temperature T in the presence of this energy $U(z)$, find $\langle z \rangle$ to lowest order
in b. In this case $U^* \gg T$ and $|W(z)| \ll T$ for all z's that matter. According to
this $U(z)$, once z exceeds z^*, it will increase to infinity. In reality z increases
only to the maximum extension r_{max} of the polymer. This maximum extension
is not accounted for in the Gaussian $p(n, r)$, as the text points out.

3.6. *Series and parallel polymers*   Two identical random-walk polymers are
fastened together end to end, to make a chain of double the length of each.
(a) What is the root-mean-squared end-to-end distance $\sqrt{\langle r^2 \rangle}$ of the two
free ends relative to that of a single polymer? (b) The same two polymers

are fastened together at both ends to make a double chain. What is the root-mean-square end-to-end distance between the two ends in this parallel configuration? *Hint:* One may construct the ensemble of double chains by letting the two chains end anywhere, and then discarding all configurations except those in which the two chains end at the same point. By this procedure, one can relate the $p(n, r)$ of the double chain to that of the two single chains.

3.3 Interior structure

For our future work we need to know much more about the polymer than its end-to-end distance. We need to characterize the whole distribution of the monomers in space. The most fundamental quantity is the average density $\bar{\rho}$ within the chain. If R is the radius of a sphere containing on average say half of the chain, then the average density is roughly n/R^3. We expect that R is of the same order as a typical end-to-end distance such as $\langle r^2 \rangle_n^{1/2}$. Using this estimate, we see that $\bar{\rho} \sim n/n^{3/2} \sim n^{-1/2} \sim R^{-1}$. Remarkably the average density becomes smaller and smaller as the chain length increases: the polymer is **tenuous**. A major issue in the pages to follow is whether such a tenuous object can have appreciable effects on the **pervaded volume**[†]. We will be concerned with effects of solvent flow and of interactions with other polymers. We shall see that these tenuous objects have remarkably strong effects. There is clearly room within the pervaded volume for arbitrarily many polymers, provided each is long enough. Thus we expect these polymers to be able to interpenetrate and entangle strongly. Another issue we face is to characterize this entangled state spatially, energetically, and dynamically.

[†] The pervaded volume of an irregular object means the spherical region enclosing that object, or the volume of that region, as noted in Chapter 1.

We may begin to see how the chain comes to have the average density $\bar{\rho}$ by examining the *local* density around a given point on the chain. For this purpose we define $\rho(r)$ as the monomer number density at r in a given configuration. For polymers passing through the origin, there is some **average local density** $\langle \rho(r) \rangle_0$ at distance r. The subscript 0 stands for the restricted average in which only chains passing through the origin are included. We expect that this local density falls off as the distance from the origin increases. Using the asymptotic behavior of $p(n, r)$ we may readily find this dependence.

If a monomer passes through r, it must be some distance i along the chain from the monomer at the origin. The probability that this monomer is at r is $p(i, r)$. The average density at r is just the sum of these probabilities for the possible monomers i: $\langle \rho(r) \rangle_0 = 2 \sum_i p(i, r)$. The factor 2 accounts for monomers i steps ahead or i steps behind the origin monomer. The r dependence may be readily found by using the scaling property of p: $p(i, r) = i^{-3/2} f(r^2/i)$ This density is finite even in an infinite chain. For large i and r where the asymptotic behavior is reached, we may replace the sum on i by an integral:

$$\langle \rho(r) \rangle_0 = 2 \int_0^\infty di \; i^{-3/2} f\left(\frac{r^2}{i}\right). \tag{3.15}$$

Defining a scaled variable $\tilde{i}$ as i/r^2, we find

$$\langle\rho(r)\rangle_0 = 2r^2 r^{-3} \int_0^\infty d\tilde{i}\, \tilde{i}^{-3/2} f(\tilde{i}^{-1}). \qquad (3.16)$$

The integral is finite: at small $\tilde{i}$, the integrand is exponentially small because f is small for large argument; at large $\tilde{i}$ f is a constant, but the integrand still falls off as an integrable power of $\tilde{i}$. The integral is simply a constant independent of r, so that $\langle\rho(r)\rangle_0 = (const)/r$.

This power-law dependence of the local density means that our random-walk polymer is a fractal set[†] in the sense of Mandelbrot [6]. That is, the average mass (number of monomers) $\langle M(r)\rangle$ within distance r of an arbitrary point of the set varies as some power r^D. Then this power is called the fractal dimension. This $\langle M(r)\rangle$ can be found for the polymer by integrating the local density: $\langle M(r)\rangle = \int_{r'<r} d^3r' \langle\rho(r')\rangle_0 = (const)r^2$. A random-walk polymer has a fractal dimension of two, like a uniform surface. Clearly, this property generalizes: any fractal with dimension D has a local density $\langle\rho(R)\rangle_0$ that falls off as R^{D-d} in any spatial dimension $d > D$. This is discussed in Problem 3.7.

[†] That is, the polymer approaches the fractal behavior as one approaches asymptotic conditions of large chains and long distances.

3.7. *Why the fractal dimension D is called a dimension* It was said in this section that in a fractal object, the number of points of the object within a given distance R of a given point grows as R^D. Show that this property holds for a straight line or a plane in three-dimensional space. (a) What is the fractal dimension D for these two objects? (b) What happens to the two D values when the line and the plane are embedded in four dimensions?

A fractal set of points necessarily contains large empty spaces. If the M particles of a fractal of radius R were uniformly dispersed in that volume, the result would be a cloud of particles of volume per particle R^d/M. The distance from an arbitrary point to the nearest particle is roughly the volume-per-particle to the $1/d$, i.e., $R/M^{1/d} \sim R^{1-D/d}$. But a fractal arrangement of points is far from uniform and the average distance from an arbitrary point to the fractal is much larger than this distance. Indeed, the average distance is of order R. This openness of fractal structures will prove very important in how they influence their environment.

This openness is an obvious property of the simplest fractal, a straight line segment. If we enclose the line in a sphere, and then choose a point within the sphere at random, the distance to the line is typically a sizeable fraction of the radius R. (The average distance is $(3\pi/16)R$ in three dimensions.) To see that this is true for general fractals, we consider the volume within distance r of a fractal with mass M. For this purpose we enclose all the points of the fractal within spheres of radius r. Each sphere contains on average $M(r)$ points. To enclose all the points requires roughly $M/M(r)$ spheres. The volume $V(r)$ enclosed by the spheres is of order $r^d(M/M(r))$. Since $M/M(r) = M(R)/M(r) = (R/r)^D$, we conclude that $V(r) \simeq R^D r^{d-D}$. The volume containing all the points lying within a distance r from the fractal is larger than $V(r)$ by only a finite factor. Thus the volume of the region that is a distance between r and $r + dr$ from the fractal is roughly $dr(dV/dr)$. Using this fact, we can express the average distance

$\langle r \rangle$ between some arbitrary point and the fractal as

$$\langle r \rangle \simeq \frac{\int_a^R dr \, r \, dV/dr}{\int_a^R dr \, dV/dr} \simeq \frac{R^{d-D+1}}{R^{d-D}} \simeq R.$$

This is what we wanted to show. Fractal structure implies tenuousness, but tenuousness does not imply fractal structure. Our cloud is as tenuous as our fractal, yet the cloud has qualitatively smaller open spaces.

The fractal property of polymers is the sign of a deeper and more powerful property called **dilation symmetry**. Appendix A of this chapter discusses this symmetry. Further geometrical implications of fractals in solution are discussed in [7].

3.3.1   Scattering

[†] The subject of scattering from condensed matter is discussed in detail in e.g. Stephen W. Lovesey **Theory of neutron scattering from condensed matter** (Oxford, Clarendon Press, 1984). The book by Jannink and Des Cloizeaux mentioned at the beginning of this chapter also has an extensive discussion of neutron and light scattering.

The behavior of the local density $\langle \rho(r) \rangle_0$ is directly observable by scattering[†]. We saw in Chapter 1 that scattering senses the size of an object: small objects diffract the incoming waves like a slit diffracts light. Scattering senses not only the overall size of a structure but also the internal distribution of matter within it, as we now show.

In a scattering experiment a plane wave or beam of scattering particles (photons, neutrons, etc.) impinges on an unknown object, such as liquid containing polymers. For definiteness we will imagine that our beam particles are neutrons: the waves are then complex scalar Schrödinger waves. The beam wave has the form $A \exp(i\vec{k} \cdot \vec{r} - i\omega t)$. The constant A is the amplitude of the wave; its square is the number of neutrons per unit volume in the beam. The angular frequency ω is proportional to the kinetic energy of the neutrons. The wave vector $\vec{k}$ is proportional to their momentum. Evidently, one can change k and ω together by varying the speed of the neutron beam. But one cannot vary k and ω independently.

When this wave encounters an atomic nucleus, such as a carbon nucleus in our solvent liquid, the great majority of it continues unperturbed. But a tiny fraction radiates outward from the nucleus in a spherical wave of the form $B \exp(i \, |k| \, r)/r$. Of course doubling the number of neutrons in the beam doubles the number scattered in this spherical wave. Thus B is proportional to A; the proportionality factor is the (complex) scattering amplitude f. A detector receives the wave far away in some particular direction (Fig. 3.6). The part of this spherical wave entering the detector is a plane wave with some wave vector $\vec{k}'$ of magnitude k. The detector measures the intensity of the wave as a function of the direction of $\vec{k}'$. The r^{-1} factor in the scattered wave is virtually the same distance for all the scatterers, since the detector is far away from the sample.

The detector senses the scattered waves from all the M scatterers j in a given volume. The amplitude from a particular scatterer at r_j is $B/r \exp(\vec{k}' \cdot (\vec{r} - \vec{r}_j))$. As we have noted, the prefactor B must be proportional to the wave function of the incident wave: $B = f A \exp(i\vec{k} \cdot \vec{r}_j - i\omega t)$. The wave entering the detector is the sum of these contributions. The **intensity** I is proportional to the absolute square of this wave. Since the scatterers in our

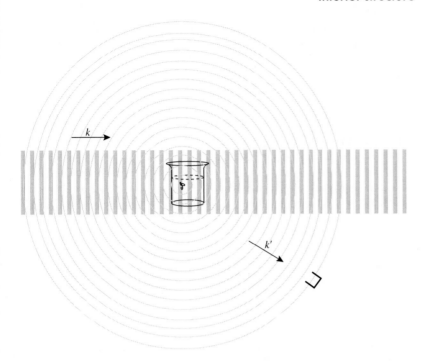

Fig. 3.6
Schematic view of neutron scattering from
a solution containing a polymer. The
detector is shown and the wavevectors $\vec{k}$
and $\vec{k}'$ are indicated.

liquid are generally in motion, the intensity I fluctuates in time. To obtain a time-independent characterization of the system, we consider the average intensity. This intensity grows with the amount of sample; to remove this uninteresting size dependence, we consider the average intensity per scatterer $\langle I \rangle / M \propto \langle |\psi|^2 \rangle / M$.

$$\frac{\langle I \rangle}{M} \propto |A\,f|^2 \frac{1}{M} \left\langle \left| \sum_j \exp(i(\vec{k}' - \vec{k}) \cdot \vec{r}_j) \right|^2 \right\rangle |\exp(i\vec{k}' \cdot \vec{r} - i\omega t)|^2 / r^2.$$

(3.17)

The last factor, with the r and t dependence, is an unimportant constant, as is the prefactor giving the dependence on beam intensity and intrinsic scattering power of the nuclei. The rest gives information about the structure. Evidently it depends on $\vec{k}$ and $\vec{k}'$ only through the difference $\vec{k}' - \vec{k} \equiv \vec{q}$[†].

This remaining factor is called the **structure factor** $S(q)$[††]. We may simplify it by expressing the squared sum as an explicit product of two sums:

$$S(q) = \frac{1}{M} \sum_{j,k} \langle \exp(i\vec{q} \cdot \vec{r}_j) \exp(-i\vec{q} \cdot \vec{r}_k) \rangle,$$

$$= \frac{1}{M} \sum_{j,k} \langle \exp(i\vec{q} \cdot (\vec{r}_j - \vec{r}_k)) \rangle.$$

(3.18)

One important property of the scattering intensity is immediately apparent from this expression. The spatial information in $S(q)$ is strongly blurred. The degree of blurring can be expressed in terms of the **scattering**

[†] Evidently q may be changed simply by varying the angle θ between the incident beam and the scattered rays that are detected. We are considering a case where the scattering particles lose no energy in the (elastic) scattering process, so that $|k'| = |k|$. Then some simple geometry shows that $|q| = 2|k|\sin(\theta/2)$.

[††] Sometimes it is necessary to consider the q dependence of the scattering amplitude f from each elementary scatterer. This q dependence reflects the distribution of matter within an elementary scatterer and is called the **form factor**. In some discussions of scattering, the elementary scatterers are taken to be an entire polyatomic structure such as a polymer. In that case, much of the information in the $S(q)$ above would appear in the form factor. The separation of the intensity of Eq. (3.17) into form and structure factors is evidently somewhat arbitrary. In our discussion we include all spatial information in the structure factor.

wavelength λ defined as $2\pi/q$. To see this, we imagine that the scatterers i were moved by arbitrary small displacements u_j. The effect would be to add a new factor $\exp(i\vec{q}\cdot(\vec{u}_j-\vec{u}_k))$ in Eq. (3.18). But if all the u_j were much smaller than the scattering wavelength λ, these new factors would all be nearly unity, and $S(q)$ would hardly change. That is, $S(q)$ cannot detect anything about the positions of the scatterers to a precision of much less than a wavelength λ. We could for example gather all the scatterers lying within a tenth of a wavelength of one another into bunches and the effect on $S(q)$ would be minor. Conversely, it is immaterial whether the scattering occurs from discrete scatterers at specific points r_j or from a smeared distribution of scattering spread over a small regions around each r_j—provided the region is much smaller than $2\pi/q$. Scattering is *myopic* within a scale $2\pi/q$. If $2\pi/q$ is larger than the entire collection of scatterers, all the phase factors in Eq. (3.18) are unity and $S(q)\to M$.

We may use our density language to simplify the expression in Eq. (3.18). The density $\rho(r)$ is the probability per unit volume that one of the scatterers is at r. If the scatterers are at positions $\{r_j\}$ the density is given by $\rho(\vec{r})=\sum_j \delta^3(\vec{r}-\vec{r}_j)$. Inserting $\int d^3 r_1 \delta^3(r_1-r_j)\ (=1)$ and $\int d^3 r_2 \delta^3(r_2-r_k)$, we may rearrange Eq. (3.18) to read

$$S(q)=\frac{1}{M}\int d^3 r_1 \int d^3 r_2 \langle \rho(r_1)\rho(r_2)\rangle \exp[i\vec{q}\cdot(\vec{r}_1-\vec{r}_2)]. \qquad (3.19)$$

The absolute location of r is clearly immaterial in the $\langle\rho\rho\rangle$. Both points may be shifted by $\vec{r}_2$ so that $\langle\rho(\vec{r}_1)\rho(\vec{r}_2)\rangle=\langle\rho(\vec{r}_1-\vec{r}_2)\rho(0)\rangle$. Now we note that the integrand depends only on $\vec{r}_1-\vec{r}_2$, which we denote by $\vec{r}$:

$$S(q)=\frac{\int d^3 r_1}{M}\int d^3 r \langle \rho(\vec{r})\rho(0)\rangle \exp[i\vec{q}\cdot\vec{r}]. \qquad (3.20)$$

The quantity $\langle\rho\rho\rangle$ is called the **density correlation function**. This $\langle\rho\rho\rangle$ is simply related to the local density defined in the previous section. We may think of it as the joint probability that a monomer is at the origin and that another is at $\vec{r}$. This is the overall probability that a monomer is at the origin times the *conditional* probability that a second monomer is at $\vec{r}$ given that one is at 0. The probability that a monomer is at the origin (or any other point in the sample) is $\langle\rho\rangle=M/(\int d^3 r)$. The conditional probability is the same as the local density $\langle\rho(r)\rangle_0$. Thus $\langle\rho(\vec{r})\rho(0)\rangle=\langle\rho\rangle\langle\rho(r)\rangle_0$[†]. The relation between $S(q)$ and the local density is simple:

$$S(q)=\int d^3 r \langle \rho(r)\rangle_0 \exp(i\vec{q}\cdot\vec{r}). \qquad (3.21)$$

The reduced scattering intensity $S(q)$ is related to the local density by a simple integral transform—the Fourier transform. This means that if one expresses the scattering density $\rho(\vec{r})$ as a sum of plane waves of all different wavevectors, the scattering intensity and $S(q)$ arise from those waves that have wavevector $\vec{q}$. If the local density is *constant* the $S(q)$ is simply zero for $q>0$.

[†] The density correlation function also contains the same information as the pair correlation function $g(r)$ introduced in Chapter 2. We recall that $g(r)$ means the average density of particles at displacement $\vec{r}$ from a particle, relative to the average density $\langle\rho\rangle$. Thus the connection is $g(r)=\langle\rho(r)\rangle_0/\langle\rho\rangle=\langle\rho(r)\rho(0)\rangle/\langle\rho\rangle^2$. Of the three quantities, the local density is most convenient for discussing fractal structure, because it avoids unimportant factors of average density.

In principle each nucleus in the fluid sample produces scattering. But in practice the scattering from a pure fluid is very weak for wavevectors q of interest here. We argued in the last chapter that structure at a given spatial scale R causes scattering at wave vectors q of order $1/R$. Thus we shall be interested in very small wave vectors relative to the size of an atom. We have seen above that the scattering from these atoms would not be appreciably changed if they were replaced by smeared-out densities several atoms in width (since such smearing is much less than $1/q$). The result in a uniform fluid is an essentially constant density. We have also seen that a perfectly constant density causes no scattering. It is thus not surprising that a simple fluid causes little scattering[†].

Any departure from this uniform density of scatterers gives rise to scattering intensity. The polymers in the fluid do this: their atoms have scattering amplitudes f that differ from those of the solvent, so that there is a contrast between the scattering per unit volume from regions that contain monomers and regions that do not. Thus, we may treat the scattering as though only the polymers were present. Whenever a liquid contains independent large objects of typical size R, the behavior of $S(q)$ depends on the size of q relative to the R. Evidently, the local density falls to zero for distances $r \gtrsim R$. When $qR \ll 1$ then $\exp(iq \cdot r) \to 1$ for all $r < R$. Thus,

$$S(q) \to \int^{R} d^3 r \langle \rho(r) \rangle_0 1 \to M. \qquad (3.22)$$

† At these small wave vectors the scattering from the fluid is independent of q and can readily be subtracted. As discussed in Problem 4.2, the amount of scattering can be shown to be proportional to the *compressibility* of the fluid relative to an ideal gas at the same density. All familiar liquids are relatively incompressible; this leads to weak scattering.

That is, the internal structure is completely invisible to $S(q)$ and the scattering is about the same as though all the scatterers were concentrated at a point. Our polymers show this behavior like any other large object. It is straightforward to see how $S(q)$ begins to vary when q grows comparable to $1/R$; this is discussed in an exercise.

In the complementary limit where $qR \gg 1$ the $S(q)$ senses the internal structure of the object. If this object is a fractal made of units of size a, as our polymer is, we may readily infer how it scatters in this limit. As discussed above, a fractal of dimension D in three-dimensional space has a local density of the form $\langle \rho(r) \rangle_0 = c\, r^{D-3}$ for $a \ll r \ll R$. For such power laws it is easy to perform the integral in Eq. (3.21) by defining $\tilde{r} \equiv |q|\vec{r}$. This yields

$$S(q) = c|q|^{-D} \int^{|q|R} d^3\tilde{r}\, \tilde{r}^{D-3} \exp(i\vec{q} \cdot \tilde{r}/|q|). \qquad (3.23)$$

The integral in $\tilde{r}$ is independent of $|q|$. It cannot depend on the direction of $\vec{q}$ since $\langle \rho(r) \rangle_0$ is independent of direction. Less obviously, it is finite for $|q|R \to \infty$ as long as $D - 3 < 0$ (it can be expressed in terms of Euler Γ functions involving D). Thus the integral is a numerical constant, and all the q dependence comes from the power in front. The same is true for any spatial dimension $d > D$. We recall that the average number of points $\langle M(r) \rangle$ within distance r of a fractal grows as r^D. Thus in scattering from a fractal object $S(q) \sim M(\lambda)$, where λ is the scattering wavelength $2\pi/q$.

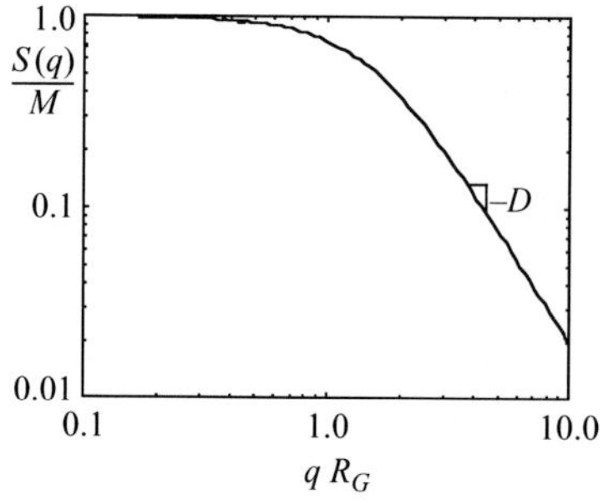

Fig. 3.7
Schematic log–log plot of $S(q)$ versus q
for a fractal set of $M \gg 1$ scatterers with
fractal dimension D. (In fact, this is the
exact structure function for a random-walk
polymer, as requested in Problem 3.7. The
wave vector q is given in the combination
qR_G, where R_G is the radius of gyration
discussed in Problem 3.11.)

The scattering is the same as though the fractal were chopped up into pieces the size of a scattering wavelength.

Figure 3.7 summarizes the expected scattering from a fractal. As q increases from $1/R$ to the inverse scatterer size $1/a$, the scattering intensity drops by a factor of M, the number of scatterers in the fractal. Remarkably, the $S(q)$ from a random-walk polymer can be calculated explicitly as a check on these ideas.

3.8. *Scattering from a random-walk polymer* It was shown in the text that the structure factor $S(q)$ of a set of n scatterers at positions r_j is $1/n \sum_{j,k=1}^{n} \langle \exp[iq \cdot (r_j - r_k)] \rangle$. Recall that the probability distribution $p(i, r)$ for a segment of length i to have its ends at displacement r can be expressed in the product form with three factors for the x, y, and z directions: $p(i, r) = A(i) \exp[-x^2/(2i\langle x^2 \rangle_1)] \times [x \to y] \times [x \to z]$, where the prefactor A assures normalization: $\int_r p(i, r) = 1$. (a) From this and the fact that $\int dx \exp(-x^2) \exp(iqx) = \pi^{1/2} \exp(-q^2/4)$, derive an expression for $S(q)$ in closed form. Note that the number of monomer pairs separated by a distance m monomers is not the same for all m. There is only one such pair for $m = n$, but there are $n - 1$ pairs when $m = 1$. If one has a sufficient range of q accessible experimentally, one may discern two **scaling regimes** in this $S(q)$: for $q^2\langle x^2 \rangle_n \ll 1$, $S(q) \sim q^0$; for $q^2\langle x^2 \rangle_n \gg 1$, $S(q) \sim q^{-2}$. To get a feeling of what range of q is needed, plot $\log(S(q))$ versus $\log(q)$ for (b) $0.3 < q\langle x^2 \rangle_n^{1/2} < 3$, (c) $0.03 < q\langle x^2 \rangle_n^{1/2} < 30$, (d) $0.001 < q\langle x^2 \rangle_n^{1/2} < 1000$. The scales should be adjusted so that the three plots are of roughly equal size. What range of q would be needed in your judgment to discriminate between the expected q^{-2} scaling and a hypothetical $q^{-1.8}$ scaling?

3.9. *Scattering from a line* The text shows that the scattering intensity at large wave vector q from a fractal object was proportional to k^{-D}. (a) Find explicitly the structure factor $S(q)$ from a line of length $2L$ oriented along the z-axis between $-L$ and L. The scattering wave vector q makes an arbitrary angle θ with the line. (b) For a line at random orientation, show that the scattering is proportional to q^{-1} for large q, as it should be for a one-dimensional fractal. Note that in three dimensions an angular average of a function $f(\theta)$ amounts to $\int_0^{-1} d(\cos\theta) f(\theta)/(4\pi)$.

3.10. *Scattering from a Gaussian Cloud* Since the monomers in a polymer chain are randomly arranged throughout a volume of limited size R, it is suggestive to approximate the polymer as a mere cloud of n monomeric scatterers confined to a region of this size. A simple realization would be a Gaussian cloud, whose local density $\rho(r)$ is given by $nCR^{-3} \exp(-(r/R)^2)$. How does the scattering intensity from such a cloud vary with wave vector q? Does it have the proper fractal behavior at large q? A useful integral is $\int_{-\infty}^{\infty} dx \exp(iqx - x^2) = \pi^{1/2} \exp(-q^2/4)$.

3.11. *Radius of gyration* The structure factor $S(q)$ of any object composed of n scatterers may be expanded in powers of the wave vector q: $S(q) = n(1 + aq + bq^2 + \cdots)$. (a) What are the coefficients a and b, expressed in terms of the positions r_j? Assume that the scatterers are isotropically arranged so that the scattering is the same in all directions. Use the power series expansion for e^x: $e^x = 1 + x + \frac{1}{2}x^2 + \cdots$. (b) The **radius of gyration** of any set of n particles is defined as the root-mean-square distance between an arbitrary pair of them: $R_G^2 \equiv n^{-2}\frac{1}{2}\sum_{jk}(\vec{r}_j - \vec{r}_k)^2$. Find an expression for R_G in terms of the scattering coefficients a and b. (c) For a random-walk polymer, how is the radius of gyration related to the mean-squared end-to-end distance $\langle r^2 \rangle_n$?

(d) How is R_G related to the root-mean-square distance of a particle from the center of mass, $n^{-1} \sum_j r_j^2$?

3.4 Self-avoidance and self-interaction

Our treatment of polymer configurations to this point has ignored the liquid environment of the molecules. A polymer coil must nestle intimately among the small molecules of its solvent. Likewise, the polymer chain must not intersect itself. The valid, self-avoiding polymers we want to study are of course a subset of the random-walk polymers of the last section. Our interest is to learn how much the properties of this subset differ from those of the full set. We shall focus on the spatial distribution of monomers, as exemplified by the mean squared end-to-end distance $\langle r^2 \rangle$. The subset of self-avoiding polymers is obtained from the full set of random-walk polymers by examining each member of the full set and discarding each configuration that has self-intersections. (This is a rather crude simplification of how equilibrium polymer configurations occur in practice. Certainly chains having two carbon atoms within 0.05 nm of each other are not present. But other closely self-approaching configurations may be partly excluded or even enhanced, depending on the details of the monomer and solvent interactions. We shall return to such refinements later. For the moment, we shall regard the self-avoidance as an all-or-nothing proposition. For example, we might exclude all configurations in which two atoms are separated by less than 0.1 nm.)

This discarding process can of course affect the end-to-end probability $p(n, r)$. The discarding must be quite dramatic in order to have a qualitative effect on the size of the chains remaining. Suppose we remove all but a fraction $t(r)$ of the chains ending at r. If t is a constant, the average size of the remaining chains is of course affected not at all: t must depend on r. If the self-avoiding chains are to be much longer than the random walks, the fraction remaining must decrease strongly as r decreases. We might imagine, for example, that the t fraction goes as a positive power of r: $t(r) \sim r^a$. Then,

$$\langle r^2 \rangle_{SA} \sim \int d^3 r \, t(r) p(n, r) r^2 \sim \int d^3 r \, r^{a+2} p(n, r). \qquad (3.24)$$

Since we know the gaussian form of $p(n, r)$, this integral can be worked out explicitly (along with the needed normalization). Since $p(n, r)$ falls off faster than any power for large r, all moments $\langle r^a \rangle^{1/a}$ are of order $\langle r^2 \rangle_{RW}^{1/2}$. (The reasoning is the same as for the $P(M)$ distribution discussed in Appendix A.) Thus any power-law dependence of the remaining fraction t would leave the average size of the self-avoiding chains basically unaffected. They would increase by only a finite factor. Conversely, if the self-avoiding chains are to be *arbitrarily* larger than the random-walk chains, t must vary more strongly than a power law. Clearly, the overall fraction remaining from the initial random walks must be indefinitely small.

Our aim in this section is to understand what types of contacts are important and how much they alter the shape of the polymer.

3.4.1   Local and global avoidance

It is instructive to look at several different types of contacts in turn, starting with *local* contacts. By local contacts I mean contacts between nearby monomers along the chain, within some fixed distance i. It is not hard to deduce the asymptotic effect of such local contacts. We imagine a random-walk chain in which only these local contacts are excluded. The asymptotic $p(n, r)$ for such chains can be found for any exclusion range i. To show what happens, we may simplify a bit further, and divide our chain into blocks of length i. We then exclude only contacts within a given block. We can immediately treat this model using Eq. (3.2). We need only build up our chain from blocks of length i rather than from single segments. Since successive blocks are statistically independent, we may again express the overall probability $p(n, r)$ in the form of Eq. (3.2). Our former reasoning holds and leads to the conclusion that $\langle r^2 \rangle \sim n$. We may also estimate how much the local contacts reduce the number of allowed configurations. The remaining fraction t_i in each block is necessarily finite. For a two-block chain any surviving configuration from the first block may be combined with any from the second, so that the surviving fraction is the product of the fractions from the two blocks. For n monomers or n/i blocks the overall remaining fraction t_n is evidently given by: $t_n = (t_i)^{n/i}$. The surviving fraction is exponentially small in n. The same conclusions hold if instead of using our self-avoiding blocks we exclude all contacts within a fixed distance i of one another. Local contacts can change the chain size by only a finite factor, and thus they cannot change the scaling behavior of chain size with chain length.

At the other extreme are *global* contacts. By these we mean contacts whose distance along the chain is a fixed fraction of the total length n. For example, we may consider the effect of contacts between the first third of the chain and the last third. These two mutually avoiding tails act to elongate the midsection of the chain. To see the potential importance of these contacts, we consider chains in which the two tails occupy the same pervaded volume: two spheres drawn around each tail interpenetrate significantly. We now ask whether the two tails are likely to have mutual contacts, leading to discarding these configurations. Contacts between two fractals such as these will be important in several contexts. In the general case we have two fractals A and B of radius R placed randomly in the same volume. A contact means a point of fractal A that lies within some small sphere of volume v surrounding some point of fractal B.

Fig. 3.8

Scheme used in Eq. (3.26) for counting contacts between two fractals in a sphere of radius R. The fractals are the first and last third of a random-walk polymer.

We may find the average number of contacts by using the local density of points within each. We pick some point from each fractal, and let one of these be at the origin. The displacement $\vec{x}$ between the two chosen points is of order R. The probability per unit volume of a contact at some point $\vec{r}$ is then $\langle \rho_A(r) \rangle_0 \, v \langle \rho_B(\vec{r} - \vec{x}) \rangle_0$, as shown in Fig. 3.8. The average number of contacts M_{AB} is then the integral of this probability over the pervaded volume. We recall that the local density $\langle \rho_A(r) \rangle_0 \sim r^{D_A - d}$, in d

dimensions. Thus,

$$M_{AB} \sim \int d^d r\, r^{D_A - d} (\vec{r} - \vec{x})^{D_B - d}. \qquad (3.25)$$

Again we simplify the integral using a scaled variable $\tilde{r} \equiv \vec{r}/R$ to give

$$M_{AB} \sim R^{D_A + D_B - d} \int_0^1 d^d \tilde{r}\, \tilde{r}^{D_A - d} (\tilde{r} - \vec{x}/R)^{D_B - d} \sim M_A M_B / R^d.$$
$$(3.26)$$

The integral is finite, since the divergence at $\tilde{r} = 0$ and $\tilde{r} = \vec{x}/R$ are integrable. It is thus a harmless constant, which we may safely ignore. Thus the average number of contacts varies as a power of the size R of the region. This power may be either positive or negative. For our two random-walk polymer tails $D_A = D_B = 2$, and the power is negative whenever these walks live in a spatial dimension d larger than four. Then the average number of contacts between two such polymers placed randomly into the same volume becomes indefinitely small as the polymers are made large. The probability of contact goes to zero. We say that two such polymers are mutually **transparent** in spatial dimensions larger than four. Evidently any two fractals are mutually transparent when $D_A + D_B < d$. One may readily confirm this result for simple fractals like lines or planes placed at random in a d-dimensional box. For our two tails in spatial dimensions greater than four, the fraction of configurations thrown away because of contacts becomes negligible, even when the two tails occupy the same volume. These contacts thus have negligible impact on the behavior of such polymers.

For $d < 4$ the two tails are mutually **opaque**; the average number of contacts grows indefinitely with R. Thus the likelihood of a contact between the two tails becomes significant[†]. We may think of each tail as a sphere of radius R which repels the other sphere. Clearly this prevents the two ends of the middle third from coming as close together as they would otherwise. Still, for a significant fraction of positions of these two ends, the tails do not overlap and their mutual repulsion does nothing. Thus the tails increase the average end-to-end distance of the midsection by a no more than a finite factor. Global contacts are not sufficient to make a qualitative change in the chain size.

In fact it is necessary to consider all types of contacts, from the most local to the most global, to account for the expansion due to self-avoidance. To get a feeling of how this comes about, we imagine a simple chain on a lattice with no self-avoidance. For convenience we take its chain length to be a power of two: $n = 2^k$. We first divide the chain into $n/2$ blocks of length two, and discard configurations with contacts within a block. Next, we group pairs of these blocks to make $n/4$ blocks of length four, and discard contacts within each of these. Then we do the same for blocks of length eight, and so on. Before each stage of this process we have a random chain of many self-avoiding blocks. We then impose mutual avoidance between each block and one of its neighbors. Considering these two blocks in isolation, the contacts involved are mostly global ones. As we have seen, these increase the size of the pair of blocks by finite factor b.

[†] The probability of contact between two mutually opaque fractals behaves in a distinctive way that is different from (a) randomly distributed clouds of particles and (b) transparent fractals or clouds. If $\tilde{r} \equiv x/R$ remains fixed as $R \to \infty$, the probability of no contacts goes to a finite limit between 0 and 1.

This distinctive behavior occurs because the points of a fractal are statistically correlated. For definiteness, we consider $\tilde{r} = \frac{1}{2}$. For any given $\tilde{r}$ and R there is a probability distribution of number of contacts: $P_{\tilde{r}}(M)$. The probability of no contacts is $P_{\tilde{r}}(0)$. The average $\langle M \rangle_{\tilde{r}}$ is $\int dM\, M\, P_{\tilde{r}}(M)$. This $\langle M \rangle_{\tilde{r}}$ can be expressed as $(1 - P_{\tilde{r}}(0)) \langle M \rangle'_{\tilde{r}}$. The new average $\langle M \rangle'_{\tilde{r}}$ is the average number of contacts for all configurations where there is at least one contact. This average can be found by choosing e.g., the contact closest to the center of the R sphere as the origin and then integrating $\langle \rho_A(r) \rangle_0 \langle \rho_B(r) \rangle_0$. This is legitimate because the tails A and B are completely arbitrary except that they touch at the origin. The result is just what we mean by $\langle M \rangle_{\tilde{r}}$ with $\tilde{r} = 0$. Thus our equation for $\langle M \rangle_{\tilde{r}}$ amounts to

$$\langle M \rangle_{\tilde{r}} = (1 - P_{\tilde{r}}(0)) \langle M \rangle_0.$$

For $\tilde{r} = 0$ the $P_{\tilde{r}}(0)$ must vanish, as the formula shows. But as $\tilde{r}$ increases towards 1, $\langle M \rangle_{\tilde{r}}$ decreases by a finite factor, while $\langle M \rangle_0$ does not. Thus, $P_{\tilde{r}}(0)$ must be finite for $\tilde{r}$ fixed and less than 1.

If the fractal particles are dispersed into two uniform clouds, there are no correlations, $\langle M \rangle'_{\tilde{r}} \to \langle M \rangle_{\tilde{r}}$ and thus $P_{\tilde{r}}(0) \to 0$ as $R \to \infty$.

When all the blocks of a chain swell by a factor b the overall size of the chain must increase by the same factor, since the chain is still a random walk of the expanded blocks. We have argued above that the expansion factor b must be finite irrespective of the number of monomers involved. Thus the expansion factors for all stages k of our process are essentially the same[†]. Therefore the self-avoiding size R_{SA} expands by a factor of order b^k relative to the random-walk size, R_{RW}. Recalling that $R_{RW} \sim n^{1/2}$ and noting that $b^k = 2^{k \log_2 b} = n^{\log_2 b}$, we conclude that $R_{SA} \sim n^{\nu}$, where $\nu = \frac{1}{2} + \log_2 b$ is a power larger than $\frac{1}{2}$. This argument says that self-avoidance alters the scaling relationship between chain length and size. The exponent ν describing the new scaling is called the **Flory swelling exponent**.

In random-walk polymers, the scaling of the overall size R reflected a pervasive dilation symmetry for the internal mass distribution. The same thing proves to be true of self-avoiding polymers. First, the end-to-end distribution function $p(n, r)$ should have a scaling property analogous to that of the random walk, since it represents a limit of indefinitely large n. The scaling form must give average end-to-end distances proportional to n^{ν}, and must preserve normalization. Thus $p(n, r) = n^{-\nu d} f(r/n^{\nu})$. If we consider a finite piece of length n within an infinite self-avoiding polymer, its mean size is of the same order as that same piece in isolation. That is, the attachment of an infinite tail to each end of the piece extends it by only a finite factor [8][††], as with the finite tails discussed above. This scaling law can be used to infer the local density $\langle \rho(r) \rangle_0$, in the same way done for the simple random-walk polymer in Eq. (3.16). The result is $\langle \rho(r) \rangle_0 = (constant) \, r^{1/\nu - 3}$. Since $\langle \rho(r) \rangle_0$ falls off as a power of r, $\langle M(r) \rangle$ is also a power of r. Repeating our reasoning for the random walk, we find $\langle M(r) \rangle = (constant) \, r^{1/\nu}$. Thus the self-avoiding polymer is a fractal whose fractal dimension $D = 1/\nu$[†††].

3.4.2 Estimating D

It is possible to calculate D systematically using sophisticated methods of renormalized field theory [9]. These methods rely on a deep correspondence between a polymer chain and a liquid at a second-order phase transition. The books of De Gennes and of Jannink and Des Cloizeaux deal extensively with this correspondence. In practice, the systematic calculations require a detailed accounting of contacts between segments of random walks, and they convey little insight about how the exponent D arises.

A much simpler argument to estimate D is based on a qualitative comparison of free energies involved in expanding a chain. It was first propounded by the eminent mid-century polymer chemist Paul Flory [10]. The Flory approach considers the work done on the system in imposing the self-avoidance constraint. If this work depends on the size R of the chain, that means the probabilities of various sizes R will be altered. For definiteness, we shall take R to be the root mean square end-to-end distance: $R = (\langle r^2 \rangle)^{1/2}$. If a chain is to expand to a size $R \gg R_{RW}$, work must be done to increase its elastic free energy. One can achieve

[†] The initial blocks, since they involve short segments, may have different expansion factors b from the majority. But the longer the chain is and the more stages k of subdivision are present, the less these initial blocks can matter.

[††] The radius of gyration grows by a factor of about 10%, compared with the same chain without tails: *cf.* [8], Chapter XV.4, Table 8.

[†††] It is natural to ask whether a self-avoiding polymer has the full dilation symmetry of the random walk, as discussed in Appendix A. The answer is yes, though the proof is too complicated to give here [8]:

$$\langle \rho(r_1), \ldots, \rho(r_k) \rangle_0$$
$$= \lambda^{-k(D-d)} \langle \rho(\lambda r_1), \ldots, \rho(\lambda r_k) \rangle_0.$$

The case $k = 1$ gives the local density we have just analyzed. Here we may set $\lambda = 1/r$ and infer $\langle \rho(r) \rangle_0 = r^{(D-d)} \langle \rho(1) \rangle_0$. Happily, this agrees with the result we just derived. The density products $\langle \rho(r_1), \ldots, \rho(r_k) \rangle_0$ behave similarly to those of a random-walk polymer. We may find the scaling properties of this product by looking at chains that pass through the labeled points in a specific order, such as the order listed. If we ignore self-avoidance except within the 0-to-r_1 segment, the r_1-to-r_2 segment, etc. we can express the density in the product form $\sum_{i_1, \ldots, i_k} p(i_1, r_1), \ldots$ used for the random walks. Using the scaling form of the p's, one finds a scaling property under dilation analogous to that for the random walk. If one now includes the remaining effects of avoidance among these segments, the densities are reduced, but only by a finite factor [8].

the needed expansion by pulling on the chain ends. Then the work S required is the free energy of extension, calculated in the last chapter: $S = \frac{3}{2}TR^2/R_{RW}^2 = \frac{3}{2}R^2/(a^2n)$, where a is the size of a monomer. Such expansion is favored because it reduces the number of contacts. To account for these contacts, we imagine that contacts are allowed, but at a cost of a small energy U per contact. That is, we replace the mutually self-avoiding monomers by weakly self-repelling ones. As we have seen, local contacts, despite their large number, are not important for the chain expansion. The number of *nonlocal* contacts may be found by imagining that the chain is a dispersed cloud of n monomers confined to the region of size R. The number of contacts is n times the probability that a given monomer has a contact. This probability is the monomer volume v times the monomer density n/R^d. Combining these facts, the overall energy V due to these contacts is given by $V \simeq vUn^2/R^d$. Asymptotically (if $d < 4$ and the chain is opaque to itself) there are indefinitely many contacts and the contact energy grows much larger than the thermal energy T; this is enough free energy to account for substantial elastic stretching.

3.12. *Simulating a self-avoiding polymer*   The structure of self-avoiding polymers is not known analytically. Often the most direct way to get quantitative information about such polymers is simply to simulate them. Because the asymptotic properties of these polymers are independent of their local construction, one can readily simulate a self-avoiding polymer using simple lattice methods. A simulation is a computational procedure in which a system is repeatedly subjected to small changes. These changes are devised so that the probability of a given configuration of the system is the same as in an equilibrium system. In this problem we develop such a simulation. The polymer will consist of a sequence of nearest-neighbor steps on a square lattice. Each monomer k occupies the lattice site $(i(k), j(k))$, where i and j are integers. To perform a simulation, one first needs to define an initial configuration that is a self-avoiding chain. We may for example use the following one: $i(k) = [[99k/100]] + 1$, $j(k) = [[(k-1)/100]] + 1$. Here $[[x]]$ means the integer part of x. Thus as the first few monomers will be at $(1, 1), (2, 1), (3, 1), \ldots, (98, 1), (99, 1), (100, 1), (100, 2), (101, 2), \ldots$ This chain is stretched in the horizontal direction with a slight negative slope. It will serve as a convenient starting configuration.

In our self-avoiding chain, any self-avoiding configuration is supposed to be as likely as any other. (This is the content of the Boltzmann Principle here, since all the allowed configurations have the same energy (namely, 0) and thus the same probability.) In our earlier simulation problem we showed, that any two configurations A and B will be equally likely if the probability to go from A to B in one step is equal to the probability to go from B to A. We shall devise a scheme that has this property.

One way to change from one self-avoiding walk to another one is to remove a monomer from one end and add it to the other. For example, we might take a monomer from the beginning of the chain ($k = 1$) and add one to the end ($k = n$) on the right side. Let us assume that this position is not blocked by another monomer. Further, let us assume that the removed monomer was just above its successor. We made our choice of where to place the end monomer by selecting one of the four possible directions at random. Thus the new monomer is placed on the right side with probability $\frac{1}{4}$. That means the probability of going from A to B is $\frac{1}{4}$ in this case. Now suppose the system is in configuration B. We again choose one of the four step directions at random, remove the monomer from the end and attach it to the beginning in the chosen direction. The probability that the upward direction was chosen is $\frac{1}{4}$. Thus the probability

A

B

to go from B to A is $\frac{1}{4}$, like the probability of going from A to B. Since this is true for all allowed moves, then all sampled configurations have equal probability.

To assure that the two probabilities are really equal, we must be a bit careful how we proceed when the chosen move is forbidden. We must select the step direction at random, test whether the chosen step is allowed, make the step if it is allowed and leave the configuration unaltered if it is not. This procedure will assure that the probability of any allowed change is the same as the probability of the reverse change. Whenever a change is made, the list $(i(1), j(1)), \ldots, (i(n), j(n))$ must be updated accordingly.

One efficient way to test for whether a candidate site is occupied is to make an image matrix $M(i, j)$. The image covers a range 1–L by 1–L which is large enough to accommodate (almost) all the polymer configurations. For every occupied site (i, j) in the lattice $M(i, j) = 1$. For the other sites $M = 0$. Thus if we wish to check whether a site (i, j) is available, we have only to test the value of one element of M, namely $M(i, j)$. We do not need to check the whole list of monomers to see if one happens to be at (i, j). The use of an image matrix does take some extra work. When the configuration is changed, $M(i, j)$ must be set to 0 for the monomer that was removed and $M(i, j)$ must be set to 1 for the site that was added. (One might hope that if we know $M(i, j)$ we can construct the list $(i(1), j(1)), \ldots, (i(n), j(n))$. This is possible on some special lattices, but not on the square lattice. For example, if the configuration of a four-monomer chain happens to be a square, there is no way to know from the occupied sites which monomers are the ends of the chain.)

The straightforward way to update the list of monomer positions when the beginning of the list is removed, is simply to shift each item in the list to the previous position. But this requires n operations for every successful move. A way to avoid this is to allow the list to start with an arbitrary element. A separate number K keeps track of which element of the list is the first monomer. Then the first monomer has position $(i(K), j(K))$. The second monomer is at $(i(K + 1), j(K + 1))$, etc. We calculate this index $mod(n)$, so that the monomer after $(i(n), j(n))$, is at $(i(1), j(1))$, continuing until the n'th monomer at $(i(K - 1), j(K - 1))$. Now when the first monomer is removed, we just change set $(i(K), j(K))$ to be the new position of the added monomer. Then we increase K by 1 to indicate that the beginning of the chain is now what was formerly the second monomer.

(a) Implement this simulation for a 64-step chain on a 100 by 100 lattice. After every 5000 steps, plot the chain to show that it is a connected chain and is self-avoiding. Do this for a dozen chains. Store the image matrix for each of these chains.

(b) Using the image matrices from part (a) calculate $M(r)$, the number of monomers within a distance r of an arbitrary monomer. This means counting the number of occupied sites within a circle of radius r of every site and taking the average. You may use a square instead of a circle; it does not change the scaling properties. Verify that $M(r) \sim r^D$. Compare the D you obtained with the $\frac{4}{3}$ calculated by the Flory argument in the text.

It is convenient to work with the average local density $M(r)/r^2$, which should vary as r^{D-2}. This density shows up the fractal behavior more clearly than $M(r)$.

3.A   Suggested experiment: *Simulated polymer properties*.   Use the simulation developed in the preceding problem to show that the fractal dimension does not depend on the degree of self-avoidance. Scale up the simulation to 128 or 256 beads. Try to obtain an accurate value for D by measuring $M(r)/r^2$, and correcting for distortions due to monomer-scale and chain-scale effects by extrapolation. Then modify the simulation so that every other bead is invisible, thus reducing the self-repulsion. (The invisible beads are freely allowed to sit at occupied sites. This reduces the average b_2 by a factor of 4.) Observe the resulting changes in $M(r)/r^2$. Determine how much the asymptotic D changes as a result. Compare the radius R with the case of full self-avoidance and compare with the prediction of the Flory formula.

The polymer is free to choose its size R. By the above estimates, the potential of mean force associated with size R is $S + V$. We thus expect that the typical size will be that which minimizes $S + V$. Since S and V are both powers of R, the minimum occurs when $S \simeq V$:

$$R^2/n \sim Un^2/R^d, \quad \text{or} \quad R^{2+d} \sim Un^3, \quad \text{or} \quad n \sim R^{(2+d)/3}. \quad (3.27)$$

We are led to the conclusion that for sufficiently large n, the fractal dimension

$$D = (2 + d)/3. \quad (3.28)$$

In three dimensions $D = \frac{5}{3}$. We note that the D value is independent of the strength U assumed for the contact interaction: for sufficiently long chains, even an arbitrarily small U still leads to a strong expansion.

This prediction for D is consistent with estimates based on systematic calculations, experiments and computer simulations. One experimental confirmation is shown in Fig. 3.9. Moreover, in two and in four dimensions, D is known exactly and it agrees with the Flory prediction. Despite the success of the Flory argument, it has not proved possible to justify it[†]. Because of the success of the Flory formula, efforts continue to find a simple way to understand the scaling of self-avoiding polymers [12].

3.13. *A lower bound on D for a self-avoiding polymer* According to the Flory argument, the constraint of self-avoidance lowers the fractal dimension D. But it must not be too much lower: if D were too low, the expanded polymer would not be opaque to itself, and the contact energy would be insignificant. (a) From this consideration, give a lower limit on D in d dimensions [13]. Is this bound consistent with the Flory formula for D?

3.14. *Triple contacts* In some situations polymers obey a modified type of self-avoidance: intersections of two monomers (of radius a) are freely allowed, but intersections involving three monomers are forbidden. (a) For n monomers in a region of radius R how does the number of triple intersections vary with n and R. Your estimate should be in the same spirit as the Flory estimate for double contacts cited above. (b) Assuming that each triple contact costs a given energy U, apply the Flory reasoning to estimate the fractal dimension of the chain in three dimensions, and in two dimensions.

3.4.3 Self-interaction and solvent quality

Up to this point we have treated the self-avoidance constraint in a simple way. We have simply removed configurations whose atoms came closer than some threshold distance. To treat this constraint more realistically, we must investigate how monomers interact in solution. As before, we shall consider our chain to be composed of segments long enough to be small polymers in their own right, yet much smaller than the overall chain. The mutual avoidance between these segments is essentially the same

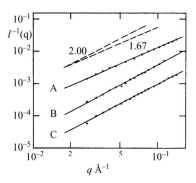

Fig. 3.9

Log–log plot of the inverse scattered intensity as a function of the scattered wave vector q, confirming the internal fractal structure, after [11, Figure 1], © American Physical Society, courtesy G. Jannink. The full points are experimental data for a polymer in a good solvent, curve A, a theta solvent, curve B, and the bulk polymer, curve C. The behavior of curves B and C will be discussed later in the chapter. The polymers used were polystyrene and deuterated polystyrene with molecular weight about 10^6. The solvents were carbon disulfide (curve A), deuterated cyclohexane at 40°C (curve B), and undeuterated polystyrene (curve C). The dashed lines are reference straight lines with slopes of 2.00 and $\frac{5}{3}$ showing the power-law behavior expected for $D = 2$ and $D = \frac{5}{3}$. The full lines are calculated curves incorporating the expected scaling.

[†] The Flory reasoning can easily lead one to wrong conclusions. For example, the argument appears to be equally justified for a chain where only the first and last third repel each other. Yet we have shown that such avoidance can only expand a chain by a factor of order unity.

whether the segments are attached to each other in sequence to form a polymer or whether the segments are floating free in solution. We thus consider two such detached segments of some nominal diameter a and ask how free they are to approach each other. We select an atom on each segment to label its position. We then examine the relative probability $g(r)$ that the two segments are separated by displacement $\vec{r}$. We introduced this $g(r)$ in Chapter 2. There we noted that this $g(r)$ is proportional to the Boltzmann-weighted sum of all solution configurations with the two segments separated by $\vec{r}$. If $r \gg a$ two segments have no influence on each other; the configurations of solvent near the one segment are the same irrespective of the position of the other. The same holds for the various configurations of the segment itself. By convention, we define this constant to be one; then $g(r)$ is called the **pair distribution function**.

As the distance r decreases, the solvent molecules around one segment may also be influenced by the other segment, so that g departs from unity. Whether g becomes larger than or smaller than one is different in different solvents. The g can be greater than one, for example, if the solvent molecules pack more easily around two nearby segments than around two distant ones. If $g > 1$ then the potential of mean force $-T \log(g)$ is negative, so that the segments experience a mutual attraction. For smaller distances $r \lesssim a$, configurations where the segments avoid each other become rare, so that the probability $g(r)$ becomes small. The same local interactions that exist for free-floating monomers also influence segments along the chain. Without this interaction, two segments which are far apart along the chain fall at displacements $\vec{r}$ with uniform probability for $r \ll R$. But when the interaction is taken into account these segments too must have a pair distribution function $g(r)$ similar to that of the free-floating segments considered above. In order to have a chain which behaves correctly, the polymer configurations must be weighted so as to reproduce the correct $g(r)$ for all pairs of segments.

To see the impact of these interactions, we consider n free-floating segments confined to a large sphere of radius R and volume Ω. Initially, we allow these to occupy any relative positions indifferently. But we must attain a final state where relative positions are governed by the proper $g(r)$ for these segments and this solvent. For definiteness, we consider a case where the true $g(r)$ is everywhere less than unity. In general work is required to accomplish this change in $g(r)$; it is this work that governs how important the interactions are. We produce the necessary change in $g(r)$ by introducing a potential of mean force $U(r)$ between our segments. First we consider the effect of introducing $U(r)$ between just one of the segments and all the others. We do this in stages, starting from the smallest r values. Thus for example, when we have reached a generic distance r_0 we turn on a potential $U(r_0)$ of sufficient strength to exclude all but a fraction $g(r_0)$ of the original segments at that distance from the chosen one. This exclusion requires work, since it reduces the volume in which the other $n - 1$ segments may live. Like any dilute gas, these segments resist compression with an effective pressure $\Pi = T (n - 1)/\Omega$. This Π is the osmotic pressure encountered in the last chapter. The volume within r_0 will be denoted v_0. To exclude segments completely from volume v_0 would require work Πv_0; to exclude a fraction $(1 - g(r_0))$ requires a proportionate amount of work, since the work is proportional to the number excluded: $(1 - g(r_0))\Pi v_0$. This process has produced the correct $g(r)$ for one segment out to a distance r_0. We may extend it to produce a correct $g(r)$ out to some slightly larger radius r_1. The new volume affected will be denoted v_1. The associated work is evidently $(1 - g(r_1))\Pi v_1$. Continuing in this way, we may impose the full $g(r)$ between the chosen segment and the others, with an associated work

$$[T(n-1)/\Omega] \sum_i v_i(1 - g(r_i)) = [T(n-1)/\Omega] \int d^3r(1 - g(r)). \quad (3.29)$$

Fig. 3.10 illustrates the first two terms of this sum.

We next choose a second segment and perform the same treatment on it, by turning on interactions between it and the $n - 2$ remaining ones. The work required is $[T(n-2)/\Omega] \int d^3r(1 - g(r))$. We proceed in this way until all segments have been chosen. The total work V is given by

$$V = T \int d^3r(1 - g(r))[1 + 2 + \cdots + n - 1]/\Omega. \quad (3.30)$$

The sum in brackets is $n(n-1)/2 \to n^2/2$. Thus when $n \gg 1$,

$$V = (Tn^2/\Omega)\left[\frac{1}{2}\int d^3r(1 - g(r))\right]. \quad (3.31)$$

We see that this V has the same form as in the Flory estimate above. But instead of assigning an *ad hoc* energy U to nearby segments, we may express the free energy associated with segment interactions in a general and rigorous way, insofar as the segments can be viewed as a dilute solution. The quantity in [...] in Eq. (3.31) arises often in solution theory. It evidently has dimensions of volume and is positive when $g(r) < 1$. It is called the **excluded volume** and we shall denote it b_2 [14]:

$$b_2 \equiv \frac{1}{2}\int d^3r(1 - g(r)). \quad (3.32)$$

This is the quantity that determines the effect of interactions between segments widely separated along the chain. Up to now we have assumed that $g(r) < 1$ so that the segment interactions must carry a positive free energy cost. Whenever b_2 is positive, this is true, and we expect the chain to expand in accord with the reasoning above. But the excluded volume may also be zero or negative if $g(r)$ is greater than 1, i.e., if $U(r)$ is attractive.

This same b_2 measures the degree of repulsion in a polymer chain. Any configuration c of a random-walk polymer has a free energy $U(c)$ owing to self-interaction. It is the potential of mean force between all the pairs of segments. To account for the interaction's effect on e.g., $\langle r^2 \rangle$, one must include the Boltzmann weight $e^{-U(c)/T}$ for each configuration. If U is sufficiently small, one may account for these interactions systematically as a

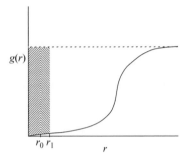

Fig. 3.10
Removal of segments near segment 1 to create the proper $g(r)$. Initially segments occur equally at all displacements $\vec{r}$. The two shaded bars show fraction of segments that must be removed from $r < r_0$ and $r_0 < r < r_1$. The volumes of these bars are the first two terms in Eq. (3.29).

power-series expansion in U. Since the important interactions are between segments that are typically separated by distances much larger than a segment, the effect of the interactions is similar to that in a dilute solution of isolated segments. Accordingly, one finds that the effects of U enter through the segment excluded volume b_2.

We now suppose that b_2 is slightly negative, so that the effective interaction between segments is attractive. Now the Flory reasoning above does not work: both the interaction energy and the elastic energy favor contraction of the chain, and no minimum-energy state can be found. To find the equilibrium size R, we must consider other effects of increasing concentration. As the concentration of segments within the chain increases, the dilute-solution formula for V must be corrected. One may include these effects as a concentration-dependent excluded volume. In terms of the concentration n/Ω, $b_2 \to b_2 + b_3 n/\Omega + \mathcal{O}(n/\Omega)^2$. In most solutions the **third virial coefficient** b_3 is positive. Thus when b_2 is negative, there is a density at which the contact energy $V = [Tn^2/\Omega](b_2 + b_3 n/\Omega)$ is minimal: $n/\Omega \simeq -b_2/b_3$. The chain elastic energy becomes irrelevant and the chain collapses to a density governed by the local segment interactions. The collapsed chain is no longer fractal: the number of segments per unit volume $n/\Omega \simeq -b_2/b_3$ is independent of chain length[†].

An intermediate case occurs when b_2 is zero. Then it is the b_3 interaction energy that must balance the elastic energy S. Setting these equal, we conclude $R^2/n \sim b_3 n^3/R^{2d}$, so that $R^{2+2d} \sim n^4$. Thus again, $n \sim R^D$, where $D = (d+1)/2$[††]. In three dimensions this is just the dimension of a random walk. We conclude that a chain with $b_2 = 0$ behaves as a simple random-walk polymer despite the remaining interaction effects of b_3, b_4, etc.[†††]

In order to verify the above predictions completely, one would need to measure changes in b_2 and observe the resulting changes in the polymer size. But it is difficult to measure b_2 explicitly. This would require measuring the osmotic pressure of a dilute solution of segments. However, the segments are altered when they are bonded together to form a chain; so even such a measurement would not give the b_2 appropriate for the chain segments. Still, we expect to see the three classes of behavior outlined above: as the effective interaction between segments is made more attractive, one passes from self-avoiding scaling with $D = \frac{5}{3}$ through random-walk scaling with $D = 2$ to the collapsed state with $D = 3$. This behavior is well known experimentally, as shown in Fig. 3.9. Polymers showing the self-avoiding scaling are said to be in **good-solvent** conditions. Those showing contraction are said to be in **poor-solvent** conditions. The intermediate state between these two is denoted the **theta** condition. A well-known theta state occurs for polystyrene in cyclohexane at about 36°C. At a few degrees higher temperature, the chain expansion is discernible. At a few degrees lower temperature, it is difficult to discern the anticipated contraction, but one sees other effects of mutual attraction, to be explored below.

As we have seen, even the smallest positive b_2 will change the scaling properties from random-walk to self-avoiding behavior if the chain length n is long enough. But for a fixed n the chain's behavior cannot change abruptly; it must change smoothly as b_2 increases from zero. We can readily

[†] In discussing this collapsed state we considered only interaction energy and neglected the work required to contract a random-walk polymer, analogous to S above. Problem 3.15 shows how to estimate this energy, and shows that it is much smaller than the energy of interaction discussed above. This is why we are justified in neglecting it.

[††] This case may be interpreted in terms of the triple-contact exclusion treated in Problem 3.14

[†††] When one examines the effect of b_3 in detail one finds subtle differences from an ideal random walk. For example, there are logarithmic corrections to the random-walk size: $R \to n^{1/2}(1 - 37/363 \times \log(n)^{-1} + \cdots)$ [15].

judge the magnitude of b_2 necessary to perturb a finite polymer significantly from its random walk state. If b_2 is small enough, the chain may be treated as a random-walk chain with radius $R \simeq R_0 \simeq an^{1/2}$ (a being the segment size). Thus the Flory interaction energy $U \simeq b_2 T n^2 / R_0^d \simeq b_2 / a^d n^{2-d/2}$. We expect the effect of U to be minor if U is much smaller than the elastic energy S. For the random-walk chain $S \simeq T$. Thus b_2 is negligible when $U \ll T$ or $b_2 \ll a^d n^{d/2-2}$. In three dimensions b_2 must be less than or about $a^3 n^{-1/2}$. As n grows bigger, the chain becomes increasingly sensitive to b_2.

For the most part the *magnitude* of b_2 is a matter of chemical detail and our scaling methods can give us little insight. But there is one important systematic effect on b_2. This occurs when the solvent molecules have a spatial extent larger than the monomers. If the solvent molecules are connected into chains of length k, this reduces the b_2 of our polymer segments by a factor k. Appendix B discusses how this reduction comes about.

From the preceding paragraphs it is clear that the interaction between monomers is more complicated than simply forbidding monomers from coming within a certain distance of each other. And yet, one type of interaction is well described by this picture, namely the good-solvent regime where b_2 is positive. One way for two monomers or segments to have a given positive b_2 is for them to be completely excluded from distances less than $(3b_2/(4\pi))^{1/3}$ and completely allowed beyond this distance. Such monomers would have the same free energy of interaction V and hence the same expansion effects as any other monomers with the same b_2. The b_2 of the real solution thus plays the role of the volume parameter v in our initial, simple model of self avoidance. We conclude that a polymer in a good solvent is well described by a self-avoiding random walk.

3.15. *Confining a polymer* Forcing a random-walk polymer to expand to a size R much larger than its natural size R_0 requires work $S \sim (R/R_0)^2$, as we have seen. It also requires work to *confine* a random-walk polymer into a volume of radius R much *less* than R_0. This work is not simply the work of bringing the two ends close together; that would just make our polymer into a big loop. We need to confine the whole chain. One strategy to convert an ideal chain into a chain of average size (say, R_G) much smaller than R_0 is to constrain a few of the random steps in the polymer. If we look at a small subsegment of our polymer, there is some length $g \ll n$ whose natural radius $R_0(g)$ is just the desired confined radius R. We divide our polymer into segments of length g, and then manipulate it so that it has the proper radius, as follows. Imagine some arbitrary configuration of the polymer. Imagine a model of it, made out of wire. Now take the end of the first segment and bend the wire at that point until the second segment overlaps the first. Now go to the end of the second segment and make a second bend so that the first three segments overlap. (We need not worry about intersections, since our random-walk polymer is invisible to itself.) We continue in this way until we have bent the wire at each of its n/g segments. The resulting wire clearly has a size of order R. This shows that we can satisfy the constraint by restricting a small fraction n/g of the random steps in the polymer. A step that was formerly free to take on all orientations is now restricted to a subset of orientations. We have restricted the configuration sum $\sum_c$ for this step by a finite factor—say K. For the whole chain, the number of configurations has been reduced by a factor $K^{n/g}$. Thus the free energy of our polymer $\mathcal{F}(R) \equiv -T \log \sum_c 1$ is $\mathcal{F}_0 + T(n/g) \log K$.

(a) Use this expression for $\mathcal{F}(R)$ to find the work to confine as a specific function of R. This estimate gives the correct scaling of the confinement energy, agreeing with more rigorous arguments. It also can be extended to the confinement of self-avoiding polymers, as the rigorous arguments cannot.

(b) Compare the work of confinement found in (a) to the interaction energy of a collapsed chain discussed in the text. Is it important or not for long chains?

3.4.4   Universal ratios

We have seen that random-walk polymers have a single asymptotic $p(n,r)$ independent of the details of how the random walk was constructed. Now we have seen that self-repelling polymers also have a common behavior independent of the details of the repulsion, provided b_2 is positive. This means that all sufficiently long polymers in good solvents are quantitatively identical to each other in the same way all random-walk polymers are. For example, they have the same function $p(n,r)$ describing their end-to-end distribution. We say that these properties are **universal**. As with random-walk polymers, and other asymptotic functions, there is some amount of arbitrary choice in the universal functions. These choices amount to scale factors for the various independent variables, e.g., the λ and μ of the "Random-Walk Polymer" section above. Thus, in a random-walk polymer, the ratio $\langle r^2 \rangle / n$ is arbitrary, since the random-walk polymer rescaling changes it by a factor of μ^2/λ. However, certain quantities characterizing the asymptotic behavior clearly have values that are independent of such scale factors. An example is $\langle r^4 \rangle / \langle r^2 \rangle^2$, encountered in a problem above. The numerator and denominator individually acquire factors of μ^4 when rescaled. But the quotient is constructed to be independent of such scale factors. Thus this quotient depends only on the universal asymptotic function $p(n,r)$ and not on the arbitrary scale factors involved with taking the asymptotic limit. This ratio (3 for a random walk polymer in one dimension) is called a **universal ratio**.

For good-solvent polymers the asymptotic behavior is also universal. Here again we seek the asymptotic spatial properties (such as the distribution function $p_s(n,r)$ in the limit $n \to \infty$). Here again we are led to use rescaled variables $\tilde{n} \equiv \lambda n$ and $\tilde{r} \equiv \mu r$ defined so that $\tilde{p}_s(\tilde{n}, \tilde{r})$ is a finite function. The specific values of μ and λ are arbitrary, as for the simple random walk. Still, one can readily construct universal quantities which are independent of these scale factors[†]. For example, the scattering intensity $I(q,n)$ can be written as $I(0,n)\tilde{S}(qR_G)$, where $R_G^2 = 3(dI(q,n)/dq^2)/I(0,n)|_{q=0}$ is the radius of gyration *cf.* Problem 3.11. Changing the arbitrary scale factors cannot change the dimensionless wave vector qR_G. Once qR_G is fixed, $\tilde{S} \equiv I(q,n)/I(0,n)$ cannot be changed by scale factors, either. Thus the function $\tilde{S}$ is the same for all asymptotic good-solvent polymers. Since good-solvent polymers are fractals, the function $\tilde{S}$ must behave at large q according to the fractal law $\tilde{S} \to h\,(qR_G)^{-D}$, as sketched in Fig. 3.7. The coefficient h gives an important quantitative connection between the short-distance structure and the overall size of a polymer. Scattering experiments give a value for h of about 1.14 ([8] using Eq. (XV–3.54)). Thus if the low-q

[†] It is natural to ask what is the reason for the strong parallels between the asymptotic scaling of the good-solvent polymers and that of the random-walk polymers. In both cases we used a scale factor for chain length and a second scale factor for geometric distance. Does this resemblance arise from a deep mathematical property of self-avoiding polymers that we have glossed over? No. It arises because we sought similar information in the two cases. We asked for the asymptotic behavior in a limit in which a single variable (n) was taken to infinity. The behavior we sought concerned only the spatial distribution of monomers on length scales that remain finite fractions of the size of the polymer when the asymptotic limit is taken. If we had asked for other information, such as the number of configurations of n monomers, additional scale factors would have been needed. The same is true if we had asked for spatial behavior on smaller scales than that of the overall polymer. For example, this analysis did not tell us that the large q (small distance) behavior of the chain was that of a fractal. We determined this from a specific analysis of how short segments of a chain resemble longer ones.

scattering is known for a given good-solvent polymer, its high q scattering intensity is completely determined. Clearly such quantities as h give great predictive power in describing polymer behavior.

3.4.5 Polyelectrolytes

It was mentioned in the first section that polymers in solution can be made with an ionic charge every few bonds along the chain. The electrostatic repulsion between these ions tends to expand the chain, and the degree of expansion proves to be more extreme than that of neutral chains. To see this, we imagine turning on the electrostatic repulsion from the most global progressively down the most local scales, inverting the procedure of the last section. We begin by considering the repulsion only between the first half of the chain and the second half. When this is done for a neutral chain, it causes the two halves to swing away from each other; this reduces the number of contacts between the halves, and thus its interaction energy, dramatically. The result of this global interaction by itself is to increase the radius by a factor of order unity. At that point the energy of interaction between the two halves is of order of the thermal energy T[†].

When we do the same thing with a polyelectrolyte chain, we must consider the electrostatic repulsion between the first and last halves. If the entire chain has n monomers, each bearing, say, one charge, the repulsion is roughly that between two charges of magnitude $en/2$ at a separation R as large as the polymer. This energy is roughly $e^2(n/2)^2/(\epsilon R)$, where ϵ is the dielectric constant of the solvent. It may be expressed in units of the thermal energy T as $T(n/2)^2(\ell/R)$. The **Bjerrum length** $\ell \equiv e^2/(\epsilon T)$ is roughly $0.7\,\mathrm{nm}$ in water at room temperature. If R were the size of a random-walk chain, so that $R \sim n^{1/2}$, this energy clearly grows indefinitely large with n. Unlike the short-ranged energy, it cannot be much reduced by simply making the two halves swing out of each other's way. This global energy would cause a large expansion of the chain—not just a finite-factor expansion as in the neutral case. Balancing this global Coulomb repulsion against the elastic energy $TR^2/(a^2n)$, the two energies are comparable when $R \sim n$. We are led to the conclusion that the swelling exponent $\nu = 1$; this is as large as ν can be for a connected chain. The repulsion is strong enough to stretch the chain to within a finite factor of its full extension, as though the chain were a rigid rod.

To confirm this picture and to see how rigid this rod is, we are led to model the chain as a row of evenly spaced charges at separation R/n, connected by random-walk polymers of unstretched length $\langle r^2 \rangle_1^{1/2} \equiv a$. The preferred distance between charges can be found by considering the tension in the polymer near the center. This tension F is equal to the repulsive force between the left and the right halves. Adding up the forces on each charge on the left half, one finds[††]

$$F \simeq T\ell(n/R)^2 \log(n). \qquad (3.33)$$

The force increases indefinitely as $n \to \infty$, even with fixed spacing n/R. This force must be balanced by the elastic tension in the chain. Since

[†] For this statement to be strictly true, we must also turn off the interactions involving monomers near the center one.

[††] Here we have approximated the sum over charges by an integral. The approximation is valid when $n \gg 1$.

random-walk chains store energy like an ideal spring, they exert a force proportional to their elongation. Specifically, $F = 3T(R/n)/a^2$. Equating these tensions leads to the preferred spacing: $R/n \simeq a(\ell \log(n)/a)^{1/3}$. If the charges were placed sparsely along the chain, $a \gg \ell$, and the spacing R/n would be much smaller than the unperturbed size a. Thus locally, the effect of the charges can be weak. Still, the overall effect is strong. Even a weakly charged chain, if long enough, would stretch the monomers out to their full extension, because of the logarithmic factor. We shall assume in what follows, though, that the charges are sparse enough so that the monomers are not strongly distorted: $R/n \ll a$.

The real chain must fluctuate around this regular rod state by stretching and by bending. If the chain stretches by a small amount γR, the associated energy must be proportional to the square of this small change. It is convenient to express the constant of proportionality as ER so that the stretching energy is $ER\gamma^2$. To estimate the **modulus** E we consider strains γ of order unity, so that the length R is doubled. Such a stretching must change both Coulomb and elastic energies by factors of order unity; hence the total energy is changed by such a factor. This means the stretching energy is of the order of the original Coulomb energy: $ER\gamma^2|_{\gamma=1} \simeq nT(\ell \log(n)/a)^{2/3}$, or $E \simeq a^{-1}(\ell \log(n)/a)^{1/3}$. Except for the weakly varying $\log(n)$ factor, the modulus E is independent of n, as in a solid, macroscopic rod. One may now estimate the amount of strain γ_T typically present because of thermal fluctuations by setting $ER\gamma_T^2 \simeq T$. Thus $\gamma_T \sim 1/R^{1/2}$ if we again neglect the $\log n$ so that E is independent of size. The $\log(n)$ factor only reduces γ_T further. As R becomes large, the thermal fluctuations in R become increasingly insignificant.

A polyelectrolyte is also rigid with respect to bending, e.g., into an arc of radius B. One can readily find the Coulomb energy needed to bend a uniformly charged rod in this way. Any small bending must cost an energy of the form $\tilde{E}R(R/B)^2$. But a bend of, say, 90° with $B \simeq R$ changes the distance between two arbitrary charges by a factor of order unity as illustrated in Fig. 3.11. Thus the Coulomb energy increases by a similar factor: $\tilde{E}R \simeq n^2\ell T/R$. The coefficient $\tilde{E}$ is of the same order of magnitude (except for the $\log(n)$) as the modulus E. The amount of thermal bending R/B_T, like the thermal strain γ_T, falls off as the square root of the chain length n. We conclude that a polyelectrolyte chain has little of the fluctuations that random-walk or self-avoiding chains have. The assumed state of a uniformly charged line is affected but little by thermal fluctuations in stretching or bending.

In this chapter we have seen that connecting arbitrary small molecules to form a flexible chain leads to remarkable structures. These structures are very different from their compact constituents. They are indefinitely tenuous and deformable. In order to think about these objects quantitatively, a new concept is needed. This is the notion of fractal scaling. Using it, we found that chains made of different constituents nevertheless converge to a common quantitative behavior, provided they are long enough. This common asymptotic behavior is embodied in scaling laws and scaling functions like the reduced scattering function $\tilde{S}(q)$ defined above.

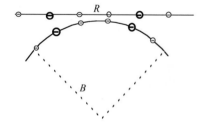

Fig. 3.11

A charged rod of length R is bent into an arc of radius of curvature $B \simeq R$, thus shortening the distance between a typical pair of charges (heavily drawn) by a finite factor.

In the next chapter we exploit and extend these scaling laws to explain the striking properties of polymer solutions: their springiness, their viscosity, and their tendency to absorb further solvent.

3.16. *Screened polyelectrolyte: the Odijk length*   In our discussion of polyelectrolytes we ignored the countercharge that neutralized the ions on the chain before it was put into solution. The interaction between the polyelectrolyte charges is modified by the countercharges. In addition, the solvent always contains further ions, not related to the polymers. We shall see in Chapter 5 that the effect of these ions is to change the interaction potential between two unit charges to the form $(e^2/r)\exp(-\kappa r)$: the potential falls off exponentially for distances larger than the **screening length** $1/\kappa$. This κ depends only on the concentration of ions and is independent of the chain length n. Thus when n is large, the interaction is relatively short-ranged: $\kappa R \gg 1$. Still, even such screened chains are remarkably rigid. By extending the methods sketched in the text, Odijk [16] found that that screened chains can be rigid on length scales much greater than the screening length $1/\kappa$.

Consider a length R of chain much longer than $1/\kappa$, bent into a gentle arc of radius $B \gg R$. As in the text, the chain is treated as a line of uniformly spaced charges separated by a distance a. (a) Find the change in Coulomb energy of a given charge in the limit of weak bending. It must be of order $1/B^2$. The total energy cost of bending the rod is evidently this energy per charge times the number of charges n. For sufficiently long R, the cost of bending the chain through a unit angle, with $B = R$, becomes of order T. This length R is called the electrostatic persistence length, or Odijk length L_0. Chains shorter than the Odijk length behave as rigid rods, even with screening; chains much longer than the Odijk length coil randomly. (b) The quantity $L_0\kappa$ is the persistence length relative to the screening length. In a limit where the distance between charges a is much smaller than the screening length $1/\kappa$, how does $L_0\kappa$ scale with κa? Is the chain flexible or rigid on the scale of $1/\kappa$?

Appendix A: Dilation symmetry

The fractal property reflects a pervasive symmetry of the polymer: a general invariance under spatial dilation. **Dilation invariance** means that these polymers are statistically essentially identical when all distances are multiplied by an arbitrary constant λ. To define this invariance operationally, we imagine a random set of points in space such as those occupied by our polymer, represented by a dot for each monomer. We study this object by making a large sample of images of its interior. The scale of these images is to be much smaller than the object as a whole, and much larger than any ultimate microstructure such as a single monomer. The images have limited resolution, so that the individual points of the set are not distinguished. Instead the object appears as a cloudy blur spanning the picture. It is natural to select images that have substantial density at the center. Similarly, it is natural to regulate the overall contrast of the images so that the set is clearly visible. Figure 3.12 shows a sample of such images. Each image represents a density $\rho(r)$ distributed in some way over a finite region of space. Dilation invariance is a property of this sample of images relative to another sample taken at a different spatial magnification. Evidently, if there is no objective way to distinguish the one set of images from the other, the random object is invariant under dilation. The figure illustrates this property: the different magnifications shown appear indistinguishable. One example of a dilation invariant set is a vertical line. The images at any magnification show a

Fig. 3.12
Eight randomly sampled images of a long self-avoiding random-walk polymer. (We show self-avoiding polymers because simple random-walk polymers are difficult to illustrate in two dimensions.) Half of the images are at a 2.5-fold magnification relative to the other half. The monomer positions are blurred to a fixed fraction of the picture size. In each case the picture was centered on a monomer. The two magnifications are interspersed randomly. The reader is invited to guess which are the magnified images. (Polymers were generated via a Monte-Carlo simulation originally written by Betsy Weatherhead and drawn by Olgica Bakajin, 1995) For the solution see the end of this section.

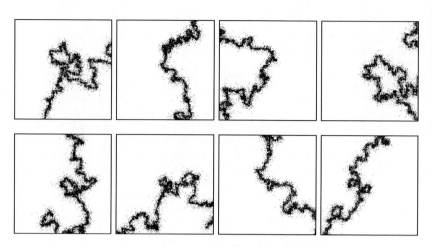

blurry line of the same darkness passing through the center. There is no way to distinguish which has the larger magnification. An example lacking dilation symmetry would be a soap froth in a pan, sampled by preparing the froth repeatedly and then photographing it. Pictures at different magnifications are clearly distinguishable because the typical bubbles have different sizes at different magnifications.

To state this invariance mathematically, we may define a set of statistics that describe the random density $\rho(r)$. One way to characterize the variations of ρ over the images is by means of the **correlation functions** $\langle\rho(r_1)\rho(r_2)\ldots\rho(r_k)\rangle_0$. The $_0$ subscript indicates that the pictures in our sample are selected to have the origin occupied. The local density discussed in the main text is the simplest example of these correlation functions. The points $r_1,\ldots,r_k$ are specific points on each image. For example, r_1 could be 1 cm inward from the upper left corner. Any of these correlation functions can be measured to arbitrary precision using a large enough sample of images. The same process performed on images magnified by a factor λ^{-1} gives $\langle\rho(\lambda r_1)\rho(\lambda r_2)\cdots\rho(\lambda r_k)\rangle_0$. Dilation invariance implies that the two statistics are the same:

$$\langle\rho(r_1)\rho(r_2)\cdots\rho(r_k)\rangle_0 = \mu_k(\lambda)\langle\rho(\lambda r_1)\rho(\lambda r_2)\cdots\rho(\lambda r_k)\rangle_0. \quad (3.34)$$

The μ prefactor accounts for a possible overall change in "contrast" upon dilation. Many of the structures that occur in complex fluids have dilation invariance of this type. An important source of recent progress in understanding these fluids is the recognition and exploitation of this symmetry.

We may show the dilation symmetry in an infinitely long random walk polymer by using again the scaling property of the end-to-end probability $p(n, r)$. We first consider a typical configuration that contributes to $\langle\rho(r_1)\cdots\rho(r_k)\rangle_0$. Any chain that contributes must pass through each of the points $\vec{r}_1,\ldots,\vec{r}_k$ as well as through the origin. The chain may pass through these points in any order, and with any arbitrary number of monomers in each segment between these points. Of these cases, we consider first the one where the points are visited in order: the chain goes from the origin to r_1, then to r_2, etc. We further suppose that the number of monomers between 0 and r_1 is i_1, and that the succeeding segments have $i_2, \ldots, i_k$ monomers. Once all this has been specified, the probability of such a chain can easily be stated: it is the product $p(i_1, r_1) p(i_2, \vec{r}_2 - \vec{r}_1) \cdots p(i_k, \vec{r}_k - \vec{r}_{k-1})$. The probability takes this simple form because each of the chain segments is statistically independent. To find the joint probability for all chains that pass through the points in this order, we must simply add all the contributions from the different possible i_1, etc.:

$$\left[\sum_{i_1} p(i_1, r_1) \right] [\ ] \cdots \left[\sum_{i_k} p(i_k, \vec{r}_k - \vec{r}_{k-1}) \right].$$

This sum is itself dilation invariant. To see this we use the scaling form $p(i, r) = i^{-3/2} f(r^2/i)$, and replace the monomer sums by integrals. The same sum dilated by a factor λ is

$$\left[\int di_1 i_1^{-3/2} f((\lambda r_1)^2/i_1) \right] [\ ] \cdots \left[\int di_k i_k^{-3/2} f(i_k, (\lambda \vec{r}_k - \lambda \vec{r}_{k-1})^2/i_k) \right]. \tag{3.35}$$

We may scale out the λ factors by defining new $\tilde{i}$ variables as $\tilde{i} \equiv \lambda^{-2} i$. Each integral may be converted to the $\tilde{i}$ variables by converting the di's and the $i^{-3/2}$ factors. This results in a factor of $\lambda^2 \lambda^{-3}$ for each integral. Thus our sum may be written

$$(\lambda^{-1})^k \left[\int d\tilde{i}_1 \tilde{i}_1^{-3/2} f((r_1)^2/\tilde{i}_1) \right] [\ ] \cdots \left[\int d\tilde{i}_k \tilde{i}_k^{-3/2} f((\vec{r}_k - \vec{r}_{k-1})^2/\tilde{i}_k) \right]. \tag{3.36}$$

Except for the λ^{-k} prefactor, this is the same as the undilated sum. Thus the sum is dilation invariant. We may repeat the reasoning for any order of passage among the points $0, \ldots, \vec{r}_k$. All the corresponding sums have dilation symmetry with the same prefactor. To obtain the overall probability of passage through these points, we must simply add these sums together for all possible orders of passage. The total evidently has the same dilation symmetry as each part. Finally the density of monomers at r is evidently the same as the probability density for a monomer to be there. That means that our joint probability is simply $\langle \rho(r_1) \cdots \rho(r_k) \rangle_0$. We have thus shown that

$$\langle \rho(r_1) \cdots \rho(r_k) \rangle_0 = \lambda^k \langle \rho(\lambda r_1) \cdots \rho(\lambda r_k) \rangle_0. \tag{3.37}$$

Comparing this result with the general definition in Eq. (3.34), we see that the μ prefactor takes a simple and suggestive form: it is a power of λ which is proportional to the number of density factors ρ. The density behaves as though it had dimensions of length^{-1} rather than its standard dimensions of length^{-3}. This latter is the scaling we would find if we multiplied *all* lengths, including the size of a monomer, by a factor λ. By contrast, our dilation invariance applies to the polymer with no change in the monomers. The case $k = 1$ shows that the polymer is a fractal object. It says

$$\langle \rho(r) \rangle_0 = \lambda \langle \rho(\lambda r) \rangle_0. \tag{3.38}$$

Setting $\lambda = \text{constant}/r$, this amounts to $\langle \rho(r) \rangle_0 \propto 1/r$: The density falls off as in a fractal with $D = 2$, confirming our previous conclusion. In general a density profile with dilation symmetry is a fractal density.

The dilation invariance amounts to a form of statistical regularity which is important in predicting properties. To illustrate this regularity, we return to the fractal analysis applied above to our polymer. We consider the mass $M(r)$ of the atoms lying within a distance r of an arbitrary monomer. We have seen that the scaling of the average M with r is that of a fractal. But the value of M obtained in a given instance fluctuates about this average. Using the dilation invariance property above, we can see how much M fluctuates. Different physical systems fluctuate by different amounts. For example, the mass in a given macroscopic volume of a *solid* object fluctuates very little: the percentage spread in the sampled M values becomes indefinitely small as the size r is made larger. In other cases, the fluctuations of M are wild. For example, if we dropped our restriction that a monomer be at the center of our sphere and studied spheres placed arbitrarily in the pervaded volume of the polymer, we would almost always obtain $M = 0$, and only rarely encounter volumes containing monomers. The distribution of M's would be heterogeneous. It is important to know how wildly M fluctuates for monomer-centered spheres. The less wild they are, the more predictable the polymer behavior is.

One way to analyze the fluctuations of M is to consider **moments** like $\langle M^k \rangle$. We can readily express such moments in terms of our $\langle \rho(r_1)\rho(r_2)\ldots\rho(r_k) \rangle_0$. Using the dilation invariance property of the $\langle \rho \ldots \rho \rangle$'s we can readily infer how the moments of M scale with the sampling size r: $M(r) = \int_{r' < r} d^3r' \rho(r')$, so that

$$\langle M(r)^k \rangle = \int_{r_1 < r} d^3 r_1 \cdots \int_{r_k < r} d^3 r_k \langle \rho(r_1)\rho(r_2) \cdots \rho(r_k) \rangle_0. \tag{3.39}$$

Using the dilation invariance property of the $\langle \rho \cdots \rho \rangle$'s we can readily infer how the M moments scale with the sampling size r: $\langle M(\lambda r)^k \rangle = \lambda^{2k} \langle M(r)^k \rangle$. This scaling implies regularity, because it says that ratios of different characteristic masses are independent of r: $[\langle M(r)^k \rangle]^{1/k} = C_k \langle M(r) \rangle$, where the constants C_k are independent of r. The mean mass, root-mean-square mass, etc. are all of the same order, so that knowing any one of these averages determines them all approximately. Another way to express this regularity is in terms of the probability distribution of sampled M values: $P_r(M)$. We may then express the normalized k-moment as

$$\langle \tilde{M}^k \rangle = \int d\tilde{M} \, \tilde{M}^k P_r(\tilde{M}), \tag{3.40}$$

where $\tilde{M} \equiv M/\langle M \rangle$. Since the moments scale the same, they are all finite for a given r, however large. Thus the $P(\tilde{M})$ must fall off faster than any power as $\tilde{M} \to \infty$. By further analysis, one may show that $P(\tilde{M})$ falls off exponentially with $\tilde{M}$. The distribution also falls off for small $\tilde{M}$, as $\exp(-1/\tilde{M})$. It thus has the form sketched in Fig. 3.13. The function $P(\tilde{M})$ must also be universal, since it can be calculated from the universal $p(n, r)$.

Compared to a solid body, the fluctuations in $M(R)$ for our random-walk polymer are large as Fig. 3.13 suggests. The relative width of the distribution, e.g., $\langle (M - \langle M \rangle)^2 \rangle / \langle M^2 \rangle$, does not grow with r, but neither does it go to zero[†]. Real, good-solvent polymers are dilation-symmetric [8], with $\mu_k(\lambda) = (\lambda^{3-D})^k$. Their $P(M)$ distributions are thought to be universal and to resemble Fig. 3.13 [17]. Other fractal objects, such as colloidal aggregates are also thought to have this form of dilation symmetry.

Appendix B: Polymeric solvents and screening

In this Appendix we discuss how linking solvent molecules together results in a weakening of interaction between segments of a polymer dissolved in that solvent. This is easiest to understand if we place the solvent and polymer on a lattice, with each segment occupying a lattice site. The solvent molecules, being larger than the segments, occupy $k > 1$ sites within some radius R_s. For concreteness, we can consider the k-site solvent molecules to be small polymers of the same species as the large polymer being studied. Thus in principle all the issues of fractal structure arise for both the solvent and the polymer, as discussed in the chapter. However, the weakening of interactions we wish to explain is independent of the details of this fractal structure. Accordingly, we will ignore questions of fractal structure here.

At first glance, our change to k-site solvent molecules seems unimportant. For example, if these solvent molecules occupied a compact cube of k sites, our system would amount to a change to a coarser lattice whose sites had a linear size of $k^{1/3}$ instead of 1. But we wish to consider tenuous solvent molecules that can interpenetrate. The k sites of each molecule pervade a volume shared by many other such molecules. This is the situation when the solvent itself consists of polymers. But it is equally true when the solvent consists of any tenuous structures, whose volume fraction decreases indefinitely as k increases.

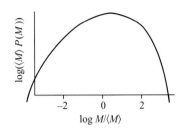

Fig. 3.13
Qualitative sketch of the mass distribution $P(M)$ as a log–log plot. The M variable has been scaled with $\langle M \rangle$; the P variable has been scaled with $\langle M \rangle^{-1}$. With this scaling the curve is universal. The downward curvature for large and for small M indicates that $P(M)$ falls off faster than any power for large or for small M.

[†] Some known sets of points called multifractals have scale invariance of a more general type. An example of a multifractal is the charge distribution on a conducting fractal that has been charged to some potential V. Multifractal densities have μ powers that are not simply proportional to the number of ρ factors. Thus multifractals have mass moments that diverge at different rates as $r \to \infty$. For a multifractal one must determine individually the scaling of all these moments in order to characterize the fluctuations of M.

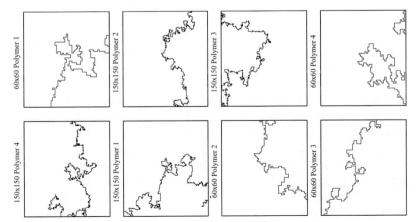

Fig. 3.14
The polymers of Fig. 3.12 without blurring.

The assumption of interpenetration makes it easier to see what happens when a long polymer occupying $N \gg k$ sites is dissolved in such a solvent. The main effect of the large solvent molecules is to reduce the pressure, as we now explain. Our system consists of many k-site solvent molecules on a lattice, such that the average number of monomers on a site is about 1. Pressure is the work required per site volume to empty a site. We used this notion of pressure above in our explanation of excluded volume. In a real lattice gas, no more than one monomer is allowed on a site: the monomers are impenetrable. This impenetrability constraint acts in a similar way whether the solvent consists of single sites or k-site molecules. For the moment we ignore it. Instead we shall allow all system configurations in which the average volume fraction ϕ is about 1. That means, for an $\mathcal{N}$ site lattice, there must be about $\mathcal{N}/k$ solvent molecules. But their positions are arbitrary. Thus our solvent forms a noninteracting, ideal gas. Now it is clear why the k-site solvent has a reduced pressure. The pressure in an ideal gas is T times the density of molecules. For our system this amounts to T/k per site-volume. The pressure is reduced by a factor k relative to that of a single-site ($k = 1$) solvent.

This reduced pressure has an important effect on the interaction between objects in the solution. We now consider two designated molecules and investigate their mutual excluded volume. To find this excluded volume, we place one of our solute molecules at the origin, and then determine the work required to place the second solute molecule into the same region. For this purpose it is convenient to identify some compact region of that contains all the sites of a given molecule. We may imagine it to be a sphere of radius R_s and volume V sites. We can suppose that all the solvent molecules have roughly the same size, so that we may take the enclosing region to be the same size for all the molecules, including our two designated molecules. The number of cells V in this region is the pervaded volume of a molecule. We shall see that the precise definition of this region is unimportant. Our assumption that the solvent molecules are tenuous implies that V is much larger than k. The molecule added at the origin causes the average volume fraction nearby to increase slightly, by an amount k/V.

We now consider the second designated molecule, and focus on configurations in which it lies in the pervaded volume of the first molecule. We compare these with configurations where the second molecule lies at some fixed position far from the first molecule. Work is required to take the second molecule from the far-away position into the pervaded volume. Since the pervaded volume has a slightly higher density, we may regard it as having slightly higher pressure. The work is the volume k of the molecule times the change of pressure Δp. This $\Delta p = (T/k)\Delta\phi = (T/k)(k/V) = T/V$. Note that this excess pressure is a factor k smaller than it would have been in a simple one-site solvent. The work W required to bring the second molecule to the origin is thus $k\Delta p = Tk/V$. We see that this work is much smaller than T. Thus there is little hindrance for the second molecule to approach the first. Nevertheless, this hindrance reduces the probability that the second molecule lies at the origin, by a factor $e^{-W/T} \to (1 - W/T)$. This reduction factor is precisely the pair distribution function for our two designated molecules: $g(r)|_{r=0}$. The same reduction factor applies roughly for typical points r in the pervaded volume: $1 - g(r)|_{r \in V} \simeq W/T$.

Now we can find the mutual excluded volume $\tilde{b}_2$ of our pair of molecules:

$$\tilde{b}_2 = \frac{1}{2} \sum_r (1 - g(r)) \simeq W/TV. \qquad (3.41)$$

Using our estimate for W, we find $\tilde{b}_2 \simeq k$. Using this result, we can infer the self-interaction energy of an n-site polymer, containing n/k of the k-site molecules. Repeating the reasoning of Eq. (3.31), the self-interaction energy V of the polymer is given by

$$V_{\text{polymer}} \simeq T(\tilde{b}_2/R^3)(n/k)^2 \simeq T b_2/R^3 n^2/k. \qquad (3.42)$$

Here b_2 is the mutual excluded volume of the original segments with $k = 1$. The interaction has been reduced by a factor k, as announced in the text. We say that the large solvent molecules have **screened** the interaction between the two solute molecules. Because of screening, the chain is but little perturbed, and $R \simeq R_0 \simeq an^{1/2}$ so long as $V_{\text{polymer}} \lesssim T$, i.e., $1 \lesssim (b_2/a^3)n^{1/2}/k$. The solvent chains can be much shorter than the polymer and still prevent it from expanding significantly.

This subtle notion of screening plays a central role in complex fluids. Screening occurs whenever two interacting objects also interact similarly with many surrounding objects that are free to move. The prime example is electrostatic screening, to be considered in Chapter 5 to follow. The profound effect of polymer screening was first recognized by Flory [10, 18], and justified by a combinatorial lattice gas argument. The unity of polymer and electrostatic screening was elucidated by Edwards [19].

The above treatment of screening used the worrisome assumption that the dense solution of solvent molecules could be treated as an ideal gas, ignoring all the packing constraints that should affect the system profoundly. In our ideal system, we constrained only the average volume fraction of the system to be some value slightly less than 1. In the real system, there can be no configuration where a site has a volume fraction exceeding one. Such a treatment is possible because we have treated the solvent molecules as arbitrarily tenuous, with the pervaded volume of each shared by indefinitely many others. Because the molecules were tenuous, we could find their mutual excluded volume by considering only the slight change of pressure required to increase the volume fraction by an arbitrarily small amount. An actual liquid made of large solvent molecules has a pressure determined by many detailed effects, including packing constraints and the mutual attraction of the atoms. However, one can create slight changes in density in a region simply by *translating* solvent molecules into the region, without appreciable changes in their local packing or attraction. Therefore it is mainly the free energy associated with the translation of the molecules that affects the small changes of concentration that are important for our system. This translational free energy can be treated without considering packing or other interactions between solvent molecules—i.e., we may consider the solvent to be an ideal gas[†].

† Another way to justify the use of our ideal system is to argue that in the tenuous limit the idealized system nearly satisfies the constraints of the real system. For definiteness, we suppose that our system has an average volume fraction $\langle \phi \rangle$ of 0.99. We consider the statistical fluctuations of ϕ within the pervaded volume V of a molecule. On average V/k molecules share this volume. In the ideal system these molecules enter and leave independently, their number fluctuates by an amount equal to the square root of the average: $(V/k)^{1/2}$. Thus in this region ϕ fluctuates over a range of $\pm (V/k)^{1/2}/(V/k)$ or $\pm (k/V)^{1/2}$. If this fluctuation amounts to much less than 1%, the occurrence of $\phi > 1$ becomes rare. To satisfy this condition would require $V > 10^4 k$. If this condition is satisfied, the ideal system nearly obeys the constraints of the real system without further intervention. Thus the ideal system is a reasonable approximation to the real system. Real liquids have an average volume fraction substantially lower than their maximum volume fraction. Thus the requirements on V/k are less stringent.

References

1. See e.g., D. W. Van Krevelen, *Properties of Polymers: Their Correlation with Chemical Structure; Their Numerical Estimation and Prediction from Additive Group Contributions*, 3rd ed. (Amsterdam: Elsevier, 1990) p. 333.
2. P. Rempp, E. Franta, and J.-E. Herz, in *Polysiloxane Copolymers/Anionic Polymerization*, Advances in Polymers Science series, no. 86 (Heidelberg: Springer, 1986), p. 147
3. G. G. Odian, *Principles of Polymerization*, 3rd ed. (New York: Wiley, 1991).
4. K. P. McGrath, M. J. Fournier, T. L. Mason, and D. A. Tirrell, *J. Am. Chem. Soc.* **114** 727 (1992).
5. H-G. Elias, *Macromolecules*, 2nd ed. (New York: Plenum Press, 1984), p. 124.
6. B. B. Mandelbrot, *The Fractal Geometry of Nature* (San Francisco: Freeman, 1982).
7. T. A. Witten, *Rev. Mod. Phys.* **70** 1531 (1998).
8. G. Jannink and J. Des Cloizeaux, *Polymers in Solution* (Oxford: Oxford, 1992).
9. J. C. Le Guillou and J. Zinn-Justin, *Phys. Rev. Lett.* **39** 95 (1977).
10. P. Flory, *Principles of Polymer Chemistry* (Ithaca, New York: Cornell, 1971), Chap. XII.
11. J. P. Cotton, D. Decker, B. Farnoux, G. Jannink, R. Ober, and C. Picot, *Phys. Rev. Lett.* **32** 1170 (1974)
12. M. A. Moore and A. J. Bray, *J. Phys. A: Math. Gen.* **11** 1353 (1978); S. F. Edwards and P. Singh, *J. Chem. Soc. Faraday Trans. 2*, **75** 1001 (1979); M. K. Kosmas and K. F. Freed, *J. Chem. Phys.* **68** 4878 (1978).
13. J. Des Cloizeaux, *J. Phys.* **31** 715 (1970).
14. R. K. Pathria, *Statistical mechanics* (Oxford: Pergamon Press, 1972).
15. B. DuPlantier, *Europhys. Lett* **1** 491 (1986); G. Jannink and J. Des Cloizeaux *op. cit.*, Chap. XIV.
16. T. Odijk, *J. Polym. Sci.* **15** 477 (1977).
17. B. M. Sterner, to be published, demonstrates numerically the universality of $P(M)$ for two-dimensional self-avoiding polymers and obtains values for several of the numerical coefficients C_k for positive and negative k.
18. M. Huggins, *J. Phys. Chem.* **46** 151 (1942); M. Huggins, *Ann. NY. Acad. Sci.* **41** 1 (1942); M. Huggins, *J. Am. Chem. Soc.* **64** 1712 (1942).
19. S. F. Edwards, *Proc. Phys. Soc.* **88** 265 (1966).

Polymer solutions

4

The previous chapter has shown how a polymer chain behaves when it is by itself in a solvent. But the properties seen in a macroscopic fluid result from many polymers interacting in the solvent. In this chapter we turn to a survey of how polymer structure, energy, and motion are altered by their mutual interactions. We first describe the energy associated with interaction in dilute solution. The following section discusses the new structural and energetic features that appear when the polymers interpenetrate. Then we turn to dynamics. The first subsection discusses the Brownian motion of a sphere, the starting point for understanding how objects move in a liquid. We show how spontaneous thermal motions are related to the work needed to create motion. We discuss the viscosity added to a liquid by inserting particles or polymers into it. Next we consider how these properties are altered when polymers interpenetrate. We describe how flow fields are modified by hydrodynamic screening. We distinguish between self-diffusion and cooperative diffusion. Finally we discuss how entanglement affects both diffusion and the relaxation of stress.

4.1 Dilute solutions

The interactions between polymers are easiest to treat when the polymers are far apart in solution and their mutual influence is weak. The main effect of this influence is on the free energy. The work required to add an additional polymer to the solution is altered by the presence of the others. To evaluate this effect, we start with an initial state in which the monomers of a given chain interact with each other but those of different chains do not. Even with no interaction between chains, one must do work to confine N polymers into a solution volume Ω. We saw this in Chapter 2 when we considered the work to compress an ideal gas. Interactions modify this work, as explained in the previous chapter under "Self-interaction and Solvent Quality." The work to change the volume available to the polymers by an amount $d\Omega$ is by definition $-\Pi\, d\Omega$, where Π is the osmotic pressure. This work and thus Π are altered when the interactions between the monomers of different chains are turned on. The work V required to turn on the interaction must have the same form as that found in the preceding chapter: $V = T(N^2/\Omega)B_2$, where B_2 is the mutual excluded volume of two polymer chains. This excluded volume must be expressible in terms of the polymers' $g_p(r)$ just as in the preceding chapter:

$$B_2 = \frac{1}{2} \int d^3 r (1 - g_p(r)). \tag{4.1}$$

The presence of V means that each increment of compression $d\Omega$ requires an extra work $-(dV/d\Omega)\,d\Omega$. Thus V adds an amount $-dV/d\Omega$ to the osmotic pressure Π:

$$\Pi = \Pi_0 - \frac{dV}{d\Omega} = \frac{NT}{\Omega}\left[1 + B_2\frac{N}{\Omega} + \mathcal{O}\left(\frac{N}{\Omega}\right)^2\right]. \tag{4.2}$$

This formula for osmotic pressure is valid irrespective of the type of solute molecule; the distinctive *polymer* features show up in how the excluded volume B_2 depends on molecular weight n and size R. We may infer how B_2 behaves by considering the pair correlation function $g_p(r)$. For $r \gtrsim R$, two polymers are out of each others' range of influence, so that $g_p(r)$ should be nearly 1. But for $r < R$ there is the possibility that g_p will be strongly influenced. To gauge the magnitude of such an influence, we consider our polymers to be in a good solvent, i.e., one in which the *monomers* have $b_2 > 0$. We have seen that the effect of the b_2 may be reproduced by excluding all configurations in which two monomers approach closer than some small distance of order $b_2^{1/3}$. To see the effect of this exclusion, we take two polymer configurations at random, move them to a separation $r < R$, and then discard the result if there are any intersections. The $g_p(r)$ is the probability that the result is retained, i.e., the probability that there are no intersections. As we saw in Eq. (3.26), the average number of intersections $\langle M_{AB}(r)\rangle \sim r^{2D-d} \sim r^{1/3\dagger}$, and the probability of no contact is significantly less than 1 for $r < R$. For example, $g_p(r) < \frac{1}{2}$ for all r smaller than some fixed fraction αR as $R \to \infty$. (This was a consequence of two polymers being mutually opaque.) This sets a lower bound on B_2:

$$B_2 = \frac{1}{2} \int d^3 r (1 - g_p(r)) > \frac{1}{2}\left[\frac{1}{2}\frac{4\pi}{3}(\alpha R)^3\right]. \tag{4.3}$$

On the other hand B_2 cannot grow faster than R^3: polymers cannot exclude each other from regions arbitrarily larger than their size. There must thus be some constant C for which $B_2 < CR^3$; this is an upper bound. In view of these two bounds, B_2/R^3 must remain finite (and greater than 0) as $R \to \infty$. The polymers increase the pressure Π as though they were mutually impenetrable spheres whose radius is roughly the geometric radius R. This is a consequence of their mutual opacity.

4.1. *Osmotic pressure of hard-sphere gas*   A dilute gas of hard spheres of radius R has a $g(r)$ which is zero for $r < 2R$ and 1 beyond. (a) What is the excluded volume B_2 for such a gas? (b) For what volume fraction $\phi = N\frac{4}{3}\pi R^3/\Omega$ of spheres is the B_2 term in the osmotic pressure equal to the ideal-gas term? (Some amusing points of comparison: a close-packed crystal has a volume fraction of 74%; random close packing has a volume fraction of 64%.)

An important further consequence of this opacity is implicit in the finiteness of B_2/R^3 as $R \to \infty$. We have argued that it is finite whenever the

---

† Here and in the following we shall use the Flory value for D, keeping in mind that the actual number may well be slightly different.

monomer excluded volume b_2 is positive. This means that there is a finite limit for B_2/R^3, when $b_2 \to 0$. More precisely, for any fixed b_2, no matter how small, some finite B_2/R^3 is ultimately reached for large enough R. The *limiting* B_2/R^3 is evidently independent of b_2. As long as b_2 is small enough, it can be varied by a large factor without changing B_2/R^3 significantly. But we have argued that all solvent effects enter only through b_2 (as long as it is positive). Thus any self-repelling flexible polymer must achieve the same value of B_2/R^3. The limiting value is *universal*. Though one can show this convincingly [1], one can only calculate its value very crudely. Thus our best estimates of the ratio come from experimental data like those of Fig. 4.1. Taking R to be the radius of gyration R_G measured in scattering experiments (see Problem 3.11), one finds $B_2/R_G^3 = 4.75 \pm 0.5$ [2]. Another way to express this universality is to define R_t as the radius of the hard sphere with the same B_2 as the polymers. $B_2 \equiv 4(\frac{4}{3})\pi R_t^3$. Then R_t/R_G is also a universal ratio. Table 4.2 summarizes this and other such ratios to be introduced below.

As the polymer concentration increases from zero, the interactions make an increasingly important contribution to the osmotic pressure Π. Ultimately the B_2 term in Eq. (4.2) becomes as large as the ideal solution pressure Π_0. The concentration $N/\Omega = 1/B_2$ where this occurs is an important one in polymer solutions. At this concentration the distance between chains $(\Omega/N)^{1/3}$ is about equal to $B_2^{-1/3}$, which as we have seen is of order R. If polymers were put at random into the solution at this concentration, a sizeable fraction of them would overlap.

The corresponding monomer concentration $nN/\Omega = n/B_2$ is called the **overlap concentration** ρ^* (pronounced "rho star"). This ρ^* is of order n/R^3, the average monomer density within a given chain volume. We have seen that this density goes as $R^{D-d} \sim R^{-4/3}$: ρ^* becomes indefinitely small as $R \to \infty$. Evidently if $\rho \ll \rho^*$, polymer interactions play a small role in the osmotic pressure or the energy of the solution. Conversely, if $\rho \gtrsim \rho^*$, interactions play an important role. The interaction energy, like Π at this overlap concentration is evidently of order of a thermal energy per chain volume: $\Pi \simeq 2\Pi_0 \simeq T R^{-3}$.

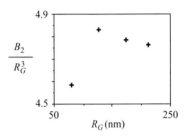

Fig. 4.1
The ratio B_2/R_G^3 versus R_G, as reported in [2], confirming that the ratio is independent of chain length. Data is for polystyrene chains of increasing molecular weight in benzene.

4.2 Semidilute solutions

For long polymers, if $\rho \simeq \rho^*$ the solution is still almost all solvent. One may thus remove some large fraction of the solvent and arrive at a solution much more concentrated than ρ^* which still has an arbitrarily small fraction of solvent. Such solutions are termed **semidilute**. There is no counterpart of this semidilute regime in small-molecule solutions. Small molecules begin to interact only at some finite density ρ. Thus one may only raise the concentration from ρ^* by a finite factor before all the solvent has been removed. In order for a semidilute regime to exist, the solute must be *tenuous* as our fractal polymers are.

As one removes more and more solvent, one arrives at concentrated solutions and then a liquid of pure polymers, with all solvent removed. This liquid is called a polymer melt, with monomer density ρ_{max}. An important measure of concentration is the density relative to that of the melt ρ/ρ_{max};

this is called the **volume fraction** and is denoted ϕ. The overlap volume fraction ϕ^* is clearly much smaller than the melt volume fraction—*viz.* 1. Volume fraction is an unambiguous way to measure concentration, since it does not depend on our definition of the monomer building block for the chain. For typical hydrocarbon polymers, with molecular weight of 10^5, ϕ^* is roughly 1%.

4.2.1 Structure

As the concentration rises above ϕ^* the spatial structure of the polymers is significantly altered. We may express this structure in terms of the local volume fraction analogous to the local density $\langle \rho \rangle_0$: $\langle \phi(r) \rangle_0 \equiv \langle \rho(r) \rangle_0 / \rho_{max}$. The r dependence of $\langle \phi(r) \rangle_0$ for $r \ll R$ can be written $(A/r)^{d-D}$; the length A depends on the type of polymer and solvent but not on its length n. Thus e.g., for polystyrene in benzene $A \simeq 1$ nm. If $\phi \gg \phi^*$, there is some distance $r \ll R$ for which $\langle \phi(r) \rangle_0 = \phi$. This distance, called the **correlation length** $\xi_\phi = A \phi^{-1/(d-D)}$, plays an important role in describing the structure of the solution. If we consider a region around a given monomer much closer than this ξ_ϕ, the local density is much greater than the average density. The average density can thus have little effect on the self-avoidance interaction. We infer that the portion of the polymer in this region is but little affected by the other chains. However, for distances larger than ξ_ϕ, the original $\langle \phi(r) \rangle_0$ is much smaller than the average volume fraction ϕ: the chain passing through the origin contributes only a tiny fraction of the overall density. Thus the other chains may alter the self-avoidance effects strongly. The segment of chain within a distance ξ_ϕ is often called a **blob**. Each chain thus may be thought of a sequence of blobs, each having an internal structure that is relatively unaffected by the other chains.

To see how self-avoidance works on scales much larger than ξ_ϕ, we consider global contacts between two blobs that are far apart along their chain. As shown in the preceding section, the swelling of a chain depends on the mutual excluded volume b_2 of its constituent monomers. Here we may take our monomers to be blobs. To find the b_2 of two blobs, we imagine a dilute admixture one-blob chains in the semidilute solution of k-blob chains. Because interpenetration of the blobs is significantly suppressed, this situation is similar to a gas of monomers in a lattice, with k-site molecules filling all the other sites. We discussed the effect of such multi-site solvents in Appendix B of the preceding chapter. There we found that the extended solvent molecules reduce the b_2 between single sites by a factor of k.

To gauge the importance of this reduced b_2, we consider the number of effective contacts within a given chain in the absence of swelling. This chain has a radius R of order $\xi k^{1/2}$ and a volume of order $(\xi k^{1/2})^3$. Each of the k blobs has an effective intersection with roughly $k b_2 / R^3$ other blobs of the same chain. Thus the number of effective intersections is $k^2 b_2 / R^3$. Accounting for the k dependence of R and b_2, this gives an effective number of contacts of $k^{2-1-3/2} \sim k^{-1/2}$. The effective interaction energy between the monomers is thus of order $T k^{-1/2}$—much smaller than the thermal energy. The interaction energy is not sufficient to expand the polymer significantly, since such expansion requires work of the order of T. We infer that a chain has the structure of an unperturbed random walk beyond the scale ξ_ϕ: it is like a random walk of blobs. The local density thus behaves as shown in Fig. 4.2. The density from the chain at the origin falls off as in a self-avoiding walk within the blob radius, and like a simple random walk beyond. Monomers from other chains have a suppressed local density near the origin. These dominate at distances beyond the blob radius. The suppression as a function of distance also obeys a power law ([1], chapter 13). Calculations suggest that its numerical value is about 0.8.

By scattering one can verify the behavior anticipated above. One may isotopically label a dilute fraction of the chains to measure the density profile of individual chains in the solution. As anticipated, on scales smaller than ξ_ϕ they have the fractal dimension of a self-avoiding chain (*viz.* $\frac{5}{3}$), while over distances much larger than ξ_ϕ they have the fractal dimension of a simple random walk (*viz.* 2) [1], as shown in Fig. 3.9. Each random-walk chain has a self density of order R^{-1}. It is arbitrarily small as the chains are made longer, with fixed total density. To achieve the given density with chains of increasing length there must be more and more chains in the volume R^3 pervaded by one chain. The chains must **interpenetrate**. They interpenetrate more and more as ϕ increases above the overlap concentration ϕ^*. This interpenetration will become important when we come to consider how this solution can flow.

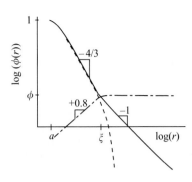

Fig. 4.2
Local volume fractions of monomer $\phi(r)$ at distance r from an arbitrary monomer in a semidilute solution of very low concentration ϕ. Solid curve: monomers from the chain going through the origin. Dot-dashed curve: volume fraction of other chains. Scales are logarithmic, so that power laws appear as straight lines. Within a blob size ξ the total volume fraction is dominated by the chain passing through the origin. Beyond the distance ξ the total volume fraction is dominated by other chains. Simple dashed curve shows excess density $\langle\phi(r)\rangle_0 - \langle\phi(\infty)\rangle_0$, which decays exponentially to zero with a decay length ξ_s.

4.2.2 Energy

The blob picture is also useful for describing the interaction energy and the osmotic pressure in the semidilute state. We saw above that interactions between pieces of a chain larger than ξ_ϕ were unimportant relative to interactions between different chains. Chemically distant parts of a given chain might as well be different chains as far as the interactions are concerned. This means that the interaction energy would be not much changed if the chains were cut into pieces the size of a blob. But we can readily estimate the energy of such cut-up chains, for these chains are at the overlap concentration ϕ^*. The distance between them is of the order of their size ξ_ϕ. As we have noted at the beginning of this section, the interaction energy at ϕ^* is of order T per cut-up chain piece—i.e., T per blob. The same is true for the osmotic pressure: $\Pi \simeq T\xi_\phi^{-3}$. The interaction energy and pressure are oblivious to the chain length in the semidilute regime.

According to this reasoning the quantity $\Pi\xi_\phi^3/T$ is predicted to be a finite number of order unity regardless of the quantitative goodness of the solvent, e.g., b_2/a^3. In this sense it is like the quantity B_2/R^3 encountered in the last section. And for similar reasons, this ratio too is universal. It has been measured via scattering experiments for several polymers and several solvents [3]. One finds $\Pi/T = (3.2\xi_\phi)^{-3}$. Recalling that $\phi = \langle\phi(\xi_\phi)\rangle_0 = (A/\xi_\phi)^{-4/3}$, one finds

$$\Pi/T = (3.2A)^{-3}\phi^{9/4}. \tag{4.4}$$

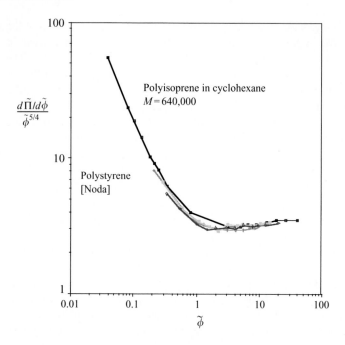

Fig. 4.3

Reduced osmotic compressibility $d\tilde{\Pi}/d\tilde{\phi}$ versus reduced volume fraction $\tilde{\phi} = \phi/\phi^*$ for various polymers and various good solvents. For each polymer sample the $d\tilde{\Pi}/d\tilde{\phi}$ values are normalized so that they approach 1 at low concentration. The $\tilde{\phi}$ values are normalized so that $d^2\tilde{\Pi}/d\tilde{\phi}^2$ approaches 2 at low concentration. Data are shown for one sample of polyisoprene in cyclohexane from [3] and three molecular weights of polystyrene from [4]. Theory [5] predicts that the pressure thus normalized should be a universal function of the normalized concentration. Theory also predicts that $d\tilde{\Pi}/d\tilde{\phi}$ should scale as $\tilde{\phi}^p$ at high concentration, where the exponent p is close to $\frac{5}{4}$. Thus $(d\tilde{\Pi}/d\tilde{\phi})/\tilde{\phi}^{5/4}$ should approach a (nearly) constant value at large $\tilde{\phi}$. Both predictions are well confirmed.

Some experimental confirmation of this universality is shown in Fig. 4.3. The figure also shows that the whole function $\Pi(\phi)$ is universal when properly scaled.

4.2. *Compressibility sum rule* When scatterers are dispersed in a liquid, the small q limit of $S(q)$ has a general interpretation, even when the scatterers are interacting strongly. In this limit, the scattered wave sees only the large-scale modulation of the scatterer density. At such large scales the scatterers can be regarded as a uniform density with slight independent random variations from point to point. As we have noted, a uniform density cannot produce scattering, but the slight variations can. According to Eq. (3.20)

$$S(q \to 0) = \int \frac{d^3r}{M} \lim_{q \to 0} \left[\int d^3r' \langle \rho(0)\rho(r') \rangle \exp(i\vec{q} \cdot \vec{r}') \right].$$

Here M is the number of scatterers in the sample. The limit as $q \to 0$ is not the same as $q = 0$. If $q = 0$ all the forward scattering along the original beam is mistakenly counted and $S(0)$ appears proportional to the sample size. To eliminate this spurious contribution, we note that as long as $q \neq 0$, the scattering cannot be affected by adding a constant to $\rho(r)$ (since $\int(constant)e^{iq \cdot r} = 0$). Accordingly, we replace $\rho(r)$ by $\rho(r) - \bar{\rho}$, where $\bar{\rho} = M/\int d^3r$ is the average density of scatterers. With this choice, the limit $q \to 0$ can be taken:

$$S(q \to 0) = \frac{1}{\bar{\rho}} \left\langle (\rho(0) - \bar{\rho}) \int d^3r' (\rho(r') - \bar{\rho}) \right\rangle.$$

We consider the scattering from some large volume Ω within the sample. The number of scatterers N in this volume is evidently $\int_\Omega d^3r \rho(r)$; its average is

$\bar{N} = \bar{\rho}\Omega$. In terms of N,

$$S(q \to 0) = 1/\bar{\rho}\langle(\rho(r) - \bar{\rho})(N - \bar{N})\rangle = \langle(N - \bar{N})^2\rangle/\bar{N}.$$

The scattering at $q \to 0$ measures the mean-squared fluctuations of the average density. (We have yet to show that this expression is independent of the volume Ω chosen!)

These fluctuations can be related to thermodynamic quantities. We recall that an average like $\langle(N - \bar{N})^2\rangle$ can be expressed

$$\langle(N - \bar{N})^2\rangle = \frac{\int dN \exp(-\mathcal{F}(N)/T)(N - \bar{N})^2}{\int dN \exp(-\mathcal{F}(N)/T)},$$

where $\mathcal{F}(N)$ is the free energy of the fluid constrained to have N particles in the volume Ω. (It includes the work of pushing particles out of Ω into the surrounding fluid. Thus it is not the free energy of an isolated system of N particles in volume Ω.) This $\mathcal{F}(N)$ must be proportional to the volume Ω: $\mathcal{F}(N) = \Omega f(\rho)$; since Ω is large, only slight fluctuations of N around $\bar{N}$ will be significant. For a small range of N around $\bar{N}$ $\mathcal{F}(N)$ can be expressed in the form $\mathcal{F}(N) = \Omega(f(\bar{\rho}) + \frac{1}{2}f''|_{\bar{\rho}}(\rho - \bar{\rho})^2)$, or equivalently, $\mathcal{F}(N) = \Omega(a + b(N - \bar{N})^2/\Omega^2)$. (There is no term proportional to f', because $\langle(N - \bar{N})\rangle = 0$.) (a) Express $\langle(N - \bar{N})^2\rangle$ in terms of the coefficient b and the temperature T. It is convenient to use the equipartition theorem from Chapter 2. (b) Express the osmotic pressure $\Pi(\rho) = -\partial\mathcal{F}/\partial\Omega$ at constant N in terms of $f(\rho)$, $f'(\rho)$, etc. for general ρ. (c) Relate $d\Pi/d\rho$ to f and its derivatives. (d) Combining the results of (a) and (c), show that $S(q \to 0) = T(d\Pi/d\rho)^{-1}$.

4.A Suggested experiment: *Turbidity of polymer solution* If a medium has a compressible density of solute, it will scatter light, as the previous problem shows. The data in Fig. 4.3 was obtained using this connection. As we have shown, the compressibility of a polymer solution goes down when the concentration goes up. This means that it should scatter less light. One can see this scattering by looking through a polymer solution. It is not completely transparent, but is a little milky or turbid. This turbidity shows up in another way when you shine the beam of a laser pointer through the solution. The beam is visible from the side. The intensity of the side-scattered light is proportional to $S(q)$, where $1/q$ is a hundred nanometers or more. Since this $1/q$ is much larger than the correlation lengths ξ in the solution, we can treat q as being essentially 0. Thus the side-scattered intensity is proportional to the osmotic compressibility of the solution. Specifically, if ρ is the number of scatterers per unit volume and Ω is the volume of the beam, the number of scatterers is $\Omega\rho$ and the scattered intensity $I \propto (\Omega\rho)S(q)$. If we shine a fixed beam through solutions of greater and greater concentration ρ, $S(q) = T/(d\Pi/d\rho) \sim \rho^{-5/4}$, as shown in Fig. 4.3. Thus $I \sim \rho^{-1/4}$: the scattered light gets *weaker* as the concentration increases.

You could use an inexpensive laser pointer and gelatin solutions of several concentrations to verify this predicted behavior. You could make up several gelatin solutions of widely varying concentrations. (Keep them warm so they do not gel. Gelation causes extra scattering not accounted for in the theory.) To measure the intensity you could use a light meter from a camera. You could also place the two samples side by side so that the laser beam shone through both of them equally. One will scatter more strongly than the other. Then reduce the beam intensity through the strongly scattering medium until the two scattered intensities look equal. You can reduce the beam by masking off a known fraction of it. Another method is to look at the light from different angles. Though $S(q)$ should be independent of angle, beam looks brighter from lower angles because you are looking at more scatterers in a given projected area. That is, the intensity seen within the beam goes as $1/\sin(\theta)$, where $\theta = 0$ is the beam direction. You could find a way to illuminate the two samples and view the two beams at different angles until their intensities looked the same.

4.2.3   Concentrated solutions and melts

As the concentration is increased further and further above ϕ^* the correlation length shrinks progressively. Ultimately there comes a point where the blobs may no longer be regarded as asymptotically long self-avoiding chains. This is the regime of concentrated solutions. If the monomer excluded volume b_2 is small, a chain must reach a certain size called ξ_T before it shows the self-avoiding properties. This size generally depends on the temperature because b_2 does; accordingly it is called [6] the **thermal blob**[†]. Even in a good solvent with large b_2, our asymptotic picture of the blob breaks down on small length scales. The number of independent steps in the random walk is no longer large. And the concentration within the blob is no longer small enough to treat the monomers as a dilute solution. At these concentrations the general and universal behavior of semidilute solutions breaks down somewhat. As the volume fraction of solvent becomes small, the osmotic pressure required to remove the remaining solvent rises dramatically, until in the melt state the liquid becomes essentially incompressible. But the lack of self-avoidance continues to hold: these chains remain simple random walks on all but the smallest length scales.

[†] The thermal blob size can be inferred from b_2 by requiring that its interaction energy be significant relative to the elastic energy of the unperturbed blob—i.e., 1: $T n^2 b_2 / \xi_T^3 \simeq T$. Here n is the number of monomers in the thermal blob: $\xi_T^2 = na^2$. The result is $\xi_T \simeq a^4/b_2$. This same criterion was used in the preceding chapter in the discussion of theta solvents to determine when the b_2 of a whole chain was sufficient to swell it appreciably.

4.3   Motion in a polymer solution

A large fraction of the distinctive practical behavior of a structured fluid involves the way it flows. It is now time to discuss the molecular motion that gives rise to these distinctive flow phenomena. We must first understand the molecular motions in a quiescent fluid in equilibrium. Before considering polymers we treat the simpler question of how a colloidal sphere of radius R moves.

4.3.1   Brownian motion of a sphere

An object in a fluid moves in response to the thermal fluctuations of the fluid around it. In this section we describe the amount of motion in terms of the viscosity of the solvent. Since the random motions of the solvent are statistically independent over long times, the long-time motion is a random walk. From our discussion of random-walk polymers, we have seen that the long-time behavior depends only on the mean-squared displacement $\langle r^2 \rangle_1$ of the individual steps. Accordingly, we are free to take these steps in a convenient, schematic way. We imagine that the particle moves in discrete steps on a cubic lattice of spacing $\sqrt{\langle r^2 \rangle_1} \equiv b$. For many purposes it is convenient to define the probability current density $\vec{j}(t, r)$: the average number of walkers crossing an imaginary surface at r per unit area in a time step Δt[††]. For definiteness we imagine a horizontal plane lying between two lattice planes. We imagine some smooth distribution

[††] We encountered this current density in the problem on Monte Carlo sampling in Chapter 2.

$b^3 p(t, r)$ giving the probability that the walker is at position r on the lattice. If the walker is just above our imaginary plane, it has a probability $\frac{1}{6}$ of passing down through the plane. The overall probability that this step happens is $\frac{1}{6} b^3 p(t, r)$. On the other hand if the walker is just below the plane, at $r - b$, the net probability that a particle moves up through the plane is $\frac{1}{6} b^3 p(t, r - b)$. The probability current $b^2 j_z$ moving through the plane at this point is the difference of these two probabilities. Since p is assumed to vary smoothly on the scale of a lattice step, we can express $b^2 j_z = -\frac{1}{6}(b^3/\Delta t) b \partial p / \partial z$. Considering the current in arbitrary directions, we have

$$\vec{j} = -\left[\tfrac{1}{6} b^2 / \Delta t\right] \vec{\nabla} p. \tag{4.5}$$

The coefficient in [ ] is called the **diffusion constant** ζ[†].

To characterize the brownian motion further, we now ask what controls the size of ζ. We can now consider the brownian motion of a simple particle, a colloidal sphere of radius R. The solvent around such a sphere may be treated as a continuum fluid with a velocity field $\vec{v}(\vec{r}, t)$. The fluid velocity on the boundary of the sphere must equal that of the sphere, but elsewhere it may fluctuate because of thermal excitation: any configuration $\vec{v}(\vec{r})$ has a probability given by the Boltzmann principle discussed in Chapter 2. Fortunately, one may determine the diffusion constant ζ without treating these fluctuations explicitly, by means of a trick discovered by Albert Einstein [7]. He noted that there is a rigorous connection between ζ of any object and the **drag coefficient** Γ describing the force required to pull it along at some small speed v[††]. To understand the Einstein relation, we imagine that our brownian particle is subjected to a weak gravitational field that exerts a force of magnitude F. As a result of this field we anticipate that the particle gradually moves downward with an average speed $\langle v \rangle$ proportional to F; this proportionality defines the drag coefficient Γ: $F = \Gamma \langle v \rangle$. Thus if the particle is at r with probability $p(r)$, this motion gives rise to a downward current density $j_\Gamma(r) = p(r)\langle v \rangle = -p(r)F/\Gamma$. After the gravitational force has acted long enough for the colloidal particles to come to equilibrium, $p(r)$ becomes a constant distribution given by the Boltzmann Principle: $p(r) = Z^{-1} \exp(-Fz/T)$. Here F represents the magnitude of the force, and z the height; the Z is a normalization constant. Since this $p(r)$ is not uniform, it gives rise to a diffusing upward current $j_\zeta = -\zeta \vec{\nabla} p$, as discussed above. If gravity were suddenly turned off, one could immediately observe this current: $j_\zeta = \zeta F/T p$. But in the equilibrium state with gravity acting, there must be no net current (otherwise $p(r)$ would not be constant in time): $j_\zeta + j_\Gamma = 0$, or $\zeta F/T p = pF/\Gamma$. We conclude that (a) the induced speed $\langle v \rangle$ is indeed proportional to F as anticipated, and (b) its magnitude is given by $\Gamma = T/\zeta$.

One may calculate the drag coefficient Γ using conventional hydrodynamics [9] and thus account for both drag and diffusion. But the basic behavior of Γ can be understood without hydrodynamic formalism. We first recall the relationship between force and velocity $\vec{v}(\vec{r})$ in a simple shear flow, as discussed in Chapter 2. Here the flow was in the x direction but varied only in the z direction. As we discussed, such flows transmit forces

[†] A comment about the connection between mean-squared displacement and ζ is in order. Our random walker takes independent steps of length b at every time interval Δt. Thus the mean-squared displacement $\langle r^2 \rangle(t) = b^2 t/(\Delta t)$. But a diffusing particle with diffusion constant ζ by definition has a mean-squared displacement in a *given* direction of $\langle x^2 \rangle(t) = 2\zeta t$. We may also recover the familiar diffusion equation for the probability $p(t, r)$, by noting that the change of p at r during a time step is the sum of probability currents j onto site r during that timestep: $\Delta p = \Delta t \nabla \cdot j = \Delta t \zeta \nabla^2 p$. This equation has the same form as the random-walk probability Eq. (3.12). See Table 4.1.

[††] This **Einstein relation** is a special case of the **fluctuation-dissipation theorem** relating any small perturbation requiring work to a fluctuating quantity in the unperturbed system in thermal equilibrium [8].

between adjacent fluid elements. We may think of a fluid element as a tiny cube of side b. A surface of a fluid element perpendicular to z experiences a force per unit area in the x direction $\sigma_{zx} = \eta_s \, \partial v_x/\partial z$. Each fluid element has momentum $b^3 \rho_s v$, where ρ_s is the mass density of the fluid. Any net force on the element causes this momentum to change: $\partial \rho_s v_x/\partial t = \partial \sigma/\partial z$. In terms of the viscosity this gives $\partial v_x/\partial t = [\eta_s/\rho_s]\partial^2 v_x/\partial z^2$. This is a diffusion law like that describing the motion of a random walker. It says that momentum **diffuses** in directions perpendicular to the momentum itself; the diffusion constant η_s/ρ_s is proportional to the viscosity and is called the **kinematic viscosity**.

4.3. *Sedimentation* Polystyrene has a density a few percent greater than that of toluene. Assume it is 5% denser. This means that a polymer in toluene is pulled downward by gravity. The point of this problem is to see how fast. (a) Write a formula for the sedimentation rate v in terms of the hydrodynamic radius R_h, the acceleration of gravity, the viscosity η_s of toluene, and the buoyant mass M_b of the polymer. You may assume, as will be shown later, that R_h is a fixed fraction of the radius of gyration R_g from Chapter 3. How does this rate scale with molecular weight? Do bigger polymers fall faster or slower? (b) Do a numerical estimate using 10-million molecular weight polymers. These each displace a volume V_d of about 17 million Å^3 or 17,000 nm^3 in the absence of solvent. Using the formula for local volume fraction $\phi(r) = (0.7 \text{ nm}/r)^{4/3}$, estimate the radius of the chain. (c) Roughly how long would it take such a chain to travel 1 cm in an ultracentrifuge where gravity is effectively multiplied by 10^5?

The transmission of momentum is more complicated when it originates at a point as with our colloidal particle. To analyze this case, we consider a small downward impulse applied at the origin at time $t = 0$. Thus the velocity field is initially a delta function in space. The momentum must propagate out to infinity. It is convenient to think of this momentum by expressing the velocity field as a sum of plane waves: $\vec{v}(r,t) = \sum_{\vec{k}} \vec{v}_{\vec{k}}(t)e^{ik\cdot r}$. For transverse components, with $\vec{v}_{\vec{k}} \perp \vec{k}$, we may use the momentum diffusion equation of the last paragraph: $\partial v_{\vec{k}}/\partial t = \eta_s/\rho_s \, k^2 v_{\vec{k}}$. However, these transverse components are not sufficient to produce our initial impulsive disturbance. Part of this disturbance is longitudinal, having pieces with $\vec{v}_k \parallel \vec{k}$. Longitudinal waves clearly cause local compression of the liquid. To know how such a disturbance spreads, we would have to know what forces result from compression. That is, such spreading depends on the **compressibility** of the fluid; viscosity is not sufficient to describe it.

But we know from elementary physics what happens when a medium is locally compressed: sound waves are generated. Thus two qualitatively different types of propagation occur in response to our impulse. The longitudinal part of the disturbance propagates out at the speed of sound. Sound waves can carry appreciable momentum if the source moves at speeds near the speed of sound. But for the gentle motions of interest here, the production of sound, and the compression of the fluid, are insignificant. The fluid may be treated as **incompressible**. The transverse part propagates by diffusion. As with all diffusion it covers a distance r in a time proportional to r^2. In the initial delta-function disturbance all wave components $\vec{v}_k$ are equal. For each k there are two transverse amplitudes and one longitudinal one, only the transverse ones survive in an incompressible fluid.

We now consider the velocity field from a particle subjected to a constant downward force F in a quiescent fluid. One can determine this field by adding all the transverse wave contributions as described in the previous paragraph. Instead, we may consider the constant force as a sequence of many impulses. The momentum from a given impulse reaches r in a time $t \simeq r^2/(\eta_s/\rho_s)$. Thus the momentum contained in a sphere of radius r is of order $Ft = Fr^2\rho_s/\eta_s$. (Over longer times the momentum begins to flow out of the sphere as fast as it is injected within it.) In terms of the velocity field, this momentum is of order $\int_{r' < r} d^3 r' \rho_s v(r')$. Taking derivatives with respect to r, we infer $Fr\rho_s/\eta_s \simeq r^2 v(r)$, or $v(r) \simeq F/(\eta_s r)$. The exact form of $v(r)$ is known as the **Oseen tensor** [5].

$$\vec{v}(r) = \frac{|F|}{8\pi \eta_s r} \left(\hat{F} + \left(\hat{F} \cdot \hat{r} \right) \hat{r} \right). \tag{4.6}$$

The $\hat{F}$ and $\hat{r}$ are unit vectors in the $\vec{F}$ and $\vec{r}$ directions. The factor in ( ) depends only on the angle the vector r makes with v and with F. It is a consequence of removing the longitudinal part of the v field. If we consider only the downward velocity in response to a downward force, it is twice as strong directly above and below the source as it is on the sides. There are also velocities directed perpendicular to the force F. The fluid flows around the particle. We shall need the Oseen tensor to analyze how the solvent flows near tenuous objects like polymers. A mathematical explanation of the Oseen tensor is given in Appendix A.

To find the drag force, we may equate the rate of work done by the force with the power P dissipated in the viscous flow. This latter has the form $\int d^3 r \eta_s \dot{\gamma}^2$. Here $\dot{\gamma}$ is the shear rate introduced in Chapter 2; in a simple shear flow, it is the derivative of the velocity. In more general flows it is a combination of derivatives of the velocity. In the Oseen field $\dot{\gamma} \simeq dv/dr \simeq F/r^2$. Thus the power density

$$\dot{w}(r) = \eta_s \dot{\gamma}^2 \simeq F^2/(\eta_s r^4).$$

The power dissipated beyond some radius R in the Oseen field is of order

$$\int_R^\infty d^3 r \dot{w}(r) \simeq F^2/(\eta_s R).$$

The power is dissipated predominantly at short distances R of the order of the object size. Since the force does work at a rate $F\langle v \rangle = F^2/\Gamma$, we infer $F^2/\Gamma = P \simeq F^2/(\eta_s R)$, so that $1/\Gamma \simeq 1/(\eta_s R)$. For a sphere, one may calculate the dissipation rate exactly to obtain $\Gamma = 6\pi \eta_s R$. Using the Einstein relation described above gives the **Stokes formula** for the diffusion constant [9]

$$\zeta = T/(6\pi \eta_s R). \tag{4.7}$$

The diffusion constant scales inversely with the sphere size. This scaling is an inevitable consequence of dimensional analysis, once we recognize that only the temperature, the viscosity, and the radius can enter the formula.

Most simple liquids have approximately the same viscosity at room temperature, as discussed in Chapter 2. Thus all particles of a given size diffuse at about the same rate in most solvents. The following benchmark is a convenient way to remember this rate. A sphere of radius $R = 11$ nm in water diffuses its own radius in 1 μs. In this time, its mean-squared displacement is R^2.

This section has shown that diffusive motion of a particle follows the same quantitative laws as a random-walk polymer. Moreover, the flow of momentum away from a source is also governed by a diffusive law, if one properly considers only flow transverse to the momentum itself. Thus each aspect of random-walk polymers that we have a studied tells a corresponding aspect of diffusive motion and of flow. I have compiled a dictionary of corresponding properties in Table 4.1.

4.3.2 Intrinsic viscosity

A large object in a liquid influences motion in another way: by increasing the viscosity of the solution. As we saw in Chapter 2, a shear flow is characterized by the shear rate $\dot{\gamma} = dv/dz$, as sketched in Fig. 4.4. With no

Table 4.1 Three types of diffusion in d dimensions

Quantity	Random-walk polymer	Diffusing particles	Diffusing momentum
Dependent variable	Probability $p(n,r)$	Particle density $\rho(r)$	Momentum density[a] $\rho_m \vec{v}$
Independent variable	Monomer number n	Time t	Time t
Material constant	$\dfrac{\langle r^2 \rangle_1}{2d}$	Diffusion constant ζ	Kinematic viscosity v
Equation	$\partial p/\partial n = (\langle r^2 \rangle_1/2d)(\nabla^2 p)$	$\partial \rho/\partial t = \zeta \nabla^2 \rho$	$\partial v_\perp/\partial t = {}^{b}v\nabla^2 v_\perp$
Mean squared distance from point source	$\langle r^2 \rangle_1 n$	$2d\,\zeta t$	$2d\,vt$
Local density ($d = 3$)	Monomer density at distance r from an arbitrary monomer $3/(\pi \langle r^2 \rangle_1 r)$	Particle density at distance r from a steady point source of i particles per unit time $i/(2\pi \zeta r)$	Velocity at distance r from particle dragged with constant force F $(\rho_m |F|/8\pi vr)[\hat{F} + (\hat{F} \cdot \hat{r})\hat{r}]$

Notes:
[a] ρ_m is mass density of the fluid.
[b] The symbol $v_\perp$ has to be interpreted using the Fourier transform method discussed in Appendix A. It is that part of $\vec{v}$ composed of wave vectors transverse to v.

Fig. 4.4
Extra viscous dissipation from a sphere in a liquid. Arrows show unperturbed velocity field, with shear rate $\dot{\gamma}_0$. Graph at right shows dissipation rate along the vertical dotted line through the sphere. Far from the sphere the rate $\dot{w}(z)$ has the unperturbed value $\dot{w}_0$. Near the sphere the shear rate and dissipation rate are enhanced. Within the sphere, they vanish. The extra dissipation due to the sphere, $\int (\dot{w}(r) - \dot{w}_0)\,d^3r$ is suggested by the shading.

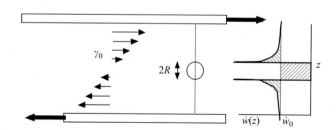

solute present, the shear has a uniform value $\dot{\gamma}_0$ everywhere and a uniform dissipation rate $\dot{w}_0 = \eta_s \dot{\gamma}_0^2$. A solid object such as a sphere perturbs this flow and increases the dissipation. With such objects present, greater dissipation occurs for a given macroscopic flow rate: the macroscopic viscosity is increased. This increase is a major way of sensing the presence of structures in a fluid. To understand how this perturbation works, we consider a single colloidal sphere of radius R in a shear flow between two horizontal sliding plates. The solvent is obliged to flow around the sphere. This increases the shear rate $\dot{\gamma}$ outside the sphere. The sphere alters the shear flow from its initial uniform state of minimum dissipation; thus it increases the dissipation[†]. The dissipation per unit volume $\dot{w}(r)$ is zero inside the sphere. But $\dot{w}(r)$ must be increased just outside the sphere. For example, the integral of the shear rate $\dot{\gamma}$ along the vertical line through the center is the difference in speed of the two plates, as for any other vertical line. Since the shear rate vanishes within the sphere, it must be greater than γ_0 above and below in order to compensate. The perturbation in velocity $v(r)$ caused by the sphere resides mostly near the sphere. It can be thought of as the sum of many Oseen tensors centered at each point of the sphere. Since these produce a velocity field that falls off as $1/r$, the resulting shear rate $\dot{\gamma}(r) = dv/dr$ falls off as $1/r^2$ as discussed above in connection with the drag force. In order to make up the deficit in shear rate within the sphere, $\dot{\gamma}/\dot{\gamma}_0$ must be significantly greater than 1 for distances $r \simeq R$. In this same region the dissipation $\dot{w}(r)/\dot{w}_0$ is also significantly greater than 1. Thus the effect of the sphere is to increase the dissipation by a finite factor in a region of the order of the sphere size. That is, the extra dissipation $\int d^3r(\dot{w}(r) - \dot{w}_0)$ is roughly $\dot{w}_0 R^3$.

By accounting for the excess dissipation, we may find the increased viscosity. On the one hand, the viscosity is defined by $\langle \dot{w} \rangle = \eta \dot{\gamma}^2$, where $\dot{\gamma}$ is the average velocity gradient. On the other hand if the fluid has volume Ω,

$$\Omega \langle \dot{w} \rangle = \dot{w}_0 \Omega + \int (\dot{w}(r) - \dot{w}_0)\, d^3r = \dot{w}_0 \left[\Omega + \int \left(\frac{\dot{w}(r)}{\dot{w}_0} - 1 \right) d^3r \right],$$

Here the first term is the dissipation due to the unperturbed flow throughout the fluid. The second term is the excess dissipation near the sphere. The integrand is significant over a volume comparable to the sphere volume $V = \frac{4}{3}\pi R^3$. The integral can be found explicitly in terms of the known velocity field around the sphere [9,11]. The result is $\int (\dot{w}(r)/\dot{w}_0 - 1)\, d^3r = \frac{5}{2} V$. The dissipation from a sphere is the same as if the dissipation were doubled over a volume equal to $\frac{5}{2}$ that of the sphere. Comparing the two expressions for the dissipation, we infer

$$\eta = \eta_s \left[1 + \frac{5}{2} \frac{V}{\Omega} \right].$$

If the solution contains N spheres far from each other, the dissipation is the sum of that due to each, so that

$$\eta = \eta_s \left[1 + \frac{5}{2} \frac{NV}{\Omega} + \mathcal{O}\left(\frac{N}{\Omega} \right)^2 \right]. \tag{4.8}$$

[†] In a viscous liquid, and many other systems where the forces are *proportional* to the velocities, the energy dissipation rate must be a quadratic function of the velocities. For such systems the steady-state velocity field is that which minimizes this energy dissipation rate. This can be proven from the equations of hydrodynamics, which amount to Newton's laws of motion applied to these linear forces ([10], Section 4.6).

The quantity in [...] is the volume-fraction of spheres ϕ. The relative increase in viscosity measures the fraction of volume in which flow is significantly perturbed. The viscosity senses only the total perturbed volume, and cannot discern the number of solute objects. The effect of a solute on the viscosity is characterized in general by the **intrinsic viscosity** $[\eta]$[†]:

$$\eta \equiv \eta_s(1 + [\eta]\phi + \cdots). \tag{4.9}$$

For an object other than a hard sphere, one may define a **viscometric radius** R_v as the radius of a sphere whose effect on the viscosity is equal to that of the object. Comparing the definition with that of R_v, we find

$$[\eta] = \tfrac{5}{2}\left(\tfrac{4}{3}\pi R_v^3/V_d\right), \tag{4.10}$$

where V_d is the volume displaced by the object (the increase in total volume upon adding one object). Thus $[\eta]$ measures the ratio of the hydrodynamically perturbed volume around an object to its displaced volume V_d[††].

4.3.3   Polymer in dilute solution: hydrodynamic opacity

Flow near a polymer arises in the same way as that near a hard sphere. But the tenuous structure and deformability of a polymer make it unclear how much the polymer flow will resemble the hard sphere flow. We find below that polymers act remarkably solid in their hydrodynamic interaction with the solvent.

A polymer, like a hard sphere, experiences a drag force if it is pulled through the solvent. If the pulling speed is slow enough, the drag force will be much smaller than the force $\simeq T/R$ required to deform it, and there will be negligible deformation. For concreteness it is helpful to model the polymer as a string of n beads, each with radius b and each separated from its predecessor by a phantom bond of length $a \gg b$. The phantom bonds do not interact with the solvent, but each bead acts like a point source of momentum, as we have discussed above. To estimate the drag force, we shall use a reference frame moving downward with the average speed of the polymer. In this frame the fluid has some asymptotic (upward) speed v_∞. We begin by supposing that the polymer is transparent to the flow, so that the flow passes *through* the polymer coil with little perturbation[†††]. Then the speed near each bead is of order v_∞ and the downward forces F transmitted into the fluid are all comparable.

Under this supposition we may readily find the amount of backflow induced by the beads. Each bead i induces a velocity field $v_i(r)$ given by the Oseen tensor. In order of magnitude $v_i(r) \simeq F/(\eta_s(r - r_i))$. The upward velocity at some point r within the polymer is thus of order

$$v_\infty - \sum_i v_i(r) \simeq v_\infty - \int d^3 r_0 \rho(r_0)\frac{F}{\eta_s(r - r_0)}.$$

[†] Here we depart slightly from the standard definition used in physical chemistry books. In the standard definition the volume fraction ϕ is replaced by the concentration in mass per unit volume.

[††] Analogous behavior occurs for a scalar diffusing field, such as charge carriers in a conductor. Here the current j is driven by the electric potential gradient E. If a perfectly conducting sphere is embedded in a slab of material, there is increased current density j and dissipation around it, and the average conductivity σ defined by $\dot{w}(r) = \sigma E^2$, increases. The overall conductivity increases by a factor which is a universal constant times the volume fraction of embedded spheres. This behavior depends on the dimension of space. In two dimensions, the effect of an embedded disk is not strongly confined to the disk region and the conductivity due to it is much larger than its area fraction. The same is true for flow in two dimensions. [The author thanks J. Wyman and I. Cohen for a useful discussion on this point.]

[†††] This assumption is called the **Rouse model** [12] for flow near a polymer.

Table 4.2 Universal ratios of length dimensions of a dilute polymer solution[a]

Ratios[b]	R_G/R_h	R'_G/R_h	R_t/R_h	R_v/R_h
Hard sphere	$(3/5)^{1/2} = 0.775$	1	1	1
Ideal chain (theory)[c]	1.48	1.91	0	1.23
Self-avoiding chain (theory)[c]	1.56	2.01	1.04	1.14
Polyisoprene in cyclohexane[d]	1.39	1.79	0.95	1.11
Polystyrene in benzene[d]	1.51	1.94	1.01	1.03
Poly(α-methylstyrene)[d]	1.55	2.0	1.05	1.11

Notes:
[a] Adapted from [14], Table VI, using their original data and earlier data they compiled from various sources.
[b] R_h is the hydrodynamic radius—the radius of a hard sphere with the same diffusion constant as the polymer. R_G is the radius of gyration, as defined in Problem 3.11. $R'_G \equiv (\frac{3}{5})^{-1/2} R_G$ is the radius of a uniform sphere with the same R_G as the polymer. R_t is the **thermodynamic radius**, that of hard spheres having the same excluded volume B_2 as the polymers. R_v is the viscometric radius, that of hard spheres having the same intrinsic viscosity as the polymers.
[c] These theories estimate the given ratios by treating $\epsilon \equiv (d - 4)$ as a small parameter and calculating the given ratio to first or to second order in ϵ.
[d] These experiments used scattering and viscosity to measure the various R values for different molecular weights. The limiting ratios for high molecular weights are reported.

If we take as our origin some bead of the polymer, then on average $\rho(r_0) = \langle \rho(r_0) \rangle_0 \sim r_0{}^{D-3}$, where D is the fractal dimension of the polymer. Thus

$$v(r) - v_\infty \sim \int^R d^3 r_0 r_0{}^{D-3} (r - r_0)^{-1}.$$

The integration is now of the same form as Eq. (3.26) for M_{AB}, the number of intersections of two fractals with dimensions D and 2 in three-dimensional space. For all fractals with $D > 1$, including our polymer, the integral goes as a positive power of R. We say that such fractals are **opaque to the flow**. We note that opacity is attained for large enough R no matter how small the individual beads were or how weakly they interacted with the fluid.

The result of this flow opacity is to invalidate our supposition that each bead feels the unperturbed speed v_∞. For this leads the conclusion that $v(r) - v_\infty$ diverges with R. In that case the backflow would be stronger than the asymptotic speed v_∞. To avoid this absurd result, we conclude that the fluid speed within the fractal is actually much smaller than v_∞, so that the forces F_i on each bead are also smaller than an isolated bead would feel. Thus, the fluid cannot move transparently through the fractal; the flow must go around it. The flow speed throughout the fractal volume is suppressed by at least a finite factor.

Because a fractal screens out the flow within itself qualitatively, it also alters the flow far away. The far-field velocity must extrapolate to a finite fraction of v_∞ at $r = R$. This means that the far field is that of a hard sphere whose radius is of order R[†]. Thus a fractal with $D > 1$ in a uniform flow causes dissipation like that of a hard sphere whose radius is of order R. Accordingly, its drag coefficient Γ or hydrodynamic radius R_h are equal to R up to a finite factor independent of the microscopic structure. A polymer must also perturb a shear flow like a hard sphere of comparable radius. Thus its effect on the viscosity is like that of a hard sphere whose radius R_v is of order R. Like other asymptotically finite ratios we have encountered above, the R_h, R_v as well as the thermodynamic R_t defined above are universal multiples of the radius of gyration R_G for any self-avoiding polymer. Table 4.2 summarizes the experimental data on these ratios. Remarkably these universal ratios are not far from 1.

[†] This hard-sphere behavior was first proposed by Zimm, and is called the **Zimm model** [13].

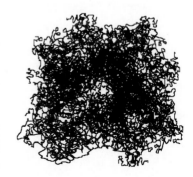

4.4. *Diffusive analog of hydrodynamic opacity* We saw in this section that tenuous fractals can interact strongly with each other by causing the surrounding solvent to move. We suggested that this interaction could be understood in terms of the mutual opacity of two fractals. This analogy is clearest if we consider the diffusive analog of hydrodynamics. Suppose a region has a uniform density of diffusing particles (random walkers), of density u in three dimensions. A fractal of radius R and dimension D is placed at the origin, as shown above. This fractal has the property of absorbing any diffusing particle that touches it. In the presence of this absorption, the density u is no longer constant: $u(r) < u(\infty)$. The picture shows the tracks of many random walkers. All tracks that intersected the fractal have been removed, leaving a depleted region around the fractal. The relative density $u(r)/u(\infty)$ is simply the probability that an (indefinitely long) random walk ending at r within the fractal does not intersect the fractal. (a) Using the formula for the intersection of two fractals, show that the relative density at the origin approaches one as $R \to \infty$ if the dimension d of space is high enough. How high must d be? For such cases, we may say that the fractal is *transparent* to a diffusing substance. In the complementary case, the relative density is reduced within the fractal by a factor of order unity (at least). In such cases, we say that the fractal is **opaque to a diffusing substance**. (b) Consider a point $r \gg R$. A walker at r has a small probability of intersecting the absorber, and thus a small probability of being removed. This probability is proportional to its local density at the origin. From this fact, find how $u(\infty) - u(r)$ varies with r.

4.3.4 Internal fluctuations

In addition to their overall diffusive motions, polymers have spontaneous internal motions. These motions can readily be seen by dynamic scattering. If one scatters from dilute polymers at a scattered wavelength $\lambda \equiv (2\pi)/q \ll \xi_\phi$, one observes a time-dependent relaxation, as we will discuss in the semidilute diffusion subsection. The characteristic relaxation time $\tau(q)$ for such scattering is roughly the time for a given scatterer to move a distance λ via internal Brownian motion. The dominant motion for scales of size λ is for the scatterer to move in concert with a segment of size λ. The time $\tau(q)$ is thus similar to the time for a sphere of size λ to diffuse its own diameter, i.e.,

$$\tau(q) \simeq \lambda^2/\zeta_\lambda \simeq \eta_s \lambda^3/T \simeq \eta_s q^{-3}/T. \tag{4.11}$$

This predicted dependence is well confirmed by experiments [15].

4.3.5 Hydrodynamic screening

The suppression of flow within a polymer coil shown above is an example of **hydrodynamic screening**. This screening is essential for describing motion in a semi-dilute polymer solution. As we have seen above, hydrodynamics describes the flow of a conserved quantity: transverse momentum. If something is present in the fluid to absorb this momentum, screening results. We now illustrate this effect with a quantitative example. We imagine a standard shear flow between two horizontal plates at separation h, when the lower one slides from left to right with a speed $v(0)$. In an ordinary fluid this leads to a horizontal velocity $v(z) = v(0)(1 - z/h)$. Now we place a number of tiny spheres in the fluid and hold them stationary by some means. The fluid is not trapped at all by these spheres, but their drag forces take momentum from the fluid. A small volume b^3 experiences a drag force proportional to $v(r)$ and to the number of spheres inside. The force can be expressed as $-b^3\Gamma\rho_2 v(r)$, where ρ_2 is the density of spheres. The other force on this volume element is the viscous force $\eta_s b^2 dv/dz$ on the bottom and the top of the element. In steady state, if the fluid is not accelerating, these forces must balance: $\eta d^2 v/dz^2 = \Gamma\rho_2 v$. This familiar equation has solutions exponentially growing with z or exponentially decreasing. The physical solution is the decreasing one[†]: $v(z) = v(0)\exp(-z/\xi_h)$, where $\xi_h^{-2} = \Gamma\rho_2/\eta_s$. The flow is now confined to a region of size ξ_h near the bottom plate. Most of the momentum is absorbed before reaching the top plate. The same hydrodynamic screening length ξ_h appears when many such one-dimensional flows are combined to make more general flows. For example the Oseen tensor describing flow around a moving sphere acquires a damping factor $\exp(-r/\xi_h)$. Thus if a small sphere is moved through a fluid containing stationary spheres, the flow around it is not much perturbed for distances $r \ll \xi_h$. But for distances $r \simeq \xi_h$, the velocity is reduced by a substantial factor.

If a small sphere is moved through a semidilute polymer solution, hydrodynamic screening also occurs, since the polymers absorb momentum being supplied by the moving sphere[††]. If the small sphere is inserted at random into the fluid, the typical distance to a monomer is the correlation length ξ_ϕ introduced above. Thus if one examines the flow at distances $r \ll \xi_\phi$, there are essentially no monomers there to impede it; accordingly, the flow is unscreened. However, at distances $r \simeq \xi_\phi$ the polymers alter the flow. If our sphere were near an isolated polymer of radius ξ_ϕ, the flow would be damped appreciably within the coil volume because of its opacity, as discussed above. The same is true for a solution of polymers of size ξ_ϕ at their overlap concentration ϕ^*. It is still true if these polymers are connected together to form a semidilute solution with the same ϕ and ξ_ϕ. At distances of order ξ_ϕ from the sphere, there is an appreciable chance that monomers are present to impede the flow speed by a finite factor. Since screening reduces the flow by a finite factor at $r \simeq \xi_\phi$, we infer that the screening length is of order ξ_ϕ. At distances much larger than ξ_ϕ, the polymer solution may be regarded as a more-or-less uniform mass of momentum-absorbing blobs. Thus exponential damping of the flow is expected, as in the last paragraph, with screening length $\xi_h \simeq \xi_\phi$.

[†] Strictly, some of the exponentially growing solution is needed to satisfy the boundary condition $v(h) = 0$. Here we restrict ourselves to a thick layer with $h \gg \xi_h$, so that this part is negligible.

[††] You may wonder what the polymers do with the momentum they have absorbed. The polymers are not attached to the container, so they cannot absorb momentum without starting to move themselves. We will consider the fate of the momentum absorbed by the polymers in Problem 4.8.

4.3.6 Semidilute diffusion

With this knowledge of hydrodynamic screening we may understand the large-scale thermal motion in a polymer solution. In a semidilute solution, one may distinguish two different kinds of motion. One is the overall relative motion between the polymers and the solvent. The other is motion of a given polymer relative to the solution. We treat the overall motion first.

Whenever the monomer concentration profile $\rho(r)$ in the solution is not uniform, a flow of polymer and solvent occurs so as to restore uniformity. This flow is described by a current density of monomers $j(r)$ analogous to the probability current defined above in the discussion of brownian motion of a sphere. If the nonuniformity is small, we expect the current to be proportional to the departure from uniformity, so that $j = -\zeta_c \nabla\rho$, again like the probability current. Thus the amount of motion is described by a diffusion constant ζ_c; this is called the **cooperative diffusion constant**. It is called cooperative since it describes the collective motion of many chains. As with the simple diffusion discussed above, the cooperative diffusion can be described in terms of the local concentration ρ or the volume fraction ϕ alone:

$$\partial\phi/\partial t = (-\nabla \cdot j)/\rho_{\max} = \zeta_c\nabla^2\phi. \tag{4.12}$$

We now ask how ζ_c should behave as concentration increases from the dilute limit. If a large-scale density gradient is produced in dilute conditions, the relaxation of the gradient occurs by the independent brownian motion of the polymers. Then cooperative diffusion is no different from the diffusion of the individual polymer chains. This remains true as long as each polymer moves independently of the others. But as ϕ approaches ϕ^* their interactions become appreciable. The osmotic pressure variation caused by the gradient is increased by the polymer interactions by a factor of order unity (in a good solvent). The interaction produces a stronger average force on each polymer away from regions of high density. This effect tends to speed up the diffusion. However, the motion of each polymer is inhibited somewhat by its neighbors. The thermal motion of the fluid around each one is damped by hydrodynamic screening from these neighbors. Still, both of these effects can only change the cooperative motion by a finite factor near ϕ^*. Thus

$$\zeta_c(\phi^*) \simeq \zeta \simeq T/(\eta_s R). \tag{4.13}$$

From this fact we can infer the behavior of ζ_c far above the concentration ϕ^*. We may attain this semidilute regime starting from a solution near ϕ^* by joining chains together, as we have done previously. Then each of the original chains becomes a blob of the solution, and each original chain's size R is now of the order of the blob radius ξ_ϕ. The joining of these polymers does not change the distribution of monomers in space very much; accordingly the solvent is free to flow around and through the blobs about as much as it could with the original solution before joining. Thus a gradient of density produces a current of monomers of about the same size whether the polymers are joined or unjoined. This means that the cooperative diffusion constant ζ_c is about the same in the two cases:

$$\zeta_c \simeq T/(\eta_s\xi_\phi) \sim \phi^{3/4}. \tag{4.14}$$

Significantly, ζ_c, like the correlation lengths and the osmotic pressure, is independent of chain length in the semidilute regime.

Cooperative diffusion can be measured by imposing a macroscopic concentration gradient and monitoring its decrease over time by some means. But it can also be observed directly by scattering from the equilibrium solution. Any instantaneous density profile $\rho(r)$ can be expressed as a superposition of plane waves $\rho(r) = \sum_q \tilde{\rho}_q e^{iq\cdot r}$, for some given complex amplitudes $\tilde{\rho}_q$. Applying the diffusion law of Eq. (4.12) to $\rho(r)$ we find an equation for the time derivative of $\tilde{\rho}(q)$: $\partial/\partial t\tilde{\rho}(q) = -q^2\zeta_c\tilde{\rho}(q)$. The amplitude evidently decays exponentially with a decay time $\tau(q) = (q^2\zeta_c)^{-1}$. As shown in the last chapter, waves scattered at wavevector q have an intensity $I(q)$ proportional to the square of this amplitude; thus,

Table 4.3 Universal semidilute length ratios inferred from polystyrene solutions[a]

Quantity	Theta solvent[b]	Carbon disulfide[c]	Toluene[d]
A (nm)[e]	0.32 ($\pm 7\%$)	0.28 ($\pm 7\%$)	0.32($\pm 7\%$)
ξ_ϕ/ξ_s[f]	0.61($\pm 9\%$)	1.23 ($\pm 8\%$)	1.23 ($\pm 9\%$)
ξ_Π/ξ_s[g]	2.97 ($\pm 5\%$)	–	3.81 ($\pm 6\%$)
ξ_f/ξ_s[h]	–	1.27($\pm 1\%$)	–
ξ_ζ/ξ_s[i]	4.29 ($\pm 10\%$)	–	1.65 ($\pm 7\%$)
r_2[j]	3 ($\pm 22\%$)	1.2 ($\pm 14\%$)	1.1 ($\pm 23\%$)
r_4	48 ($\pm 28\%$)	4.5 ($\pm 21\%$)	4.3 ($\pm 29\%$)

$I(q, t) \propto |\rho(q, t)|^2 = I(q, 0)e^{-2t/\tau(q)}$. The decay time is roughly the time for the density to diffuse a wavelength $2\pi/q$.

If many initial states were prepared at random with proper Boltzmann-weighted probabilities, and the intensity $I(q, t)$ monitored for each, each experiment would show a different initial intensity, but the same decay time $\tau(q)/2$. One may measure this decay conveniently by measuring the **correlation function** $\langle I(q, t + t')I(q, t')\rangle_{t'}$, where $\langle \ \rangle_{t'}$ is the average over initial times t'. Under the diffusive law, this correlation function has the form [16]

$$\langle I(q, t + t')I(q, t')\rangle_{t'} = \langle (I(q) - \langle I(q)\rangle)^2\rangle \, e^{-2t/\tau(q)} + \langle I(q)\rangle^2. \quad (4.15)$$

The scattering from an equilibrium fluid behaves in this way, as well. The random thermal fluctuations have the effect of preparing initial states of random amplitudes, which then decay according to the diffusion equation. At any given moment the intensity is the cumulative effect of many partially decayed random fluctuations.

Careful scattering measurements have been done to test the predicted behavior of $\tau(q)$. The data [17] show good consistency with the ϕ dependence predicted above. They lead to the estimate $\zeta_c = T/[6\pi\eta_s(\alpha\xi_\phi)]$, where $\alpha = 1.3 \pm 10\%$, as inferred from Table 4.3.

4.5. *Cooperative diffusion and permeability*   The cooperative diffusion constant ζ_c tells the flow of concentration in response to a concentration gradient. One can readily infer from this the flow of fluid in response to an osmotic *pressure* gradient. The ratio of fluid velocity to a pressure gradient is defined as the **permeability** P: $\vec{v} = \vec{j}/\rho = -(P/\eta)\vec{\nabla}p$. The permeability is defined for flow of any fluid; no polymers or other solute need be involved. Here η is the viscosity of the fluid and p is the pressure, j is the current density of fluid and ρ is the particle density of the fluid. Evidently the permeability has the dimensions of a length squared. (a) Find the permeability P relating the average velocity to the pressure gradient in a long circular pipe of radius R. Recall that the velocity profile in the pipe is parabolic while the pressure is constant over the cross-section of the pipe. You can find the pressure gradient by equating the power required to maintain the pressure with the viscous energy dissipation in the fluid. (b) If monomers with number density ρ_n are flowing to the right with a current density j_n, what is the average speed of monomers with respect to solvent? Assume monomers have the same volume as solvent molecules. Considering the monomers as fixed, what is the solvent current j_s in terms of j_n and volume fraction $\phi = \rho_n/(\rho_n + \rho_s)$? (c) A semidilute solution has a monomer density ρ_n and solvent viscosity η_s. Relate its permeability P to its cooperative diffusion constant ζ_c and the osmotic compressibility $d\Pi/d\rho_n$.

Notes:
[a] These ratios were inferred by Huang and Witten [18] using scattering and cooperative-diffusion data from the literature.
[b] These values were inferred from data on polystyrene in the theta solvent cyclohexane at 36 °C [19–22].
[c] These values were inferred from data on perdeuterated polystyrene in the good solvent carbon disulfide at 20 °C [23–25].
[d] These values were inferred from data on ordinary polystyrene in the good solvent toluene at 25 °C [26–28].
[e] Fractal amplitude defined by $\langle \phi(r)\rangle_0 = (A/r)^{D-d}$, as in the subsection "Semidilute Solutions" above.
[f] ξ_s is the correlation length inferred from scattering: $S(q) = const.(1 - (q\xi_s)^2 + \mathcal{O}(q^4))$. ξ_ϕ is the distance from a monomer at which the local volume fraction $\langle \phi(r)\rangle_0$ of a large single chain inferred from high-q scattering is equal to the overall volume fraction ϕ: $\langle \phi(r)\rangle_0|_{r=\xi_\phi} = \phi$, as described in the subsection "Semidilute Solutions."
[g] ξ_Π is the length scale implicit in the osmotic pressure Π: $\Pi = T\xi_\Pi^{-3}$.
[h] ξ_f is defined by $S(q) = S(0)(q\xi_f)^{-D}$ for q in the fractal regime.
[i] ξ_ζ is the radius of a sphere whose stokes diffusion constant is the same as the cooperative diffusion constant. It is rigorously related to ξ_Π and the permeability length ξ_p, as discussed in the text.
[j] r_n is the reduced moment of the concentration profile defined by $r_n \equiv \int_0^\infty dx x^n(\langle \phi(x\xi_\phi)\rangle_0/\phi - 1)$. The r_2 and r_4 determine the ratios $\xi_s : \xi_\phi : \xi_\Pi : \xi_f$, as explained in [18].

(d) Recognizing that Π is a power of ρ_n and can be related to the correlation length ξ_Π by T/ξ_Π^3, relate the permeability to ξ_Π and the hydrodynamic correlation length ξ_h defined by $\zeta_c = T/(6\pi\eta_s\xi_c)$. (e) What diameter pipe has the same permeability as a semidilute solution whose scattering correlation length is ξ_s? For this estimate, you may assume that the various ξ's are equal: $\xi_h \simeq \xi_\Pi \simeq \xi_s$. You may ignore numerical factors.

The solution is given at the end of the chapter.

Though the diffusive law for ρ is valid over sufficiently long times and distances, other effects can modify the relaxation of ρ over shorter times. The chief one is elastic stress in the polymer chains. We shall see below that such stress relaxes in a time that is independent of the length scale of observation. Thus for sufficiently great observation scales, this stress relaxation is always much faster than diffusion. Then the stress relaxation effects have a negligible effect on the observed density relaxation.

4.3.7 Semidilute self-diffusion without entanglement

The diffusion of a single chain behaves quite differently from that of the overall density. As the concentration rises towards ϕ^* the chain's motion becomes inhibited by its neighbors and it diffuses slower. Again the **self-diffusion** constant ζ_s changes by a finite factor at $\phi \simeq \phi^*$. And again we can use the ϕ^* behavior to infer that for semidilute chains. We begin by selecting k chains at random at ϕ^* and considering the diffusion of their center of mass. Each chain diffuses because of the random motions of the fluid around it. Different chains are typically several hydrodynamic screening lengths from each other because of hydrodynamic screening[†]. The random fluid motion is separate and independent for such chains. Accordingly their brownian motions are independent. After a time t the center of mass displacement $\bar{r}(t)$ can be expressed in terms of the displacements $r_i(t)$ of the chains i: $k\bar{r}(t) = \sum_{i=1}^{k} r_i(t)$. The mean-square displacement $\langle \bar{r}^2 \rangle$ can be written

$$k^2 \langle \bar{r}^2 \rangle = \left\langle \left[\sum_{i=1}^{k} r_i(t) \right]^2 \right\rangle$$

$$= \sum_{i,j}^{k} \langle r_i(t) r_j(t) \rangle.$$

Insofar as the individual chains move independently, there is no correlation between r_i and r_j, so that $\langle r_i r_j \rangle = 0$ unless $i = j$. Also, when $i = j$, $\langle r_i r_i \rangle = \langle r^2 \rangle$ is independent of the chain chosen. Thus $k^2 \langle \bar{r}^2 \rangle = k\langle r^2 \rangle$. Thus the diffusion constant $\zeta_{cm} = \langle \bar{r}^2(t) \rangle / t \sim \zeta/k$, where ζ is the diffusion constant of one of the chains.

[†] This is a good starting assumption, but it is not rigorously valid, as explored in Problem 4.8.

4.6. *Damping of jello*   In Problem 2.7 we calculated the damping expected for a jiggling cube of Jello owing to the viscous dissipation of the water. We assumed that the flow was just that of pure water undergoing the overall motion of the block of Jello. We found that this dissipation appeared too little to account for the damping of real Jello. In the previous section we have recognized that relative flow of solvent through a polymer solution like Jello produces an extra dissipation. This problem revisits the Jello question to see whether we can account better for its damping. If a block of Jello were stretched to double its natural length, it would eventually expel a finite fraction of the water inside. The stretched water wants to conserve its volume when stretched, while the polymer network wants to increase its volume. Thus when the Jello is distorted, there is a current of water through the polymer network. From this we can estimate the dissipation of Jello. From the oscillation frequency and density of Jello, we know its modulus G is roughly 4000 J/M^3 (or 40,000 erg/cm^3). This modulus is of the same order as the osmotic pressure. (a) From this information, estimate the blob size ξ. A block of Jello is rapidly subjected to a shear strain of magnitude γ. This strain produces an osmotic pressure difference of order $G\gamma$ between the center and the outside. (b) Estimate the rate of energy dissipation per unit volume caused by the resulting flow in a cube of side b. (c) Assuming this dissipation has a small effect on the motion, estimate the damping rate for a jiggling cube of Jello 5 cm on a side.

4.B Suggested experiment: *Jello elasticity*   The theory of polymer elasticity makes some clearcut predictions about Jello that you can check. There is a clear predicted dependence of the modulus and hence the oscillation frequency as one changes the concentration. Try checking this by making up some Jello samples at different strengths, ranging from half strength to quadruple strength. It is probably better to use unflavored gelatin instead of Jello. Measure the oscillation frequencies and damping rates for samples of the same size and shape but different concentrations. You could gauge the effect of sliding motion against the bottom by using bottoms of different consistency, including a thin layer of more concentrated Jello. From these measurements infer the change of modulus. Does the modulus change with concentration as predicted for a semidilute solution? Note that the modulus depends on cooling time, so cool for a long time.

We now form a semidilute solution from the original ϕ^* solution by joining up all the chains in groups of k chains to make long, interpenetrating ones. We compare the motion of the center of mass of one k chain with the center of mass motion of the last paragraph. The motion of this connected chain is now different in many ways from that of the disconnected chains; for example, the constituent chains cannot now move far apart. These differences arise from the forces between the k connected subunits of the k chain, formerly absent. However, such internal forces cannot affect the motion of the center of mass. Accordingly, we infer that the long chain diffuses like the separated subchains; its diffusion constant is reduced by a factor k relative to that of one subchain. The subchains, defined to be at their ϕ^* have size $\simeq \xi_\phi$ and diffusion constants ζ roughly as large as they would have been in dilute solution. This ζ is that of cooperative diffusion, as we have just seen. The number of monomers $M(\xi)$ in one subchain (or blob) is of order ξ^D. The molecular weight M is evidently $kM(\xi)$. Combining, we may obtain the scaling of the self-diffusion constant ζ_s:

$$\zeta_s \simeq \zeta_c/k \sim M^{-1}\xi^{D+2-d} \sim M^{-1}\phi^{(D+2-d)/(D-d)}. \qquad (4.16)$$

For polymers in a good solvent $\zeta_s \sim \xi^{2/3}/M \sim \phi^{-1/2}/M$.

4.3.8 Motion with entanglements

The above picture reduces the motion of a chain in solution to that of an object whose parts are subjected to independent random forces. This simple model of motion is called the **Rouse model** [5]. The Rouse model describes self-diffusion reasonably well when the concentration is not too far above ϕ^*. But for solutions far above ϕ^* we must consider another difference between our original solution of subchains and the final solution of connected k-chains: *entanglement*. The forbidding task of accounting for entanglements quantitatively was reduced to a simple and intuitive model by Edwards and by deGennes [5, 10] in the 1970s. Our discussion below recounts these ideas.

Intuitively, entanglements between chains impose constraints that go beyond mere hydrodynamic drag. We may begin to anticipate how they influence a given chain by replacing the other chains with fixed, regular obstacles, such as a regular lattice of bars of spacing ξ. If a random walk polymer threads through such a "jungle-gym" lattice, it certainly cannot diffuse according to the Rouse model discussed above. If a random force pushes the chain against one of the bars, the bar exerts a restoring force that prevents the chain from crossing it. The chain is confined. If each end of the chain were to be fixed, this confinement would be permanent; the chain must forever follow the same path through the jungle gym that it followed initially. But if the ends are free to move, there is one type of motion it can perform without a restoring force from the bars. This is for each monomer to move in the direction of its successor along the chain. The result of such a motion is equivalent to removing a bit of the chain at one end and adding it on to the other end. This motion is called **reptation** [5].

Of all the different small random motions of our chain only a reptational motion results in no restoring force from the bars. Thus for this motion the random thermal forces are free to act unopposed. This means that the chain is free to do random small steps along its own path: it performs curvilinear brownian motion. The displacement along the path $s(t)$ obeys $\langle s^2(t) \rangle \propto t$. The constant of proportionality is the curvilinear or tube diffusion constant ζ_t. This diffusion motion is the same as it would be if the contorted sequence of blobs were straightened out into a straight path of length $L \simeq \xi k$. Each section of the chain of size ξ is subjected to independent random forces, as in the Rouse model above. Accordingly the diffusion constant for the overall motion of a chain containing k such sections is reduced by a factor of k. The curvilinear diffusion constant is simply the Rouse diffusion constant derived above, *viz.* $\zeta_c(M(\xi)/M)$. Under this diffusion, the chain ultimately moves a curvilinear distance equal to its path length L. At this point the chain is no longer constrained by its original entanglements. The time τ_{rep} required is given by $L^2/\tau_{\text{rep}} \simeq \zeta_c(M(\xi)/M)$. This τ_{rep} plays a central role in many phenomena of entangled polymers. Using $L = \xi(M/M(\xi))$, we may infer the scaling of τ_{rep}:

$$\tau_{\text{rep}} \sim M^3 \xi^{d-3D} \sim M^3 \phi^{(3D-d)/(d-D)}. \tag{4.17}$$

For polymers in a good solvent $D \simeq \frac{5}{3}$ and $\tau_{\text{rep}} \sim M^3 \phi^{3/2}$. The reptation picture predicts the self diffusion of entangled polymers. In a time τ_{rep} the chain has moved a spatial (as distinguished from curvilinear) distance of the order of its own size R. Each subsequent interval τ_{rep} leads to another independent translation of order R. Thus the long-time motion is diffusive, with a diffusion constant $\zeta_{\text{rep}} \simeq R^2/\tau_{\text{rep}}$. Using $R^2 \simeq \xi^2(M/M(\xi))$ one finds in good solvents that

$$\zeta_{\text{rep}} \sim M^{-2}\phi^{-7/4}. \qquad (4.18)$$

It is wise to verify this important result.

Real entanglements between polymers differ from our jungle-gym entanglements in two basic ways. First, the polymers are not a regular array of lines, but are random coils. This means, for instance that the test chain encounters a given chain at qualitatively more places than it would encounter a given bar in the jungle gym. Nevertheless, experiments and computer simulations of the diffusion of a mobile chain among frozen or immobilized chains confirm the reptation picture [10]. They support the view that the entanglement constraints confine each chain to its original path. A second defect of the jungle-gym model is that it neglects the motion of the other polymers. This motion raises the possibility that the chains can avoid the entanglement constraints by some type of collective motion. While we cannot rule out this possibility, we can argue that the reptation hypothesis is self-consistent even when the other chains move. If all the chains move by reptation, a test chain may release an entanglement constraint imposed by a given bar in two ways. The first way, treated above, is to reptate away from the constraining bar. The second is for the chain represented by that bar to reptate so that it no longer constrains the test chain. The time required for a given constraint to be released in this second way is the time for the constraining chain's end to move to the constraining point. Since the constraining chain is also assumed to move by reptation, this time is roughly the reptation time τ_{rep}. Thus in the time τ_{rep} required for the test chain to escape by reptation, a finite fraction of the entanglements have also been released by their own reptation. This allows the test chain to diffuse slightly further, but does not alter our simple scaling picture.

The reptation model for diffusion has recently been stringently tested by experiments and simulations [29]. In addition to measuring the reptation time, one may also check further predicted details of the motion and check predicted changes in the diffusion when different chain lengths coexist in the solvent. Overall these tests give strong confirmation to the reptation model and its recent refinements.

4.7. *Diffusion in a theta solvent*   If polymers are dissolved in a theta solvent in the semidilute regime, the cooperative and self-diffusion constants are modified because the blobs are random walks with $D = 2$ instead of self-avoiding walks with $D = \frac{5}{3}$. How do ζ_c, ζ_s, and ζ_{rep} scale with molecular weight M and volume fraction ϕ in theta solvents? That is, do they change faster or slower than with good solvents?

4.3.9 Stress relaxation and viscosity

Entanglement also controls how elastic energy is stored and dissipated in a polymer solution or melt. We have studied the work required to *compress* polymers into a smaller volume of solution; it is given by the osmotic pressure Π. The polymers also store energy when the solution is distorted, even with no change of volume. We discussed how this energy is measured in the Experimental Probes section of Chapter 2. There we told how rheometers apply a known, oscillating shear strain and measure the stress that oscillates in phase or in quadrature with that strain.

In strongly interpenetrating polymer solutions, this stress occurs because of entanglement. To understand the effect, we return to our jungle-gym model. We impose a step shear strain like that shown in Fig. 2.3. This amounts to tilting the jungle-gym lattice so that each square of the lattice distorts (affinely) into a parallelogram, like the sample as a whole. This distorts the shape of the polymer and its internal density profile. If we consider the end-to-end vectors of each blob of the chain, they are initially isotropically distributed. But after the distortion, the vectors become anisotropic. They are more concentrated along the long diagonal of the sheared square and less concentrated along the short diagonal. The anisotropy has a strength of order unity under one unit of shear. Each of these blob distortions costs energy. To estimate this energy, we imagine that the anisotropic distribution of vectors were produced by an external potential $U(\theta)$. The distribution of angles $f(\theta) = \exp(-U(\theta)/T)$. In order to produce an $f(\theta)$ that varies over a factor of order unity, $U(\theta)$ must be of order T. The work required to turn on this potential must also be of order T, since the energy U affects most of the possible angles θ. This means that the work required to make the shear distortion is of order T for every cell of the lattice, or for every blob of the chain.

4.C Suggested experiment: *Rubber elasticity and temperature* The elastic restoring force of rubber is essentially the restoring force of an elongated random walk. This means the force should be proportional to the absolute temperature, provided the basic local structure of the rubber polymers does not depend on temperature. If this is true, one should observe a roughly 30% increase in the spring constant as the temperature is changed from $0\,°C\,(273\,K)$ to $100\,°C\,(373\,K)$. Devise an experiment with rubber bands to test this prediction. You might use weights, a ruler, boiling water, and ice water.

Since a unit of shear strain stores energy at a density T/ξ^3, the solution has a step-strain modulus G_0 of this order, owing to entanglements. This G_0 is independent of molecular weight and is of the order of the osmotic pressure $\Pi^\dagger$. The solution retains this stored energy until the entanglement constraints are released. If each chain is prevented from reptating, by immobilizing its ends or by crosslinking, the stored energy lasts indefinitely and the sample is an elastic solid—a **gel**. Similarly a polymer melt immobilized by a few crosslinks makes an elastic solid—a **rubber**. The order of magnitude of this modulus has been mentioned in Problems 2.6 and 2.7. The highest-modulus rubbers are those with the highest density of entanglements. Empirically the largest moduli for a weakly crosslinked rubber are of the order of 1 atm, or $10^5\,J/m^3$.

$\dagger$ The ratio of G_0 to Π is predicted to be universal in the semidilute regime in good solvents. But this universality has not been verified experimentally. Instead, Π appears to vary more strongly with solvent quality than G_0 does. As concentration increases above the semidilute regime or solvent quality diminishes, the predicted universality is expected to be compromised. These limitations may account for why the universality has not been observed.

4.8. *Stokes diffusion in semidilute solution* In treating the self-diffusion of a polymer in semidilute solution, we have pretended that the neighboring chains completely remove any momentum in the fluid, so that the flow around a point source of momentum is completely screened out. But this can not be completely true, since a chain that absorbs the momentum can only give it to the adjacent solution; the chains do not remove the momentum entirely. This means that if

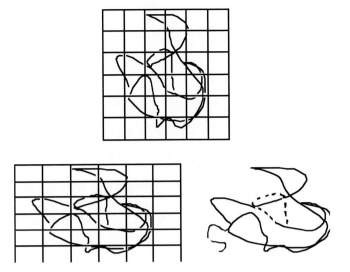

Fig. 4.5
How reptation leads to stress relaxation.
Top picture shows a chain confined in a
jungle-gym representing the other chains.
Bottom left picture shows the effect of a
horizontal stretching: the whole structure
is distorted and each section of the chain
becomes anisotropic. Bottom right picture
shows the chain after some time has
passed. The heavy section on the left
shows the part of the chain that has
reptated out of the initial constraints. This
part is isotropic and holds no stress. The
dashed section on the right shows the part
of the initial chain that has reptated away.
This section is equal in length to the heavy
section on the left.

a chain is gradually dragged through a semidilute solution, the drag coefficient Γ must obey Stokes law valid for any viscous liquid: $\Gamma = 6\pi\eta R_h$, where η is the viscosity of the solution. The hydrodynamic radius R_h must be of the order of the geometric radius R of the chain or smaller. Thus the corresponding self-diffusion constant $\zeta_s = T/\Gamma$ must not be too small: $\zeta_s \gtrsim T/(6\pi\eta R)$. In the text we have derived scaling laws for both the viscosity η and the self-diffusion constant ζ_s, without regard to this Stokes Law constraint. This problem explores whether the two approaches are mutually consistent. (a) Using the scaling of η and R for an entangled, semidilute solution, find the scaling expected for the self-diffusion constant according to Stokes' law. Compare with the scaling for self-diffusion according to the reptation model.

If the chains are free to reptate, the stored energy can relax (Fig. 4.5). In our jungle-gym model, a given cell of the lattice continues to store energy until the chain end has reached that cell and relaxed the constraint. The same is true for a real entangled solution. The time required to relax a finite fraction of these constraints is evidently the reptation time τ_{rep}. By extending this reasoning, one can make detailed predictions of the fraction of the initial stress relaxed as a function of time. These predictions agree well with experiment [10] (Fig. 4.6).

4.9. *Partial stress relaxation function in polymer melt*   If the volume fraction of polymer is increased above the semidilute regime, the correlation length shrinks to the size of a monomer. This is the concentrated solution regime. When all the solvent is removed, the polymer liquid is called a melt. In this regime most of what we derived about the semidilute regime remains valid. The point of this problem is to calculate how the stress $G(t)\gamma$ begins to fall below its initial value $G(0)\gamma$ in a strongly entangled solution or melt. The stress falls over time because part of each chain has reptated out of its initial (distorted) tube and into a new (isotropic and stress-free) set of entanglements, as shown in Fig. 4.5. In effect, part of the initial distorted tube has disappeared. The lost stress is just the initial stress times the fraction of initial tube which has disappeared. This disappearance occurs as the chain executes a random walk forward and backward in its tube. At a given time t the amount of tube removed from the left is the maximum curvilinear distance the chain has traveled to the right in that time. This maximum is roughly the root mean square curvilinear distance

Fig. 4.6

Storage modulus G' versus oscillation frequency ω for narrow-distribution polystyrene melts, reprinted with permission from [30, Fig. 2]. © 1987 American Chemical Society. Molecular weight ranges from 8.9×10^3 (curve L9) to 5.8×10^5 (curve L18). The modulus drops off below a characteristic frequency on each curve. This frequency is the inverse of the stress relaxation time. The high molecular weight samples have a large range of frequencies where the modulus is independent of frequency. Over this **plateau region**, the liquid behaves like an elastic solid. The associated modulus is the step-strain modulus discussed in the text. (To achieve the huge reported range of oscillation frequencies, the experimenters use a trick. They cool the short polymers to slow their motion and heat the long ones to speed up their motion. The temperature effect can be separately quantified. The reported frequencies have been multiplied by a factor a_T, determined separately for each sample, to show what the moduli would be at a fixed temperature of 160 °C.)

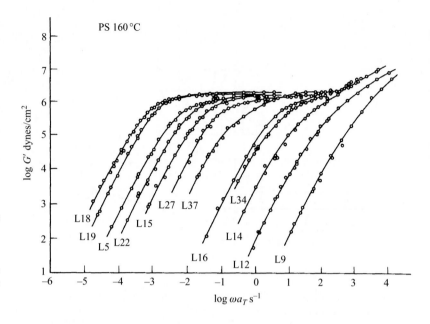

traveled. In a given time, about the same amount is lost on each end. (a) Derive a function for the fractional stress lost as a function of time t after the initial strain in terms of the curvilinear diffusion constant ζ_t. Do not worry about late times when the amount of tube lost is nearly the entire tube. And do not worry about numerical prefactors.

Knowing the step-strain modulus and the relaxation time, we can determine the viscosity η of the entangled solution. As we saw in Chapter 2, the viscosity scales as the modulus times the relaxation time. If one increases molecular weight M at a given concentration ϕ, the reptation time increases as M^3, while the modulus remains constant. Thus the viscosity should increase as M^3. Experimentally, the viscosity agrees better with $M^{3.4}$ than with M^3. After many years of puzzlement it appears that this disagreement is not fundamental. It appears that in the case of viscosity it is particularly difficult to attain the asymptotic $M \gg M(\xi)$ behavior. Several effects alter the viscosity when M is not asymptotically large. One may estimate these effects quantitatively and they seem to explain why an apparent $M^{3.4}$ behavior is observed in the experimental range of M [10].

If a liquid undergoes step strains that are more frequent than the relaxation rate, the stress from each step progressively builds up. Thus, if a steady shear rate exceeds the inverse relaxation time, substantial elastic stress can build up. In particular, **normal stress** builds up as the polymer chains stretch along the streamlines of the flow. This stress from stretched polymer chains is the origin of the unusual flow behavior pictured in Fig. 1.6. The appearance of steady-state deformation and its attendant change in the stress at shear rates exceeding the relaxation rate is known as the Cox–Merz rule [31]. Cox–Merz behavior is observed in many structured fluids; this is natural since the large structures found in such fluids make for long relaxation times. Thus the Cox–Merz criterion for shear-rate dependent viscosity can be met at moderate shear rates.

4.4 Conclusion

In this chapter we have recounted the basic behavior of large, flexible chain molecules in solution. The qualitative elements of randomness, self-repulsion, and entanglement combine to make distinctive spatial structure, energy storage, and dynamic response. We have seen that many of these properties can be understood using primitive geometric notions such as the number of contacts between two fractals. These geometric properties allow one to predict all the asymptotic scaling behavior of long polymers as a function of molecular weight and concentration. In many cases the common geometric basis of these polymer properties allows one to predict one property like the diffusion constant in terms of a completely different property like the osmotic pressure, as exemplified in e.g., the universal ratios of Table 4.2. In the coming chapters we explore to what degree such generality can be extended to other forms of structured fluid.

Appendix A: Origin of the Oseen tensor

The velocity field from an external force-per-unit-volume $\vec{f}(r,t)$ obeys Newton's second law. The component α of the velocity field obeys

$$\rho \partial v_\alpha / \partial t - \partial_\beta \sigma_{\beta\alpha} = f_\alpha,$$

for the small velocities considered here. (Repeated indices are to be summed.) The ∂_β means the spatial derivative in the β direction. The stress σ arises from two origins. The first is viscous stress proportional to the velocity gradient tensor $(\partial_\beta v_\alpha + \partial_\alpha v_\beta)$. The second is the stress that arises from compressive forces. These forces generate a scalar pressure field p.

$$\sigma_{\beta\alpha} = \eta(\partial_\beta v_\alpha + \partial_\alpha v_\beta) - \delta_{\beta\alpha} p$$

For an incompressible fluid the pressure p must be such as to prevent v_β from having any compressional part, i.e., $\partial_\beta v_\beta = 0$. Since we wish to consider steady velocity fields, the time derivative will vanish. Our equation for v_β now reads

$$-\eta \partial_\beta \partial_\beta v_\alpha + \partial_\alpha p = f_\alpha.$$

(A term in $\partial_\beta \partial_\alpha v_\beta$ has dropped away because $\partial_\beta v_\beta = 0$.)

To find v_β around a point source, we Fourier transform in space, defining $\vec{v}(r) = \sum_k \vec{\tilde{v}}_k \exp(ik \cdot r)$. The incompressibility condition becomes $\vec{k} \cdot \vec{\tilde{v}}_k = 0$. The equation for the $\tilde{v}$ is thus

$$\eta k^2 \tilde{v}_\alpha + i k_\alpha \tilde{p} = \tilde{f}_\alpha.$$

Thus

$$\eta \tilde{v}_\alpha + i k_\alpha \tilde{p}/k^2 = \tilde{f}_\alpha/k^2.$$

We can find $\tilde{p}$ by imposing the incompressibility condition $k_\alpha \tilde{v}_\alpha = 0$:

$$0 + i\tilde{p} = k_\alpha \tilde{f}_\alpha / k^2.$$

Thus

$$\eta \tilde{v}_\alpha = \frac{1}{k^2}(f_\alpha - k_\alpha(k_\beta \tilde{f}_\beta / k^2)).$$

For a localized force $\vec{F}$ at the origin $\vec{\tilde{f}}_k = \vec{F}/(4\pi)$. The desired Oseen tensor is thus given by

$$\vec{v}(r) = \frac{|F|}{4\pi\eta} \sum_k \frac{1}{k^2}[\hat{F} - \hat{k}(\hat{k} \cdot \hat{F})]e^{ik \cdot r}.$$

The first term must integrate to $\vec{F}/(\eta r)$ up to a numerical factor. The second term must be in the radial $\hat{r}$ direction and must be such that $\partial_\beta v_\beta = 0$. The result is the Oseen tensor given in the text.

Solution to Problem 4.5 (Deriving permeability)

(a) The speed at distance r from the axis is given by $v(r) = v_0(1 - (r/R)^2)$. The average speed $\bar{v}$ is $\frac{1}{2}v_0$, as can be readily found by integrating. The source of the pressure does work in a small length L at the rate $\dot{W} = $ force $\times$ velocity, or $\dot{W} = -L\bar{v}(\pi R^2)\vec{\nabla}p = -\frac{1}{2}Lv_0(\pi R^2)\vec{\nabla}p$. Here p is the pressure and z is the coordinate along the axis. This work goes into viscous dissipation. The local rate of dissipation per unit volume $\dot{w}$ is $\eta(dv/dr)^2 = (\eta v_0^2/R^4)(2r)^2$. Integrating,

$$\dot{W} = L \int \pi dr^2 \dot{w}(r) = (4\pi)L(\eta v_0^2/R^4)\int_0^{R^2} dr^2 r^2 = L\eta v_0^2(4\pi)\left(\frac{1}{2}\right).$$

Combining, we find $\frac{1}{2}\vec{\nabla}pv_0\pi R^2 = \eta v_0^2(2\pi)$, or $R^2\vec{\nabla}p = 4\eta v_0$. Now we can find the permeability P, using $\bar{v} = -(P/\eta)\vec{\nabla}p$, or $v_0 = -2(P/\eta)\vec{\nabla}p$. Comparing with our other relation between v_0 and $\vec{\nabla}p$, we conclude $P = R^2/8$.

(b) Monomers with density ρ_n and speed v have a current density $j_n = \rho_n v$ so $v = j_n/\rho_n$. Viewing the monomers as fixed, we then have solvent flowing at speed v, and current $j_s = v\rho_s = j_n\rho_s/\rho_n = j_n(1 - \phi)/\phi$.

(c) The cooperative diffusion constant ζ_c relates the monomer current j_n to the gradient of monomer density $\vec{\nabla}\rho_n$: $j_n = -\zeta_c\vec{\nabla}\rho_n$. The permeability P gives a like relation between solvent current j_s and solvent density ρ_s: $j_s = \rho_s P\vec{\nabla}p/\eta_s$. The pressure we are considering here is osmotic pressure Π, so we shall use Π for p henceforth. To compare these we must relate changes of $\vec{\nabla}\rho_n$ to $\vec{\nabla}\Pi$. We can say $\vec{\nabla}\rho_n = (d\rho_n/d\Pi)\vec{\nabla}\Pi$. Combining, we have

$$j_n = -\zeta_c\left(\frac{d\rho_n}{d\Pi}\right)\vec{\nabla}\Pi = j_s\left(\frac{\rho_n}{\rho_s}\right) = -\left(\frac{\rho_n}{\rho_s}\right)\frac{P}{\eta_s}\rho_s\vec{\nabla}\Pi,$$

so that

$$\zeta_c\left(\frac{d\rho_n}{d\Pi}\right) = \rho_n\frac{P}{\eta_s}; \quad P = \eta_s\zeta_c\frac{1}{\rho_n}\left(\frac{d\rho_n}{d\Pi}\right) = \eta_s\zeta_c\frac{1}{\phi}\left(\frac{d\phi}{d\Pi}\right).$$

(d) Π varies as a power of ϕ: $\Pi = \phi^q$, where the exponent $q = 3/(3 - D)$. Thus $d\Pi/d\phi = q\Pi/\phi$ and

$$P = \eta_s \zeta_c/(q\Pi) = \eta_s \zeta_c \xi_\Pi^3/(qT).$$

Expressing ζ_c in terms of $\xi_\zeta \equiv T/(6\pi \eta_s \xi_\zeta)$, we have

$$P = \frac{1}{6\pi \xi_\zeta} q \xi_\Pi^3 = \frac{3 - D}{18\pi}\left(\frac{\xi_\Pi^3}{\xi_\zeta}\right).$$

Thus the solvent flows through the polymer network as though it were a bunch of pipes of radius $R \simeq \xi$.

(e) Specifically, $R^2/8 = (3 - D/18\pi)\xi^2$ or using $D \simeq \frac{5}{3}$, $R \simeq 0.4\xi$. We can get a more refined estimate using the data in Table 4.3: $\xi_\Pi = 3.81\xi_s$; $\xi_\zeta = 1.65\xi_s$. Then

$$R \simeq (2.5 \pm 10\%)\xi_s,$$

where ξ_s is the scattering correlation length defined by $S(q) = S(0)(1 - (q\xi_s)^2 + \cdots)$.

References

1. G. Jannink and J. Des Cloizeaux, *Polymers in Solution* (Oxford, UK: Oxford University Press, 1992).
2. N. Nemoto, Y. Makita, Y. Tsunashima, and M. Kurata, *Macromolecules* **17** 425 (1984).
3. M. Adam, L. J. Fetters, W. W. Graessley, and T. A. Witten, *Macromolecules* **24** 2434 (1991).
4. I. Noda, N. Kato, T. Kitano, and M. Nagasawa, *Macromolecules* **14** 668 (1981).
5. P. G. deGennes, *Scaling Concepts in Polymer Physics* (Ithaca, NY: Cornell, 1979).
6. M. Daoud and G. Jannink, *J. Physique* **37** 973 (1976).
7. A. Einstein, *Ann. Physik* **17** 549 (1905); **19** 371 (1906).
8. R. K. Pathria, *Statistical Mechanics* (Oxford: Pergamon Press, 1972), chap. 13.
9. J. Happel and H. Brenner, *Low Reynolds Number Hydrodynamics With Special Applications to Particulate Media*, 2nd rev. ed. (Leiden: Noordhoff International Publishing, 1973).
10. M. Doi and S. F. Edwards, *The Theory of Polymer Dynamics* (Oxford: Oxford University Press, 1986).
11. A. Einstein, *Ann. Phys.* **34** 591 (1911).
12. P. E. Rouse, *J. Chem. Phys.* **21** 1273 (1953).
13. B. Zimm, *J. Chem. Phys.* **24** 269 (1956).
14. N. S. Davidson, L. J. Fetters, W.G. Funk, N. Hadjichristidis, and W. W. Graessley, *Macromolecules* **20** 2614 (1987).
15. B. Ewen and D. Richter, *Adv. Polymer. Sci.* **134** 1–129 (1997).
16. See e.g. B. J. Berne and R. Pecora, *Dynamic Light Scattering: With Applications to Chemistry, Biology, and Physics* (New York: Wiley, 1976).
17. A. Z. Akcasu, G. C. Summerfield, C. C. Han, C. Y. Kim, and H. Yu, *J. of Polym. Sci. Part B Polym. Phys.* **18** 863 (1980).
18. J. R. Huang and T. A. Witten, *Macromolecules* **35** 10225 (2002).
19. M. Adam and M. Delsanti, *Macromolecules* **18** 1760 (1985).
20. P. Stepanek, R. Perzynski, M. Delsanti, and M. Adam, *Macromolecules* **17** 2340 (1984).
21. J. P. Cotton, M. Nierlich, F. Boué, M. Daoud, B. Farnoux, G. Jannink, R. Duplessix, and C. Picot, *J. Chem. Phys.* **65** 1101 (1976).
22. J. Roots and B. Nyström, *Macromolecules* **13** 1595 (1980).

23. M. Rawiso, R. Duplessix, and C. Picot, *Macromolecules* **20** 630 (1987).
24. J. des Cloizeaux and G. Jannink, *Polymers in Solution, Their Modelling and Structure* (Oxford: Oxford University Press, 1990).
25. B. Farnoux, *Ann. Phys.* **1** 73 (1976).
26. Y. Higo, N. Ueno, and I. Noda, *Polym. J.* **15** 367 (1983).
27. F. Hamada, S. Kinugasa, H. Hayashi, and A. Nakajima, *Macromolecules* **18** 2290 (1985).
28. D. W. Schaefer and C. C. Han, in *Dynamic Light Scattering: Application of Photon Correlation Spectroscopy*, ed. R. Pecora (New York: Plenum Press, 1985).
29. K. Kremer, G. S. Grest, *J. Chem. Phys.* **92** 5057 (1990); D. Richter, B. Farago, L. J. Fetters, J. S. Huang, B. Ewen, and C. Lartigue, *Physical Rev. Lett.* **64** 1389 (1990).
30. S. Onogi, T. Masuda, and K. Kitagawa, *Macromolecules* **3** 109 (1970).
31. W. P. Cox and E. H. Merz, *J. Polymer Sci.* **28** 619 (1958).

Colloids*

5

Colloidal dispersions[†], are homogeneous suspensions of solid particles in a fluid. Most inks and paints are colloidal dispersions. The particle size ranges from nanometers to microns. Such a particle is large enough to approach its bulk properties on the atomic scale but small enough that its thermal energy dominates its gravitational energy[††]. As discussed in Chapter 1, colloidal suspensions or **dispersions** have a variety of direct applications as well as serving as intermediates in various processing technologies, such as ceramics. Colloidal particles tend to attract each other and thus aggregate together. If this aggregation grows unchecked the aggregates eventually migrate to the top or bottom of the container, thus destroying the desired dispersed state. Much of colloid science is concerned with maintaining the dispersed state. This is the problem of colloidal stabilization. Stabilization is achieved by modifying the particles' surfaces to prevent aggregation. Thus understanding and control of interfacial forces is central to colloid science and to this chapter.

In this respect, colloidal dispersions resemble emulsions and foams. As noted in Chapter 1, **emulsions** differ from colloids in that the dispersed objects are liquid droplets rather than solid particles. In a **foam** the dispersed objects are in the gas phase. Most of our reasoning about colloidal stability applies equally to emulsions and foams.

In the next section we discuss the main causes of the attractive interactions that compromise the stability of colloidal dispersions. The following section deals with means of combating these attractions via repulsive forces. The last section discusses consequences of strong colloidal interaction: the particles take on various forms of spatial organization: crystals, liquid crystals, fractal aggregates, and magnetic states. Appendix A treats one of the major mechanisms of attraction in detail. Appendix B explains the various forms of fractal aggregation mentioned in Chapter 1.

How are colloids made? There are two general approaches to the production of colloidal-sized particles: **dispersion**[†††] and **condensation**. These are reviewed in Everett's [2] book on colloid science. **Dispersion** means the breakup of larger-sized particles into smaller ones, usually by some type of rough mechanical treatment, such as grinding or shaking. It is difficult by such means to apply sufficiently strong stresses to achieve the lower range of colloidal dimensions. More importantly, dispersion generally results in

[†] Historically colloids referred to any sticky, gelatinous material. Here, we follow the more recent usage.

[††] This means that gravity is not strong enough to drive the particles to the bottom or top of the container, as discussed in Problem 2.2.

[†††] "Dispersion" is used to denote both a process for making colloids and the resulting state in which particles are dispersed in a liquid.

*This chapter is heavily based on an earlier draft by Phillip A. Pincus, itself based on [1]. The organization of that draft, some text, and several figures survive in the present version.

a broad distribution of sizes, shapes, and compositions; this is awkward for quantitative experiments and applications.

Condensation refers to various methods where dispersed molecules are induced to come together to form the colloidal particle. These often involve a nucleation process[†] in the carrier fluid. If all the nucleation can be initiated in a short time period, then subsequent growth proceeds in a fairly uniform manner, leading to a narrow distribution of sizes. Silver halide suspensions for photographic applications and polymer latex systems used in paints may be prepared this way.

[†] A familiar form of nucleation is the formation of a cloud of water droplets when moist air is cooled.

5.1 Attractive forces: why colloids are sticky

Colloidal particles have a strong tendency to aggregate into large clusters, which may be compact or tenuous (Fig. 1.5), under the influence of ubiquitous attractive forces, which become increasingly potent as the particles become larger. **Stabilization** refers to processes whereby this precipitation is prevented from occurring, resulting in a homogeneous distribution of the particles throughout the volume of the fluid. Colloidal stabilization generally requires the modification of the particle surfaces in order to create repulsive interactions which may compensate the attractive forces. The subject of surface forces is reviewed in detail in Israelachvili's [3] book. In this section we discuss some of these interactions.

5.1.1 Induced-dipole interactions

[††] Induced-dipole interactions are variously called van der Waals interactions, Keesom forces, charge-fluctuation interactions, London interactions, and dispersion forces. These different names arose from considering different mechanisms for charge polarization to be induced. The name "dispersion forces" does not refer to colloidal dispersions, but rather to optical dispersion, the slowing of light waves in matter. The electronic susceptibility of the material causes both this slowing and induced attraction between its molecules. In this chapter we shall deal only with the basic features common to all these mechanisms, and will think of them as generalized van der Waals interactions.

Van der Waals or induced-dipole[††] interactions are ubiquitous interactions which have important consequences in condensed matter. They exist between all atoms and molecules and do not depend on whether the species are electrically charged. Indeed, the shape-specific forces which govern enzymatic processes in biochemistry rely on these interactions. The van der Waals interaction is, in detail, rather complicated and arises from fluctuating atomic and molecular electrical dipoles. It is treated in depth in the excellent book by Mohanty and Ninham [4] and also in the briefer discussion by Israelachvili [3]. We present a simple classical argument which demonstrates the origin of these interactions.

The interaction energy of an atom (or molecule) of electric dipole moment $\vec{\mu}$, with an electric field $\vec{E}$ is

$$U = -\vec{\mu} \cdot \vec{E}. \tag{5.1}$$

Even if the atom does not have a permanent dipole moment, it must nevertheless have a nonzero μ owing to its **polarizability** α: $\mu = \alpha E$. Then the interaction energy between the atom and the field is given by

$$U = -\int \vec{\mu}(E) \cdot \vec{d}E = -\frac{1}{2}\alpha E^2, \tag{5.2}$$

where the factor of one half arises because the electric field must first polarize the electronic structure of the atom before the two fields can couple. Suppose now that in the absence of any external $\vec{E}$ field, there are two identical polarizable atoms separated by a displacement $\vec{r}$ which is large compared to atomic dimensions, so that we can treat them as point dipoles. At some instant there may be a thermal or quantum fluctuation causing one of the atoms (say, atom 1) to have a nonzero moment μ_1. If $\vec{\mu}_1$ is pointing toward atom 2, i.e., parallel to $\vec{r}$, the electric field at atom 2 is $\vec{E}_2 = 2\vec{\mu}_1 r^{-3}$. Then, using Eq. (5.2), there is an interaction between the two atoms given by

$$U_{1,2} = -2\alpha\mu_1^2 r^{-6}. \tag{5.3}$$

5.1. *Fluctuating dipoles*   We assumed above that the instantaneous dipole was directed parallel to $\vec{r}$. Recalculate $U_{1,2}$, relaxing this assumption.

This result merits some discussion. The stability of the unpolarized atom implies $\alpha \geq 0$. Then the interaction energy $U_{1,2}$ is negative definite; the interaction is attractive independently of whether the fluctuating moment $\vec{\mu}_1$ is pointing toward or away from atom 2. We can now average $U_{1,2}$ over all fluctuations of atom 1; in Eq. (5.3), the factor μ_1^2 is then replaced by its average $\langle\mu_1^2\rangle$, which is nonzero even though $\langle\vec{\mu}_1\rangle = 0$. The r^{-6} force law is characteristic of the unretarded van der Waals interaction [4]. Let us try to estimate the order of magnitude of $U_{1,2}$. The mean square value of the fluctuating dipole moment, $\langle\mu_1^2\rangle$, is related to the elementary atomic dipole given by the electronic charge e multiplied by an atomic length, a, i.e., $\langle\mu_1^2\rangle \sim (ea)^2$; the atomic polarizability scales with the atomic volume, $\alpha \sim a^3$ [3]. The interatomic coupling may be generally expressed as

$$U_{1,2} = -C(a/r_{1,2})^6, \tag{5.4}$$

where C is a constant with dimensions of energy and magnitude comparable to the binding energy of an electron—several electron volts. The length a is an atomic size scale and is thus a few tenths of a nanometer in magnitude. As $r_{1,2}$ decreases from infinity, the van der Waals attraction increases, but it is ultimately overwhelmed by short-range repulsion when the electron clouds of the two atoms begin to overlap. The repulsion becomes dominant when $r_{1,2}$ is 2–3 times a. Thus at the most favorable separation, $U_{1,2}$ is only a small fraction of C. It is typically a few percent of the atomic ionization energy and is thus comparable to the thermal energy T at room temperature. This $U_{1,2}$ is reduced further when the two atoms are not in free space but in a homogeneous, isotropic medium. If the medium consists of atoms just like atoms 1 and 2, the attractive force between 1 and 2 vanishes, since the attraction is the same in all directions. Thus it is not surprising that when the two atoms are embedded in a solvent, the constant C grows with the *difference* of polarizabilities between the atoms and the solvent. This solvent effect may reduce C from its vacuum value significantly. However, it is not generally possible to adjust the solvent to cancel out the van der Waals attraction completely. Even if the solvent had the same static polarizability as the atoms, this would not cancel out the attractions. To achieve this cancellation, the solvent would have to respond like the two atoms at all frequencies where fluctuations occur, since fluctuations at all frequencies contribute to the van der Waals attraction.

5.1.2 Solid bodies

Using Eq. (5.4) and assuming binary interactions, the van der Waals interaction between bodies containing many atoms can be calculated. The mutual energy V of two thick slabs 1 and 2 at separation h is then given by $V = \sum_{r_1} \sum_{r_2} C(a/r_{1,2})^6$. Here r_1 is the position of an atom in slab 1 and r_2 is the position of an atom in slab 2. Supposing that each atom occupies a volume Ω, this sum can be approximated by an integral:

$$V(h) = Ca^6\Omega^{-2} \int dx_1 dy_1$$

$$\times \left[\int_{-\infty}^{0} dz_1 \int dxdy \int_{h}^{\infty} dz_2 \, [x^2 + y^2 + (z_1 - z_2)^2]^{-3} \right],$$
$$(5.5)$$

where x and y are the transverse coordinates of atom 2 relative to atom 1. Now, the integral in [. . .] is clearly independent of x_1 and y_1, so that the x_1 and y_1 integrals give simply the surface area A. The remaining [. . .] integrals are finite and depend only on the separation h. For dimensional consistency, [. . .] must be a constant times h^{-2}. This is conventionally written

$$V(h) = -HA(12\pi h^2)^{-1}, \qquad (5.6)$$

where the coefficient H is called the **Hamaker constant**. The Hamaker constant is an energy characteristic of the material and the intervening medium.

5.2. *Van der Waals interaction between layers*   (a) Find the Hamaker constant H for two thick slabs consisting of identical atoms interacting via Eq. (5.4), expressing H in terms of the energy C, the atomic length a and the volume per atom Ω. (b) How does $V(h)$ depend on h if the two infinite slabs are replaced by films of thickness d in the limit $d \ll h$?

Using Eq. (5.6), it can be shown (Problem 5.3) that the mutual interaction energy of two spheres has the form

$$V_b(h) \simeq -kH(b/h), \qquad (5.7)$$

where b is the sphere radius and $b \gg h$. This is an example of the Derjaguin Approximation [3]. It is important to note that when two particles touch, i.e., h in Eq. (5.7) becomes a molecular dimension a, the interaction energy between the particles becomes arbitrarily larger than the Hamaker constant and thus becomes arbitrarily large compared to T if the particle size b is large enough. We may estimate the time τ for two initially contacting spheres separated by a fraction of a nanometer to escape from each other using the Eyring formula from Chapter 2 in the section on Approach to Equilibrium: $\tau \simeq \tau_0 \exp(\Delta U/T)$. Since U is large, the typical time for escape τ may become long even on laboratory timescales. In effect, the attraction is irreversible. As time goes on, the particles aggregate. The aggregates may eventually become large and dense enough so that gravity may cause settling (sedimentation) to the bottom of the container or floating to the top (**creaming**).

For simplicity we have discussed the van der Waals interaction in the context of single atoms and atomic media. But this mechanism of attraction is much more general. It arises simply because materials are polarizable and the charge within them fluctuates. Since all condensed matter has these properties, van der Waals attractions operate in all condensed matter, atomic or molecular, insulating or metallic. Because the van der Waals energy has the same distance dependence for all materials, all *like* surfaces separated by another medium have characteristic Hamaker constants, and attractive interactions of the form of Eq. (5.6). But these attractions are just a special case of an even more general attractive mechanism, to be discussed next.

5.3. *Derjaguin approximation*   The Derjaguin approximation relates the force between curved bodies to the interactions between flat surfaces composed of the same materials. This approximation can be used to find the force between a sphere and a plane, equivalent to the interaction between two spheres. Consider a sphere of radius b at a distance of closest approach h from a half space of the same material, with $b \gg h$. Suppose that the interaction energy per unit area between to half spaces separated by a distance z is known to be $v(z)$. (a) Show that if $v(z)$ decays sufficiently rapidly with z, then the force between the sphere and the plane is $F(h) = 2\pi b v(h)$. (b) Using the results of Problem 5.2, show that the van der Waals interaction potential energy between a sphere and a plane is of the form $V_b(h) = -kH(b/h)$. What is the constant k?

5.1.3   Perturbation-Attraction Theorem

Most attractive colloidal interactions arise from the property of the surrounding fluid that its equilibrium structure may be altered in response to a disturbance, in this case the presence of a colloidal particle. The van der Waals force, discussed in the previous paragraph, may be thought of as resulting from a distortion of the fluctuating electric and magnetic fields in the space between the particles. The following simple argument demonstrates why these interactions are generally attractive. Consider the insertion of a particle into the solvent. The solvent may respond by a redistribution of the structure of the internal degrees of freedom, such as a distortion of the average packing density of the solvent (*cf.* Fig. 2.1) or concentration variations in a mixed solvent. If the particle alters at least one such internal degree of freedom, a work W is required to deform the internal degree of freedom away from its equilibrium value in the absence of the particle. Fig. 5.1 illustrates such a deformation in the simple case of an elastic medium such as a slab of foam rubber. A heavy object placed on the slab produces a deformation of the surface nearby. The displacement of the surface $u(r)$ at a distance r from the object dies off exponentially with distance: $u(r) \sim e^{-r/\xi}$ at large distances from the object. The decay length ξ of the exponential depends only on the properties of the medium, and not on the perturbing object. A particle inserted into a fluid often perturbs the fluid in an analogous way. The analogous decay lengths may range from a few tenths of a nanometer up to microns[†].

Now let us try to insert another particle, identical to the first one, into the medium, as illustrated in Fig. 5.1. There are two situations, depending on whether the distance between the particles r is less than or greater than the

[†] The case of van der Waals forces is somewhat different, since these have no exponential cutoff: $\xi \to \infty$. Instead they show a power law decay of the perturbation rather than an exponential fall off. This difference is associated with the fact that photons, the quanta of electromagnetic radiation, are massless.

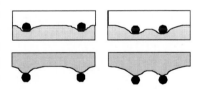

Fig. 5.1
The Perturbation-Attraction Theorem. Top left: two perturbing particles in a deformable medium, illustrated as a slab compressed by the particles' weight. Top right: two particles which share the same deformed region make a deeper depression, and thereby lower the energy of the system. Bottom left and right: particles' weight now expands the slab. Still the deformation is enhanced when they are close; interaction is again attractive.

decay length ξ. If $r \gg \xi$, the two particles are effectively decoupled and the work required is equal to that of the first: *viz. W*. However if $r \leq \xi$, the second particle profits, because some of the needed deformation has already been done. Thus the deformation energy required to add the second particle ($r \leq \xi$) is less than W. This results in an effective *attractive* interaction between the two particles, of range comparable to ξ. Remarkably, the two perturbing objects attract each other regardless of the direction of deformation, as illustrated in the bottom half of Fig. 5.1.

This reasoning suggests that whenever our inserted particle perturbs its surroundings as described above, two similar particles experience an induced attraction. We refer to this statement as the **Perturbation-Attraction Theorem**. We can verify the theorem under certain conditions: (i) the solvent has sufficient time to reach thermal equilibrium in response to the perturbing particle i.e., the solvent is *annealed*; (ii) the solvent deformation may be described as a scalar field; and (iii) the perturbation of the solvent falls off monotonically with distance from a perturbing particle. We now confirm this conclusion with a mathematical argument for the case where the perturbations are weak. We suppose that the medium is characterized by some (scalar) quantity $\psi(r)$ whose equilibrium value is zero. For example, ψ could be the concentration of some solute in the solution measured from its equilibrium value. We wish to characterize the work required to insert two particles at separation x. We imagine inserting these two particles in sequence, first at the origin, and then at the point x. The first insertion requires a work denoted W_1.

We may think of W_1 as the sum of three pieces. The first piece is the work required if ψ were fixed at its equilibrium value of 0. We call this piece W_0. However, by hypothesis, ψ does not stay fixed. This is because there is an interaction between the particle and the medium that depends on ψ. This interaction (free) energy alters W_1. It adds an energy of the form $W'\psi$. Here we have assumed that the perturbation of ψ is small and have thus expanded the interaction energy to lowest order in ψ. The coefficient W' may be of either sign. Finally, perturbing ψ at the origin alters the medium, and this alteration requires work. The (free) energy of the medium is increased if ψ changes from its equilibrium value. We can express this change as a function of the perturbed value at the particle: $\psi(0)$. If the perturbation is small, we may again express this energy to lowest order in the perturbation: $\frac{1}{2}\chi\psi(0)^2$. Since the medium's (free) energy must be minimum when $\psi(0) = 0$, the coefficient χ must be positive. Summarizing, the insertion energy W_1 has the form

$$W_1(\psi(0)) = W_0 + W'\psi(0) + \tfrac{1}{2}\chi\psi(0)^2. \tag{5.8}$$

When we insert the particle, $\psi(0)$ adjusts itself to minimize W_1. One readily verifies that $\psi(0) = -W'/\chi$ and $W_1 = W_0 - \frac{1}{2}W'^2/\chi$.

Our first particle has perturbed the quantity ψ in the whole vicinity of the origin. The perturbed ψ at x can be expressed $f(x)\psi(0)$. The simplest and most common behavior is for $f(x)$ to decay monotonically to zero as x increases, as in our foam-rubber example above. We shall assume that $f(x)$ behaves in this way. Later we shall identify conditions when this monotonic decrease occurs. In Problem 5.4 we will note a contrasting case where $f(x)$ oscillates instead. This $f(x)$ can be related to the net work $W_1(\psi(0), \psi(x))$ to change both $\psi(0)$ and $\psi(x)$. This W must be minimal when $\psi(x) = f(x)\psi(0)$. Also, since ψ is assumed small, we take W to be quadratic in both ψ's. One can easily deduce a $W_1(\psi(0), \psi(x))$ that is consistent with Eq. (5.8) and which gives $\psi(x) = f(x)\psi(0)$ when $\partial W_1/\partial \psi(x) = 0$:

$$W_1(\psi(0), \psi(x))$$
$$= W_0 + W'\psi(0) + \tfrac{1}{2}\chi\big(\psi(0)^2 + \psi(x)^2 - 2f(x)\psi(0)\psi(x)\big).$$

We now add the second particle at x. This adds a work $W_0 + W'\psi(x)$ to the total work W:

$$W(\psi(0), \psi(x)) = 2W_0 + W'\psi(0) + W'\psi(x)$$
$$+ \tfrac{1}{2}\chi\big(\psi(0)^2 + \psi(x)^2 - 2f(x)\psi(0)\psi(x)\big). \quad (5.9)$$

The free energy must now accommodate to both particles:

$$0 = \frac{\partial W}{\partial \psi(0)} = W' + \chi(\psi(0) - f(x)\psi(x)),$$

and

$$0 = \frac{\partial W}{\partial \psi(x)} = W' + \chi(\psi(x) - f(x)\psi(0)).$$

Since the two ψ's play equivalent roles in these linear equations, they have equal values:

$$\psi(0) = \psi(x) = \frac{-W'}{\chi(1 - f(x))}.$$

Since f was positive, both ψ's increase owing to their proximity. Moreover, at this equilibrium value of the ψ's, The free energy W is given by

$$W = W_{\min} = 2W_0 - \frac{W'^2}{\chi(1 - f)}. \quad (5.10)$$

The nonlocal response f lowers the energy and thus induces an attractive interaction between the two particles, as claimed.

Our proof covers the case where the perturbation on the medium is weak. But perturbation leads to attraction in some cases of strong perturbation, provided further conditions are met. Appendix A treats an important category of strong perturbations.

One further category of perturbation not treated in our discussion is the case where the perturbed quantity ψ is not a scalar. An important example is the case where the particles induce a vector dipole moment in the molecules around the particle. Such polarization often occurs when the solvent is water. We suppose that the polarization vectors tend to point *towards* particle 1 and then estimate the energy to add particle 2. The dipoles nearest particle 1

† 2,6-Lutidine or 2,6-dimethyl pyridine is a small organic molecule whose chemical formula is C_7H_9N. Lutidine and water mixtures undergo a continuous phase transition into a lutidine-rich and a water-rich phase at a convenient composition and temperature. At such a transition fluctuations of composition occur that have spatial extents arbitrarily larger than the size of a molecule. Lutidine–water mixtures are often used for studying the effects of phase separation in a solvent.

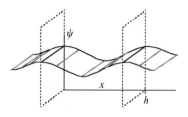

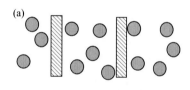

(a)

(b)

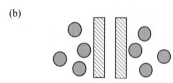

Fig. 5.2
Depletion attractions. Two plates immersed in a suspension of small particles. (a) When the separation between the plates exceeds the particle diameter, particles may enter the interstitial region with only minor modifications of the particle distribution. Thus the internal and external osmotic pressures balance and there is no interplanar force induced by the particles. (b) The separation is smaller than a particle diameter; particles cannot enter. The external osmotic pressure is not balanced and there is a net attractive force per unit area between the plates which is equal to Π, the osmotic pressure of the reservoir.

are now oriented *away* from where particle 2 is to be inserted. This adds to the W_2 rather than diminishing it. Thus non-scalar perturbations need not produce attraction.

An example of the Perturbation-Attraction Theorem occurs when colloidal particles are suspended in a mixed solvent, composed of a miscible binary mixture such as lutidine and water†. Normally the particle surface prefers one of the solvents; this preference alters the concentration near the particle. The concentration gradients extend a correlation length ξ from the particle; when the composition and temperature approach the conditions of phase separation, ξ becomes large. We refer the reader to a text on phase transitions such as the book by H. E. Stanley [5]. Other particles are then attracted to the regions of high favorable solvent concentration. As the Theorem suggests, colloidal particles in mixed solvents tend to aggregate when the solvent is near the threshold of demixing [6].

5.4. *When perturbation causes repulsion*   Generally two like objects that perturb a medium experience a mutual attraction. But sometimes they do not. Here is a simple and important counterexample. The medium is a membrane with a bending rigidity: the membrane resists curvature. Such a membrane whose height above some reference plane is ψ has a free energy $W[\psi]$ of the form $W = \int dx\, A\psi^2 + B(d^2\psi/dx^2)^2$ provided ψ varies only with x. If $d\psi/dx$ is not too large, the second term is simply the square of the curvature. This is a quadratic free energy and can be readily solved, though there is an awkward integral to do. One finds that a perturbation ψ_0 induced by a phantom vertical plane at $x = 0$ with interaction energy $V = \lambda\psi(0)$ has the form $\psi_0(x) = C\lambda e^{-\kappa z}[\cos(kx) + \delta \sin(kx)]$ for positive x. Evidently, $\psi_0(-x) = \psi_0(x)$.

(a) Find the prefactor C and the coefficients δ, κ and k, that minimize $W[\psi] + V(\psi(0))$. Note that $d/dx\,\psi|_{x=0} = 0$ by symmetry.
(b) Now a second plane is inserted at distance $x = h$ from the first. Using the discussion in the text, find the energy W of the medium with $\psi(0)$ and $\psi(h)$ fixed.
(c) How does the total energy $V + W$ vary with separation h? You can ignore the overall magnitude of this energy. Are there separations where $W + V$ is larger than at $h = \infty$? If so, where?

5.1.4 Depletion forces

A further example of the Perturbation-Attraction Theorem is provided by **depletion attractions** between colloidal particles. Such attraction occurs in situations like that pictured in Fig. 2.1. The large colloidal particles are in a solvent containing smaller particles that cannot penetrate the colloidal particles. This is an important case of the mixed-solvent situation we just discussed. The theorem tells us that attractive interaction is expected when the separation between the surfaces is small enough to sense the structure of the solvent, *viz.* the diameter of the small particles. For this case, we can confirm this prediction by a more primitive argument, that directly gives the magnitude of the attraction.

We first consider Fig. 5.2, which depicts two surfaces immersed in a suspension of smaller colloidal particles. These planes may be viewed as an approximation to the surfaces of much larger colloidal particles immersed in a suspension of smaller ones as in Fig. 2.1. When the particles are not excluded from the interplanar region, the pressures on each wall are balanced. However, when the particles are excluded by their size from entering the channel between the plates, the osmotic pressure Π on the outer surfaces which is associated with the suspended particles is not counterbalanced, leading to an effective attractive force per unit area between the plates of magnitude Π. The range of this interaction is approximately the particle diameter.

What happens when the colloidal particles of Fig. 5.2 are replaced by flexible polymer coils? There are two distinct cases: the polymers are either attracted to or repelled from the solid substrate. Which case is operative in any particular situation depends upon the three media—polymer, substrate, and solvent—and the interactions between them. Polymer adsorption due to attraction to the surfaces often leads to repulsive colloidal forces and will be discussed in the next section. The non-adsorbing situation leads to depletion attractions. Indeed, when the solution is dilute and the polymer coils are well separated from one another, the system is quite analogous to that depicted in Fig. 5.2. In fact the osmotic pressure which drives the plates together at separations smaller than the polymer radius of gyration, R_G, is simply $\Pi = (c/M)T$, where c is the monomer concentration of chains having degree of polymerization M. Of course, polymer coils are different from colloidal particles in that polymers are deformable. Thus the chains may distort and squeeze into the interplanar space even if the separation, h, is small, $h \leq R_G$, as in Fig. 5.3. The free energy associated with this deformation can be estimated (in a good solvent) as T per blob, as studied in Problem 3.15. We discuss this confinement energy more carefully in Chapter 6 to follow. Totalling over all the blobs in the channel yields a free energy per polymer $\Delta\mathcal{F}$ given by

$$\Delta\mathcal{F} \simeq (M/M(h))T \simeq M(a/h)^{5/3}T. \qquad (5.11)$$

As in Chapter 4 $M(h)$ is the number of monomer units per blob of size h, while a is the size of a monomer. The number of coils in the interstitial region is suppressed by the Boltzmann factor $e^{-\Delta\mathcal{F}/T}$; this decreases exponentially with $1/h$ and the attractive force between plates will approach $\Pi = (c/M)T$ exponentially with $1/h$. This occurs for $h \simeq R_G$. Thus there is only a quantitative, not qualitative, difference between the polymer and small rigid particle situations.

When the polymer solution is semidilute, the coils interpenetrate and form the transient network discussed in Chapter 4. The solution of volume fraction ϕ may be considered as a melt of blobs of size ξ, with $M(\xi) \simeq (\xi/a)^{5/3}$ monomers per blob. When $h \geq \xi$ and there are many blobs between the surfaces, the polymer-induced interaction is weak. However, as the plates approach one another and $h < \xi$, the polymers are squeezed out of the channel and as $h \to 0$, there is an osmotic pressure $\Pi \simeq \phi^{9/4}a^{-3}T$ pushing the surfaces together. Therefore, even in a semidilute solution there exists an attractive force between surfaces induced by depletion effects. In that case, the range of the negative **disjoining pressure** is the correlation length, ξ. Note that as the polymer solution becomes increasingly concentrated, (i) the range of the disjoining pressure decreases

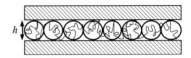

Fig. 5.3
Two dimensional representation of a polymer coil confined to the region between two plates for $h < R_G$. The circles denote blobs of diameter h.

from R_G in dilute solutions to $\xi \simeq a\phi^{-3/4}$ in semi-dilute solutions; (ii) the strength of the attractive force increases as $\phi^{9/4}$ in the semidilute range.

5.2 Repulsive forces

We have seen that there are many mechanisms which induce attraction and potential aggregation of colloidal particles. These may be van der Waals forces which are parametrized by the Hamaker constant H or indirect forces associated with nonlocal deformation of the solvent matrix. Furthermore, as the colloidal particles become larger, the binding energy between touching dimers increases. Therefore, in the absence of opposing forces, Brownian colloidal particles will eventually aggregate into sufficiently large clusters that they sediment or cream out of suspension. Colloid stabilization then requires one or more mechanisms to provide repulsive forces to counter-act the attractive interactions. These repulsions are generally supplied by elastic effects; usually some kind of deformable "bumper" is attached to the particles. The bumper distorts as two particles approach one another. This deformation results in an elastic restoring force, tending to keep the particles apart. In practice, it is often the case that the bumper elasticity is of entropic origin, so that its modulus scales like T. This is indeed the situation when the stabilization is carried out (a) by flexible macromolecules, via **steric stabilization** or (b) by electric charge, via **electrostatic stabilization**. Below we discuss each of these cases in turn.

Both of these methods give repulsive forces that have some characteristic range R and that fall off rapidly for larger distances. Thus at very long distances, the van der Waals attraction dominates, and the potential of interaction resembles that of Fig. 5.5. Under these conditions we can already deduce some general requirements for colloidal stability. Since one cannot compensate for the van der Waals attraction at all distances, one must at least assure that it is weak enough to be harmless in those regions where it dominates. As we showed in Chapter 2, two particles have a substantially increased likelihood of being together whenever their attraction has strength of T or more. If two isolated particles occupy a volume Ω, and have attraction of strength V_0 over volume ω their net probability of being together is of order $\exp|V_0/T| \, (\omega/\Omega)$. Thus when $V_0 \gtrsim T \log(\Omega/\omega)$, the particles are likely to be together. In a solution of many such particles, any particle is likely paired under the same conditions, where now Ω is the volume per particle. This strong association creates phase separation.

To avoid this phase separation we must have $|V_0| \lesssim T$. (Here we have ignored the weak $\log \Omega$ dependence.) Since the van der Waals energy is a substantial part of the total near this minimum, we may use Eq. (5.7) to estimate V_0. If h_0 is the separation at the minimum, stability requires that $-kH(b/h_0) \lesssim T$, or

$$h_0 \gtrsim b(H/T).$$

For many particles and solvents, H happens to be of order T. This tells us that we must provide a repulsion sufficient to push the attractive minimum out to a separation h_0 of the order of the particle radius b. Accordingly, the range of the repulsion must be of order b or more. A shorter-ranged repulsion will not do.

5.5. *Depletion binding of colloidal particles* Consider two identical spherical colloidal particles of radius b which are just touching each other while immersed in a semi-dilute, good solvent polymer solution of volume fraction ϕ and correlation length ξ. The polymers do not adsorb on the particles. Thus the concentration is depleted at small distances z from the surface: $\phi(z) \rightarrow 0$ for $z \ll \xi$ and $\phi(z) \rightarrow \phi$ for $z \gtrsim \xi$. If $b \gg \xi$, estimate the binding energy of the dimer arising from the depletion forces i.e., the work required to separate the particles. [Hint: Use the concept of the Derjaguin approximation of Problem 5.3.]

5.2.1 Steric stabilization

We have seen in the last section that polymers which do not adhere to surfaces immersed in a solution induce attractive forces between them. This interaction arises from the depletion of polymers (compared to the bulk solution) in the channel between the colloidal grains. If the polymers were prevented from escaping from the interstitial region, their compression as the particles approach each other would lead to an osmotic *repulsion*. This may be seen by again referring to Fig. 5.3 and the associated discussion for dilute polymer solutions. The free energy cost in deforming a polymer coil trapped between two surfaces is given by Eq. (5.11). If there exist N of these chains per unit area of surface (where N is the number of nonoverlapping trapped polymers per unit area), the corresponding free energy penalty per unit area ΔF is given by $\Delta F = N \Delta \mathcal{F}$, and there is a disjoining pressure,

$$\Pi_d = -\frac{d\Delta F}{dh} \simeq \frac{5}{3} M \, N \left(\frac{a}{h}\right)^{5/3} h^{-1} T, \qquad (5.12)$$

if the chains are not permitted to leave the interstitial region. This trapping may be engendered by the attachment of the polymers to the surface by, e.g., formation of covalent chemical bonds between an end monomer and the surface, as in Fig. 5.4.

Such bonds are generally sufficiently strong to prevent the polymers from leaving the surface except under the harshest treatments. Is this disjoining pressure strong enough to prevent binding of the particles under the influence of van der Waals forces? This question may be addressed by combining Eqs. (5.6) and (5.11). The form of the interaction free energy between the surfaces is sketched in Fig. 5.5. The absolute minimum is at contact between the surfaces, but there exists a barrier to overcome in order for the two surfaces to feel the strong binding at contact. If the barrier height is much greater than T, the adhesive contact is effectively unattainable during any laboratory experiment. Note that there is a **secondary minimum** at larger separations ($h \geq R_G$), where the polymeric bumper is not deformed but there still remains van der Waals attraction. If the depth of this minimum is several T, the particles may aggregate in this minimum, making loose aggregates called **flocs**. The maxima and minima are determined by the disjoining free energy per unit area multiplied by an appropriate area; for example, for the flat plates this is the area of the surfaces. For spheres of radius b, large compared to the nearest interparticle separation, in the Derjaguin Approximation this area is proportional to b.

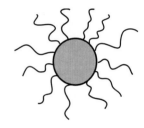

Fig. 5.4
Sketch of a colloidal particle with chemically end-bonded polymers which are non-adsorbing. The polymer hairs act as bumpers opposing aggregation.

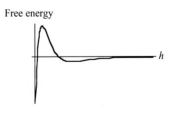

Free energy

h

Fig. 5.5
Sketch of the disjoining free energy versus the interplanar separation taking into account the repulsions from dilute grafted hairs and van der Waals attraction. The absolute minimum is at contact with a barrier and a secondary minimum is at larger separations.

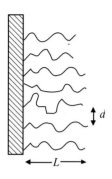

Fig. 5.6
Sketch of a polymer brush grafted to a solid substrate. The height of the brush is L; the average distance d between grafting points is less than the radius of gyration, R_G, of isolated coils. Thus in good solvents, adjacent chains are strongly interacting.

The strength of the polymer-induced interactions depends on how much polymer is in the system under study. For example, if the barrier (Fig. 5.5) is not sufficiently high to prevent aggregation with dilute hairs, the concentration of polymer that is grafted to the surfaces may be increased (with some chemical ingenuity) into the regime where their radii of gyration start to overlap; this is called the **brush regime** and is depicted for a planar surface in Fig. 5.6.

In order to analyze the contribution of the polymer brush to the disjoining pressure, we first give a basic description of the height L (Fig. 5.6) of the isolated brush due to Alexander and deGennes [7]. The excluded volume repulsions between the overlapping polymer coils provides an internal pressure, pushing the chains into the solution so as to reduce the internal monomer concentration.

The equilibrium height is then determined by a balance between the excluded-volume repulsions and the stretching elasticity of the polymers. To estimate these, we may use the blob scaling arguments of the Chapter 4. Chains of M monomers of size a are grafted to a surface at separation $d \gg a$ but substantially smaller than the chain size. In that case the local environment is semidilute, and we may describe it in terms of blobs of size ξ, with k blobs per chain. Now we can express the interaction energy U of each chain (the energy required to insert it against the local osmotic pressure) in the form $U \simeq Tk$. This energy favors few blobs per chain and thus dilute conditions. The opposing stretching energy S may be found by treating the chain as a random walk of blobs. Each chain must stretch to a height of order of the thickness of the polymer layer L. Thus $S \simeq T(L/\xi)^2(1/k)$. The thickness L in turn determines the concentration of blobs: $1/\xi^3 \simeq k/(d^2L)$. Thus $L \simeq k\xi^3/d^2$. We may now express the stretching energy S in terms of k and ξ:

$$S \simeq T(k\xi^2/d^2)^2(1/k) \simeq Tk(\xi/d)^4.$$

We see that both energies U and S are proportional to k; thus the optimal volume fraction is independent of M. At the optimal volume fraction $S \simeq U$, so that $1 \simeq (\xi/d)^4$. Thus $\xi \simeq d$. The blob size and concentration are independent of the chain length M; thus the height of the layer is simply proportional to M; the chains become stretched as anticipated. The force per unit area needed to compress the layer is of the order of its osmotic pressure:

$$\Pi \simeq T/d^3. \tag{5.13}$$

The effective barrier against van der Waals-induced aggregation may thus be augmented by increasing the grafting density, which decreases d in Eq. (5.13). Within this **Alexander–de Gennes approximation** for the grafted brush, the effective disjoining pressure is given by Eq. (5.13) for $h \leq L$ and drops abruptly to zero for $h \geq L$. In fact the dropoff is gentle, because not all the chains stretch to the same height L. The chains gain both stretching and osmotic free energy if some stretch more and some less. The optimal concentration profile, is close to a parabola [7]. The resulting disjoining pressure has been calculated [8] and is shown in Fig. 5.7. The Alexander–de Gennes scaling form for the brush height is preserved but the detailed concentration profile eliminates the spurious abrupt drop-off.

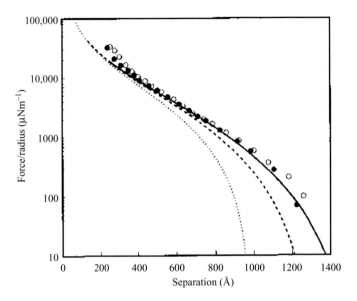

Fig. 5.7
Measured forces between two surfaces coated with grafted polystyrene chains about 1400 monomers long in toluene, after [9]. The curves show predicted force using free-solution properties of polystyrene in toluene, together with the parabolic-profile theory described in the text, after [8].

In these last paragraphs, we have discussed how polymers which are intrinsically repelled by solid surfaces may stabilize colloidal dispersions when the chains are firmly end attached to the substrates. However it would be more economical to use homopolymers, since no special chemistry would need to be performed to get the polymers to attach to the surfaces. Indeed, this can be accomplished by using polymers which adsorb onto the surfaces of the solids to be stabilized. A cartoon of this situation is given in Fig. 5.8. Then the polymer loops can act as bumpers keeping the particles out of adhesive contact. This scheme may indeed work, but it is quite delicate [1].

In fact several conditions must be satisfied for homopolymer stabilization. These include: (i) **irreversible adsorption**; (ii) sufficient polymer to saturate the surfaces; (iii) a sufficiently thick polymer **corona**. These conditions will now be discussed quantitatively.

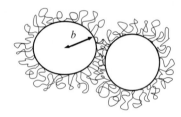

Fig. 5.8
A sketch of two interacting colloidal particles with adsorbed polymer. Note the deformation of the polymer distribution in the interstitial region.

Let us consider the picture, Fig. 5.8, in more detail. The "bumper" view is that the coronae, which extend a finite distance, L, into the solvent in a series of **loops and tails** when compressed by two particles approaching one another, cause an osmotic repulsion, as discussed in the first paragraph of this section. The range of this repulsion, L, must be such that the van der Waals attractive energy between the particles when the coronae are just touching is less than about T. Thus the minimal stabilizing corona thickness L should be proportional to the particle size, $L \propto b$, as noted at the beginning of this section. However, for the adsorbed polymer to produce an osmotic repulsive force between two surfaces, the free energy of the interstitial chain must be increased. (This is because the disjoining pressure is $\Pi_d = -\partial F/\partial h$. Eq. (5.12).) Such an increase in free energy per chain would take these polymers out of thermodynamic equilibrium with the bulk solution. Thus the polymers would desorb, reducing any repulsion. This is another example of the Perturbation-Attraction Theorem. Consequently, to achieve the strongest repulsive forces, the polymers should

Fig. 5.9
Sketch of a polymer chain adsorbing on a flat surface. Note that the coil is constrained in the direction normal to the surface and spreads out along the surface.

be *out of equilibrium* with the polymer dissolved in the solvent†. We discuss below why this might indeed occur for adsorbing polymers. Condition (ii), concerning surface saturation, arises from the fact that the polymers are adsorbing; when the two surfaces are within a distance that can be spanned by one coil, a given chain can adsorb on both surfaces to form a **bridge**. Such bridging will contribute a negative term to the polymer free energy and thus an attractive component to the force between surfaces. If, however, the surface is effectively covered by polymer, adsorbing sites will not be available and this bridging will be screened out. Thus, effective stabilization requires well-coated surfaces.

Some of the energetics of adsorbed polymers may be understood by a simple argument given by de Gennes [11] for the geometry of a single adsorbed chain. Referring to Fig. 5.9, the adsorption free energy of the polymer may be approximated as

$$\mathcal{F}/T \simeq (R_0/L)^2 - \epsilon Mx, \tag{5.14}$$

where $R_0^2 = Na^2$ is the characteristic size of a Gaussian chain, ϵT is the adsorption energy gained per segment in contact with the surface, and x is the fraction of segments which contact the surface. The first term is the free energy required to confine the chain to a region of thickness L. This confinement energy was first encountered in Problem 3.15. The second term is the local gain in free energy when monomers touch the surface, or adsorption energy. Making the simple assumptions that the monomers are uniformly distributed throughout the slab and that $x \simeq a/L$, and minimizing f with respect to L, we obtain $a/L \simeq \epsilon$. Thus, provided $\epsilon > M^{-1/2}$, the polymer is strongly deformed from its Gaussian conformation. Typically, $\epsilon \approx 1$. The inequality is easily fulfilled for $M \geq 100$. (In the presence of excluded volume interactions, each term in Eq. (5.14) is modified but the result for L remains essentially unchanged. We discuss adsorption in more detail in Chapter 6.) The adsorption free energy per chain is $M\epsilon^2 T$ which may be substantially greater than T. This is essentially the magnitude of the energy barrier in the Eyring formula in Chapter 2 for the kinetics of polymer desorption. If $M\epsilon^2 \gg 1$, the desorption rate decreases exponentially in time; this means that often, especially for high molecular weight polymers, the polymers cannot jump off the surface during an experimental observation time. Under such conditions, the chains may be considered to be irreversibly adsorbed, i.e., out of thermal equilibrium with the dissolved unadsorbed coils. This explains why the adsorbed polymers may be practically considered to be trapped on the surfaces and the Perturbation-Attraction Theorem fails to be applicable.

Thus adsorbing high molecular weight polymers are often irreversibly adsorbed because the overall barrier height for desorption is proportional to M. However, the layer thickness, L, is only a few nanometers (for $\epsilon \simeq \frac{1}{10}$) and independent of molecular weight. This would not be a thick enough layer to stabilize the colloidal particles against van der Waals-induced aggregation. Fortunately, one may readily achieve thicker layers whose surfaces are saturated with polymers. If the surfaces are not saturated with polymers these gain several T in free energy by adsorbing. Accordingly their probability of being desorbed in solution is exponentially small. Thus, even for quite dilute solutions, adsorbing surfaces can be saturated and each adsorbed chain will have many chains nearby. Then, even for dilute solutions, the local concentration in the interfacial region may be sufficiently high that it is in the semidilute regime. Different chains will be competing for adsorption sites on the surface, leading to a reduced number of surface contacts per polymer. This, in turn, induces chain swelling and a corresponding thickening of the surface layer relative to the single chain situation.

How thick will the layer become? In Chapter 4, it was shown that the characteristic length over which there are significant concentration variations in a polymer solution is the correlation length, ξ_ϕ, which varies from the polymer radius, R, in dilute solutions to $\xi_\phi \propto \phi^{-3/4}$ in semidilute solutions. This suggests that, in solutions with sufficient polymer to saturate the surfaces, the coating thickness corresponds to ξ_ϕ. In Chapter 6, it will be shown that this result indeed follows from a more careful calculation of the adsorption profile. Then the interaction range, L, should be $L \simeq \xi_\phi$. In dilute solutions where $\xi_\phi \simeq R$, this implies that the optimal polymer molecular weight for stabilization of colloidal particles of size b should be given by $M \simeq (b/a)^D$ where D is the fractal dimension relating chain size to chain length, $R \simeq M^{1/D}a$. The molecular weight of adsorbed polymers which is required to stabilize colloidal particles increases with nearly the square of the particle dimension. This is in contrast to grafted chains, for which the molecular weight only grows linearly with particle size. Thus more polymer is required in the adsorption case, but less chemistry is necessary to synthesize specific grafting bonds. Notice that increasing the polymer concentration in the solution is deleterious to colloid stabilization. This is because the increased osmotic pressure of the solution squeezes the adsorbing layer and reduces the thickness from R to ξ_ϕ as the concentration increases. Thus the corona thickness L becomes molecular weight independent and decreases with increasing concentration as $\phi^{-3/4}$. Too much polymer destabilizes the suspension! Thus colloidal control with adsorbing homopolymers is delicate and requires attention to the molecular weight and concentration for the given particulate system to be controlled.

5.2.2 Electrostatic stabilization

In polar solvents where hydrocarbon polymers are rather difficult to dissolve, colloidal particles are often stabilized by attaching electrically charged groups to the surfaces, creating locally unbalanced charges and consequently Coulomb repulsion between the particles. The high dielectric constant of such solvents permits charge separation and ionization, as explained in Chapter 7. Nevertheless global neutrality is generally maintained, and the nature of the electrostatic interactions must be considered in this context.

Let us first consider a salt solution in which monovalent salt molecules, e.g., NaCl are dissolved. Assuming that all the molecules are dissociated, there is then a concentration, c, of ions, half Na$^+$ cations and half Cl$^-$ anions. What is the effective potential experienced by a test charge at $\vec{r}$

arising from, e.g., a Na^+ ion at the origin? The Na^+ ion attracts Cl^- ions and repels other Na^+ ions. Therefore if the test charge is at a large distance (compared to the average interionic spacing $d \equiv c^{-1/3}$) from the origin, Gauss' Theorem would yield a reduction in the effective electrostatic potential: $\Phi(\vec{r})$. This reduction is called **Debye–Hückel screening**. The form of the Debye–Hückel screening may be derived directly from Maxwell's electrostatic equation (written in cgs units)

$$\epsilon \nabla^2 \Phi(\vec{r}) = -4\pi\rho(\vec{r}), \qquad (5.15)$$

where ϵ is the dielectric constant (80 for water), Φ is the electrostatic potential and $\rho(\vec{r}) = \rho_+(\vec{r}) - \rho_-(\vec{r})$ is the charge density at $\vec{r}$; $\rho_\pm(\vec{r})$ are the anion and cation concentrations. In a solution, the charges are mobile and may adjust to the local forces that they experience; the position of each ion is governed by its Boltzmann probability owing to the local potential that it experiences.

$$\rho_\pm(\vec{r}) \simeq \rho_\pm^0 \exp(\mp e\Phi(\vec{r})/T), \qquad (5.16)$$

where $\rho_\pm^0$ are constants that adjust the total number of ions of each species to correspond to the concentration of ions dissolved, i.e., $c\Omega = \int(\rho_+ + \rho_-)(\vec{r}) \, d^3r$, where Ω is the total volume of solution. We now suppose that the potential seen by an ion is that arising from the mean charge density $\rho(r)$ of Eq. (5.15). Combining Eqs. (5.15) and (5.16) generates a nonlinear second order partial differential equation for the self-consistent electrostatic potential, $\Phi(\vec{r})$:

$$\epsilon \nabla^2 \Phi = 4\pi \left[-\rho_+^0 \exp\left(\frac{-e\Phi(r)}{T}\right) + \rho_-^0 \exp\left(\frac{e\Phi(r)}{T}\right) \right]. \qquad (5.17)$$

This is called the **Boltzmann–Poisson Equation**. It is not exact, because it neglects correlations between the ions, as well as their fluctuations. Each ion is taken to move independently in the *average* potential generated by the other ions.

If the electrostatic potential energy range is smaller than T, the Boltzmann–Poisson equation may be linearized by expanding the exponentials to first order in a Taylor series, yielding the **Debye–Hückel equation**:

$$\nabla^2 \Phi(\vec{r}) = \kappa^2 \Phi(\vec{r}), \qquad (5.18)$$

where κ^{-1} is the **Debye screening length** and is given by

$$\kappa^2 = (4\pi e/\epsilon T)(\rho_+^0 + \rho_-^0). \qquad (5.19)$$

For the monovalent salt solutions, $\rho_+^0 = \rho_-^0 = \frac{1}{2}ec$ and the Debye length may be expressed as $\kappa^2 = 4\pi\ell c$, where $\ell \equiv e^2/\epsilon T$ is the **Bjerrum length** encountered in our discussion of polyelectrolytes ($\ell \simeq 0.7$ nm for water). A 1 mM salt solution corresponds to a Debye length of approximately 10 nm. If a positive ion is assumed to be at the origin, the solution of Eq (5.18) is the screened Coulomb potential,

$$e\Phi(r) = T\ell/re^{-\kappa r}, \tag{5.20}$$

which is of the same form as the Yukawa potential that is familiar from nuclear physics. This is a reasonable approximation if there are many screening ions in a screening length, which corresponds to $\ell/d \ll 1$, as may be verified using the formula for κ.

5.6. *Debye–Hückel screening* (a) Show that the condition that there are many screening ions in a Debye length κ^{-1} is equivalent to $\ell/d \ll 1$ for monovalent salts, where d is the mean spacing between ions in solution and ℓ is the Bjerrum length. (b) What is the corresponding result for multivalent salts? (c) For a CaCl$_2$ solution in water, what is the range of concentrations for which the Debye–Hückel approximation is valid?

A natural extension of Eq. (5.20) to the case of a colloidal particle of size b and having Z ionizable groups firmly anchored to the surface is

$$e\Phi(r) = ZT\,\frac{\ell}{r}\,e^{-\kappa(r-b)}\,\frac{1}{\kappa b + 1}. \tag{5.21}$$

However this is typically an overestimate of the strength of the repulsive potential between charged colloidal particles. This can be seen by considering the potential energy of a negative counterion at the surface of the particle, $e\Phi(b) = -4\pi b\ell\sigma T$, where σ is the number of anchored charges per unit area. For $b \simeq 1\,\mu$ and $\sigma \simeq 10^{-4}/\,\text{nm}^2$, this gives $|e\Phi(b)| \simeq 10^4 T$. There is therefore a strong attractive force driving the **counterions** to recondense on the particles, so that the particle is incompletely ionized. This arises because the strong, generally covalent anchoring of the surface charges leads to very high local electric fields. This results in a reduction of the **effective charge**, Z^*, of the particle as experienced by a distant test charge. We can readily estimate Z^* for colloidal particles whose radius b is much larger than the screening length κ^{-1}, the typical case. The surface charge gives an enhanced density of counterions near the surface. The Boltzmann enhancement factor is $\exp(e\Phi(b)/T)$, where $e\Phi(b) \simeq Z^*T\ell/b$ is the surface potential. The enhancement acts over a volume of order b^2/κ. The probability that the charge is in the enhanced region is the Boltzmann-weighted ratio of the enhanced volume to the volume per ion $1/c$. This ratio is about 1 when $e\phi(b) \simeq -T\ln(c\kappa/(b^2))$ or

$$Z^* \simeq -(b/\ell)\log(c\kappa/b^2). \tag{5.22}$$

5.7. *Debye–Hückel potential of a sphere* Derive Eq. (5.21) from the Debye–Hückel equation (5.18).

For the numerical example above, this gives $Z^*/Z \simeq 10^{-3}$; the effective charge is only about 0.1% of the nominal charge. The effective electrostatic potential is then given by Eq. (5.21) with Z replaced by Z^*. There exists some evidence for this reduction in effective charge in mixed micellar systems [12] and charged colloids [13].

Is the electrostatic repulsion between charged colloidal particles sufficient to stabilize them against the flocculation induced by the van der Waals attractions of Eq. (5.7)? In terms of the effective charge, the Coulomb repulsion dominates the van der Waals attraction (at short distances) when $Z^{*2} \gtrsim (b/\ell)(H/T)$. However, at interparticle separations exceeding about κ^{-1} the van der Waals force, which only decays as a power law, wins, giving a force curve qualitatively similar to Fig. 5.5. Flocculation in the secondary minimum is prevented if $T \gtrsim H(\kappa b)$. If we consider the Debye length, κ^{-1}, to be the corona thickness in this case, as with polymer stabilization, the corona thickness must grow in proportion to the particle size to prevent destabilization. If too much salt is added to a charge stabilized suspension, it separates; this phenomenon is called **salting out**.

5.A   Suggested experiment: *Salting out*   Skim milk is a colloid whose largest particles are spherical casein micelles about 100–200 nm in diameter [14]. These casein micelles are stabilized by a short brush of charged polymers. According to the discussion in the text, it should be possible to make the micelles aggregate by adding salt so that the screening length is only a small fraction of the radius. The project consists of testing this hypothesis. A suggested approach is first to dilute the milk with distilled water so that the micelles are far apart and so that the solution is just vaguely cloudy. Try to determine the micelle size by shining a laser through the suspension. As implied in the Chapter 4 under "semidilute diffusion" the flickering time is the time for a micelle to diffuse a distance λ/θ where θ is the (small) angle between the beam and the observed speckle point. Determine the amount of salt needed to make the Debye screening length equal to the radius. Make salt solutions with different amounts of salt added, ranging from two orders of magnitude too much to two orders of magnitude too little. Make eight samples, each $\sqrt{10}$ more dilute than the last. Then add 5–10% full-strength skim milk to each sample and mix. (Adding milk to salt water is better than adding concentrated saltwater to milk. You do not want the milk to see saltier water than it should, even for a moment before it gets mixed.) Shine the laser pointer through each sample and look for signs of scattering intensity increasing with time, indicating aggregation. Leave the samples in the refrigerator over night and see whether any samples undergo macroscopic precipitation. If so, play with the precipitate and try to determine whether the particles are stuck together tightly or loosely. Weigh the wet precipitates, let them dry and weigh them again, to determine what fraction was water. A smaller Debye screening length should lead to a denser precipitate.

The reduction in the effective charge is sometimes called **charge renormalization**. When it is expressed as an effective surface charge density, σ^*, we find $\sigma^* \simeq \ln\phi/(4\pi b\ell)$. Note that $\sigma^* \to 0$ as $b \to \infty$. This implies that all the counterions are bound for large particles, and therefore the use of the Debye–Hückel approximation is suspect. This result can be understood fairly easily in terms of the nonlinear Boltzmann–Poisson equation. Consider the special case of a flat half space with a surface charge density σ (of immobilized charges) that is in contact with pure water, i.e., there is no dissolved salt. The only ions in solution are the counterions associated with the surface charged groups. If the surface is negatively charged, the Boltzmann–Poisson equation can be written

$$-\frac{d^2\phi}{dz^2} = (2\lambda^2)^{-1}e^{-\phi}, \qquad (5.23)$$

where z is the coordinate perpendicular to the surface, $\phi(z) \equiv e\Phi(z)/T$ is the potential measured from the surface and λ is given by $\lambda^2 \equiv \epsilon T/(8\pi e\rho^0)$.

With this definition of λ, ρ^0 is the free charge density adjacent to the surface. One readily verifies this by evaluating Eq. (5.23) at the surface, and recognizing Poisson's equation with charge density ρ^0. We must solve the equation in order to find this ρ^0. This one dimensional Boltzmann–Poisson equation has the solution,

$$\phi = 2\ln[1 + (z/2\lambda)]. \tag{5.24}$$

This may easily be verified by inserting (5.24) into (5.23). We note that $\phi \to \infty$ as $z \to \infty$. The boundary condition which is satisfied at the surface is given by Gauss' Theorem and is $\partial\phi/\partial z|_{z=0} = 4\pi\sigma\ell$. The corresponding charge distribution is

$$\rho = \rho^0[1 + (z/2\lambda)]^{-2}. \tag{5.25}$$

The constant amplitude ρ^0 is computed from the condition of charge neutrality, $e\sigma = \int_0^\infty \rho(z)\,dz$. This results in $\lambda = (4\pi\sigma\ell)^{-1}$. The result that $\rho(\infty) = 0$ implies that all the counterions are bound to the surface and there is no ionization, in agreement with the charge renormalization argument given above. The majority of the counterions are contained in a layer of thickness λ adjacent to the interface.

The length scale λ, known as the **double-layer thickness** or the **Gouy–Chapman length** is an important measure of surface charge. It is evidently the distance from a surface of charge density σ in empty space whose potential energy ϕ differs from that at the surface by an amount T. Alternatively, it may readily be seen to be the Debye screening length associated with a slab in which the ions are self-consistently confined, i.e., $\lambda^{-2} \simeq 4\pi\rho^0\ell/e$, with $\rho^0 = e\sigma\lambda$. Thus the charge renormalization discussed above for spherical particles arises from the nonlinearity of the Boltzmann–Poisson equation, which is neglected in the Debye–Hückel approximation. For a typical charged surface with an area per charge of approximately $0.5\,\text{nm}^2$, the double layer thickness, λ, is of order $1\,\text{nm}$. This small value for the characteristic length scale over which the counterions are confined, brings into question the validity of the continuum Boltzmann–Poisson equation. Numerical simulations [15] have shown that for monovalent counterions these results are relatively accurate. However for multivalent ions (see Problem 5.8), λ is much smaller and the Boltzmann–Poisson approximation is even qualitatively inaccurate [16][†]. Here we restrict our attention to the monovalent case for simplicity, bearing in mind that the important situations with polyvalent counterions must be treated more microscopically.

5.8. *Multivalent counterions* (a) Determine the Debye screening length κ^{-1} for a solution of m different ionic species each having a valency z_m and concentration ρ_m. (b) What is the double layer λ thickness when the counterions have valency z? Estimate λ for a surface charge density corresponding to an area per charge of $5\,\text{nm}^2$.

The disjoining pressure between two identical charged surfaces in the absence of added salt is considered in Problem 5.9. However, an excellent estimate may be obtained from the single surface solutions to the Boltzmann–Poisson equation treated above. If the separation between the

[†] When ions have a valency of more than one, the Coulomb energy of two adjacent ions is often substantially greater than T, and the ionic positions become highly correlated. One consequence of these correlations is **overcharging**: Counterions continue to be attracted to the surface even after it has been neutralized. Overcharging can thus reverse the net charge on a surface. It can cause two like-charged surfaces to be bound together, with a layer of counterions in between. Ions with a valency of two or three appear to play an important role in binding macro-ions like DNA together.

two surfaces is $2h$, then by symmetry there is no electric field on the mid-plane, i.e., $\partial\phi/\partial z|_{z=h} = 0$. Then the only force that can be transmitted is associated with the osmotic pressure Π of the counterion "gas":

$$\Pi = T\rho(h)/e. \qquad (5.26)$$

Within the Boltzmann–Poisson approximation, the calculation of the disjoining pressure is then reduced to the determination of the counterion concentration on the mid-plane. For $h \gg \lambda$, this may be approximated by twice the value arising from a single surface (the sum of the contributions from each interface) concentration at h, i.e., $\rho(h)/e \simeq (\pi h^2 \ell)^{-1}$, and a disjoining pressure,

$$\Pi \simeq T(\pi h^2 \ell)^{-1}. \qquad (5.27)$$

In the absence of added salt, the repulsive electrostatic force between the surfaces decays through a power law. With added electrolyte, the pressure decays asymptotically as $e^{-\kappa h}$. For pure pH 7 water without added electrolyte, the Debye length, κ^{-1}, is a few microns; this sets a practical upper limit for the effective range of the electrostatic interactions in aqueous solvents. Quantitative experiments, e.g., using the surface force apparatus [3] have validated the basic ideas embodied in this treatment.

5.9. *Disjoining pressure between charged interfaces* Consider two identical charged surfaces of charge densities σ separated by a distance $2h$ in water. There is no added salt and the neutralizing counterions are monovalent. (a) By solving the Boltzmann–Poisson equation, determine the counterion distribution, i.e., $\rho(z)$. (b) Find an expression for the disjoining pressure and compare to Eq. (5.27).

5.3 Organized states

The interactions between colloidal particles may be sufficiently strong to induce various forms of long range ordering analogous to self-organization of atoms and molecules into crystals or rigid nonsymmetric molecular liquids into liquid crystals. In this section, we discuss briefly several such cases.

5.3.1 Colloidal crystals

One of the best studied of the ordered colloidal systems is that of charged latex particles in low ionic-strength water [17]. These are spherical plastic particles generally in the range of from 10 nm to tens of microns in diameter. They are often prepared by polymerization of species such as styrene in emulsion droplets which are charge stabilized in water by sulfonic acid containing surfactants [18]. This results in quite uniformly sized spherical particles with surfaces decorated with ionizable groups. These particles interact principally via a repulsive Yukawa potential, Eq. (5.21), which may be strong enough to induce crystallization. This occurs when the interaction energy gained by localizing the particles on a lattice (and hence maximally far from one another) overwhelms the translational free energy associated with the particle delocalization into a liquid. Assuming highly charged particles, so that the charge renormalization condition is applicable, the scaling form for the freezing may be written as $b^2/\ell a > k$, where k is a constant of order 10 and a is the mean interparticle spacing. Note that this implies a minimum particle size $kb \simeq 5\,\text{nm}$, below which there can be no freezing under the action of electrostatic forces. Generally freezing occurs when the Debye screening length is close to the interparticle separation. A more complete phase diagram has been given by [19]. Often the colloidal crystals have lattice constants which are comparable to the particle dimensions, which is often in the range of the wavelength of visible light. Thus diffraction may be easily observed. These colloidal crystals are quite soft, and because of the ease of imagining the building units, they are often studied as a paradigm for real crystals and their properties, such as defect structures and other metallurgical aspects. The motion is, however, rather different because the "atoms" are embedded in a viscous fluid.

5.3.2 Lyotropic liquid crystals

Suppose the colloidal particles are not spherical but, e.g., rod-like, with length L and radius b, with $L \gg b$, then there is an additional degree of freedom, namely the orientations of the rods. (An example of such relatively rigid rods is provided by various viruses such as Tobacco Mosaic Virus which is 300 nm long and 18 nm in diameter.) This allows the possibility of another state of order in which the rods throughout the liquid tend to point in a common direction. Such an arrangement, similar to logs in a log jam, is called a **nematic liquid crystal**. The liquid spontaneously chooses an alignment direction, and thus not all directions are equivalent as they are in an ordinary liquid. This loss of directional equivalence amounts to a **spontaneously broken symmetry**.

The reason for nematic symmetry breaking was explained by Onsager [20]. In Chapter 4, the excluded volume parameter, b_2, was defined, and in Problem 5.10 it is determined that for an isotropic distribution of rigid rods, $b_2 \propto bL^2$, which is L/b larger than the actual rod volume. This result suggests the possibility of a phase transition to a state of parallel rods where the effective excluded volume is simply the geometric volume. This provides the mechanism for the formation of lyotropic liquid crystals as a function of increasing concentration of rods; they can simply pack more efficiently. The disordered and nematically ordered states are shown in Fig. 5.10. With ever increasing concentration even more types of order may develop. For example, a **smectic liquid crystal** is a layered structure in which there is no center of mass order in the layers, as illustrated in Fig. 5.11. Thus each layer is a two dimensional fluid. An excellent, detailed treatment of liquid crystals is given in the book by de Gennes and Prost [21].

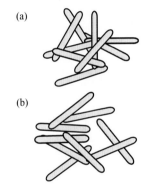

Fig. 5.10
(a) Sketch of an isotropic solution of rods.
(b) Weakly nematically ordered rods.

Fig. 5.11
A smectic liquid crystal.

5.10. *Rod excluded volume*   A slender rod of length L and radius $b \ll L$ is centered at the origin and pointed along the z axis. A second identical rod lies along the x axis at a distance x from the first. Its orientation is arbitrary.

(a) What is the probability that the second rod will intersect the first? Ignore numerical prefactors and consider the dependence on L, b, and x. How small must x become in order for the probability to be of order unity?

(b) Using the results of (a), estimate the mutual excluded volume of two such rods if they are dispersed in a solution. Again, ignore numerical prefactors.

(c) Repeat parts (a) and (b) with the rods replaced by disks of diameter L and thickness b. Compare with the result for the rods. Compare with the result for hard spheres of diameter L found in Chapter 4.

5.3.3   Fractal aggregates

Much of the discussion of this chapter has centered on the methods to prevent colloidal aggregation and suspension instability. But there are cases where aggregation is desirable. Here the strong tendency to aggregate can be turned to advantage to produce distinctive and useful aggregated structures like those of Fig. 1.5. These aggregates are different from ordinary precipitates that form when small molecules attract one another in a solvent. The difference arises because the attraction between the particles of Fig. 1.5 is very strong. It is so strong that particles once joined cannot move or unstick. The randomness in these aggregates is thus of a different type than the thermal randomness we have encountered up to now. Accordingly, a different approach is needed to analyze it. These aggregates, like polymers, are fractal structures. Their fractal dimension depends on the conditions of aggregation. When aggregation occurs as fast as possible and the rate is limited only by the Brownian movement of the aggregates, the process is said to be **diffusion limited**, and the fractal dimension D is observed to be about 1.7. When the joining of particles is strongly inhibited by an activation barrier (as discussed in Chapter 2) the aggregation is said to be **reaction limited**. Then the fractal dimension D is observed to be about 2.1.

Many properties of fractal aggregates are analogous to those of polymers. For example, fractal aggregates have hydrodynamic radii comparable to their geometric sizes. Some properties of these aggregates are qualitatively different from those of polymers, because aggregates are rigid structures that hold their initial shape. If an external force distorts an aggregate, it responds with an elastic restoring force. This force arises from the joining interaction that holds the constituent particles together. The scaling of this restoring force with the aggregate size can be derived in terms of geometric scaling properties: *viz.* the fractal dimension D and another geometrically defined exponent. In Appendix B we lay the groundwork for a quantitative description of fractal aggregates and their properties.

5.3.4   Anisotropic interactions

In the previous sections, the forces which exist between spherical colloidal particles have been assumed to depend only on the distance between them. Anisotropic effects, e.g., liquid crystalline ordering, were associated with nonspherical objects. However, even spherical particles may have angular dependences in their interactions if there is an internal direction defined within the particles. For example, the particles may be **ferromagnetic**, having a permanent magnetic dipole moment, or they may be electrically or magnetically polarizable, and respond to external electric or magnetic fields. In these cases, the interaction between the particles depends on the angle that the interparticle vector makes with the external field. Suspensions of ferromagnetic particles, like magnetite, iron, and cobalt, are called **ferrofluids**, and have applications to fluid seals, printing, etc. [22]. **Electrorheological fluids**, e.g., starch suspensions, are dispersions of particles having large electric polarizabilities, which develop electric dipole–dipole interactions in the presence of externally applied electric fields [23]. These systems are being investigated for clutches and self-adjusting suspensions in automotive applications.

The new feature to be considered in a fluid of dipoles is the dipolar energy. We consider a suspension of spherical particles which possess identical net dipole moments, μ. For concreteness, they will be assumed to be ferromagnetic with magnetization M and of radius b. In a typical ferrofluid, $M \simeq 100\,\mathrm{Oe}$ and $b \simeq 10\,\mathrm{nm}$, so that the dipole moment $\mu \simeq Mb^3 \simeq 4 \times 10^{-16}$ cgs units. The dipolar interaction energy between two particles separated by a vector, $\vec{r}$ ($r > b$), is given by

$$U_{1,2} = r^{-3}[\vec{\mu}_1 \cdot \vec{\mu}_2 - 3(\vec{\mu}_1 \cdot \vec{r})(\vec{\mu}_2 \cdot \vec{r})r^{-2}]. \tag{5.28}$$

The strength of the dipolar interaction between particles may be expressed in terms of a dimensionless coupling constant, $\lambda \equiv \mu^2/4b^3T$. At ambient temperatures, $\lambda \simeq 1$, which implies that the dipolar interaction may significantly influence the organization of the colloidal suspension. When $\lambda \geq 1$, the anisotropic nature of the dipolar energy favors a head-to-tail configuration of the dipoles and may lead to chain formation, If the strength of the dipolar interaction is sufficient, some form of phase separation might be expected, but the complete phase behavior in the zero external field has yet to be determined.

5.4   Colloidal motion

A colloidal suspension shows the same kinds of time-dependent properties as a polymer solution. The particles diffuse and sediment. The fluid as a whole flows in response to stress. Colloidal particles are essentially solid spheres. The motion of hard spheres was explained in Chapter 4. Thus we need only recall those previous results here.

In the dilute state, colloidal particles undergo Brownian motion, characterized by a self-diffusion constant. This is the Stokes diffusion constant of Eq. (4.7), varying inversely with the particle size. This same diffusion constant tells the speed of motion in response to a given force, such as sedimentation under gravity, according to the Einstein relation. Particles with a stabilizing corona of polymers have a hydrodynamic radius large enough to include most of the corona, since this polymer layer is hydrodynamically opaque. This is true whether the corona is a brush or an adsorbed layer. Colloids stabilized by charge may in principle suffer an analogous effect. The countercharge in the Debye screening cloud around a moving charged particle feels electrostatic forces tending to drag it along with the particle. These forces in turn tend to entrain the fluid in the Debye zone. These effects are subtle and cannot readily be summarized. In any case, they slow the diffusion by at most a factor of order unity. Such "electroviscous effects" are treated, e.g., in [24]. When colloidal particles repel each other strongly,

Brownian motion is inhibited, as with polymers. Each particle becomes entrapped in the cage formed by its neighbors. This glassy effect occurs with small molecules as well, and we mentioned it in our discussion of viscosity in Chapter 2. In such systems, as with polymer solutions, diffusion is characterized by a slowed self-diffusion and a speeded-up cooperative diffusion.

When strong interaction leads to marked changes in structure such as colloidal crystallization, it must also lead to storage of mechanical energy. Thus colloidal crystals must have elastic moduli and associated viscosity. Strong crystalline order occurs when the interaction energy per particle is a few hundred T. Micron-scale colloidal particles thus have an energy density measured in T per μm^3. The corresponding energy density in a semidilute polymer solution is T per blob volume, where the blob size is a few nanometers. Evidently the colloidal modulus is weaker than the typical polymer modulus. One generally cannot sense the mechanical moduli without special apparatus. Analogous interaction effects occur in emulsions and foams, as discussed in Chapter 7. There the underlying forces come from interfacial tension, and the resulting forces are stronger.

5.4.1 Electrophoresis

Charged colloidal particles move in a distinctive way when an electric field acts on them. The contrast with conventional forcing is greatest when there are many free ions and the screening length is short. In Chapter 4, we analyzed the response to external forces by tracing the added momentum as it flowed outward from the forced particle. The long-ranged flow field of the surrounding fluid leads to hydrodynamic drag. However, when a charged particle is forced by an electric field, the situation is radically different. In any unit of time, the momentum added to the particle is balanced by opposite momentum added to the counterions in the oppositely charged screening layer. For a uniformly charged body that is smooth on the scale of a screening length, the two momenta cancel nearly completely, and virtually no momentum is transferred to the fluid. Thus no long-range flow occurs. The particle crawls through the liquid rather than swimming through it. All the dissipation leading to the retarding force occurs in the narrow screening zone. Thus the velocity of each bit of surface is determined independently of the rest. The **electrophoretic mobility** depends on the surface charge density and the screening length, but not on the size of the particle.

When the charge on a particle is nonuniform, the phenomenon of electrophoresis becomes startlingly rich and complex, as recent discoveries have shown [25]. If two regions on the particle have different charge densities, the electrophoretic force wants to pull them at different speeds. Since the two regions are obliged to move at the same speed, one region must exert a force on the other. This means that each region experiences a net unbalanced force, which it must give to the surrounding fluid, producing hydrodynamic backflow. An opposing momentum is injected at the other region. A net momentum current comparable to the internal force is injected into the fluid, and long-range Stokes flow occurs. One may control the nature of the induced motion in striking ways by controlling the shape of the particle and the placement of charge on it. One may induce motion at right angles to the electric field or rotation with no translation, for example [25].

5.4.2   Soret effect

Other perturbing influences create motion analogous to simple electrophoresis. A particle inserted into a fluid acquires interfacial energy because of the interaction of the surface with adjacent fluid molecules, as we will learn in Chapter 6 to come. If the fluid is inhomogeneous, the interfacial energy can be different on different parts of the particle. For example, a temperature gradient induces a gradient in the interfacial energy. Typically it is reduced at higher temperature. Thus by translating to a region of higher temperature, the particle can lower its interfacial energy. The gradient of this energy is a force. It causes motion of the particle. This motion in response to a gradient of interfacial energy is called the **Soret effect** [26]. As in the case of simple electrophoresis, this motion does not add momentum to the fluid. Thus it does not produce hydrodynamic drag. One may imagine inhomogeneous surfaces that produce unbalanced Soret forces analogous to the unbalanced electrophoretic forces discussed above. It would seem that analogous drag phenomena would occur.

Appendix A: Perturbation attraction in a square-gradient medium

As we have suggested above, two like objects that perturb a fluid tend to attract each other. In this appendix we show that this attraction is *inevitable* under a wide range of conditions [27]. We consider two identical, parallel surfaces of area L^2 in a fluid at some small separation $2h$. We suppose that each surface perturbs the fluid by altering some scalar property ψ of the fluid immediately adjacent to the surface. This ψ may be e.g., the local concentration of polymer or colloid, or the chemical composition, as in the examples in the text. Naturally, if ψ is perturbed at the surfaces, it is perturbed nearby as well. Thus there is a perturbed profile $\psi(z)$ at all distances z from the surfaces. To describe this profile quantitatively, we must consider the free-energy cost of perturbing ψ in the liquid. This cost depends on the whole profile $\psi(z)$: it is a functional of the function ψ. We denote it as $W[\psi]$. We may also treat the effect of the each surface as a contribution to the free energy: $V(\psi_s)$. Here ψ_s is the value of ψ at the surfaces.

The $\psi(z)$ profile is that which minimizes the total free energy functional $\mathcal{F}/L^2 \equiv W[\psi(z)] + 2V(\psi_s)$. The first term is the work required to alter ψ from its equilibrium value ψ_0. When $\psi(z) = \psi_0$, we shall set $W = 0$. Any change of ψ from ψ_0 requires positive work. Naturally the free energy $W[\psi] + 2V(\psi_s)$ depends on the separation of the surfaces $2h$. The work required to separate the surfaces is precisely the change in this free energy.

For many fluids, the free energy W for a nonuniform profile has a simple form, provided $\psi(z)$ varies gradually enough. Then only the difference between $\psi(z)$ and ψ in *adjacent* regions matters. That is, the cost for each region depends only on the gradient of ψ there. Except for this gradient effect, the work to perturb ψ in a region is the same as though ψ were constant in space. In other words, $W[\psi]$ has the form

$$W[\psi] = 2 \int_0^h dz \, w(\psi(z)) + \frac{1}{2} m(\psi(z)) \left(\frac{d\psi}{dz}\right)^2 . \qquad (5.29)$$

Here the $w(\psi)$ is the free energy per unit volume required to make a *uniform* change from ψ_0 to ψ. We have restricted the integration to the left half of the system, noting that the contribution from the right half must be identical. We shall suppose that the variation of ψ is gentle, so that we may neglect higher powers of the gradient $d\psi/dz$[†]. Its coefficient m may depend in general on the local value of ψ[††].

Two identical surfaces immersed in such a square-gradient medium must attract each other at long distances. Further, the attractive force per unit area is simply the local energy density at the midplane w_m. To see this, we separate the surfaces by a small amount $2\Delta h$ and examine the change in free energy $\Delta\mathcal{F}$. We perform the separation in two stages. First we separate the surfaces without allowing a change in the ψ profile. We simply extend its midpoint ψ_m over the extra interval Δh on each side of the system. Second we allow the ψ field to relax to minimize $\mathcal{F}$.

In the first stage $\mathcal{F}$ changes by an amount $2\Delta h L^2 w_m$. Since ψ is constant here, the gradient energy is zero. In the second stage, ψ changes in the pre-existing region and in the newly created gap region. The resulting change of $\mathcal{F}$ in the gap region is second order; it is the product of the small change $\Delta\psi_m$ and the small interval $2\Delta h$. The profile in the old region from 0 to h was that which minimized the free energy $\mathcal{F}$ with the constraint that $d\psi/dz$ vanish at h. This derivative vanished because h was at the midpoint of the full system. To examine how this profile changes, we first determine the optimal profile between 0 and h without constraints. We allow $d\psi/dz(h)$ to vary freely. But the optimal value of $d\psi/dz(h)$ is zero in a square-gradient medium. We may check this by dividing the interval up into many equal, discrete intervals ending at $z_1, z_2, \ldots, z_k = h$. The corresponding ψ values are $\psi_1, \ldots, \psi_k$. The free energy $\mathcal{F}$ is the sum of contributions from each interval. To minimize $\mathcal{F}$ in the i'th interval we need only adjust ψ_i and ψ_{i-1}. The only interval that involves ψ_k is the last one. The free energy contribution to it is

$$\frac{h}{k} w(\psi_{k-1}) + \frac{h}{k} m(\psi_{k-1})(\psi_k - \psi_{k-1})^2 \left(\frac{h}{k}\right)^{-2} .$$

Since ψ_k appears only in the gradient term, the optimal value for it is that which makes the gradient vanish. Minimizing $\mathcal{F}$ *automatically* makes the gradient vanish at h. The vanishing of the gradient is not an additional constraint[†††]. When we allow $\psi(z)$ to relax to the new, wider interval, there is no proportionate change in $\mathcal{F}$. The original $\psi(z)$ already minimized $\mathcal{F}$ in the old region 0–h with respect to arbitrary variations. This means any small change $\Delta\psi$ can make at most a second-order change in $\mathcal{F}$ in that region. Thus the only change in $\mathcal{F}$ is the contribution from the gap region,

[†] There is no piece linear in this gradient. Such a piece would not be invariant under a change of coordinate system that replaces z by $-z$; it is thus ruled out by the requirement that the free energy density must be independent of coordinate system.

[††] As in the polymer solution, any fluid with a free energy of this form has spatial correlations in the fluctuating equilibrium $\psi(z)$. These correlations die off exponentially in space with a decay length ξ given by $\xi^2 = d^2w/d\psi^2|_0/m(\psi_0)$.

[†††] The author is grateful to Alexei Tkachenko for pointing this redundancy out to him.

proportional to w_h. The increase in $\mathcal{F}$ implies an attractive force per unit area

$$\frac{d(\mathcal{F}/L^2)}{d(2h)} = w_h.$$

Though we have assumed our surfaces to be identical and flat, this force law holds more generally. If two surfaces are sufficiently similar, there must be a point between them where the gradient of ψ vanishes. Then the above argument can be extended to show that the force per unit area at that point is attractive and equal to w.

While this perturbation attraction occurs quite broadly, there are important cases where the mechanism does not apply. One is the case of surfaces with *grafted* polymers treated in the chapter. Here polymers are chosen which are attracted to the surfaces only at one end. The attraction results in a polymer-enriched layer near each surface. When the two surfaces are brought together so that the layers interpenetrate, a repulsion occurs, not the attraction implied by the Perturbation-Attraction Theorem. Why is the theorem not applicable? The reason is that the free energy does not have the square-gradient form, as in Eq. (5.29). Instead, the energy cost of a nonuniform concentration is nonlocal. The cost at position z does not depend merely on ψ at z and on ψ at neighboring z. It depends on ψ throughout the layer. Thus e.g., if some chains were removed from the layer, the free energy at the midpoint h would be affected—even if some external agent forced ψ to be fixed in the intervening layer. Thus the free energy cost of nonuniformity cannot be found by adding up gradient contributions from the whole layer. Nonlocal differences such as $\psi_s - \psi(h)$ also contribute directly to the free energy.

Another case where perturbation does not lead to attraction is the deformation of a membrane with bending rigidity, as treated in Problem 5.4.

Appendix B: Colloidal aggregates

We have discussed several mechanisms of colloidal attraction and several ways of countering such attractions with compensating repulsions. Repulsions of sufficient strength can induce the colloidal particles to form ordered states: the particles form a periodic lattice in the solution. Attractions of sufficient strength can also induce self-organization. The simplest effect of attraction is the precipitation or creaming mentioned above. Rather than dispersing throughout the solution, precipitating colloidal particles prefer to assemble into compact masses, in which their mutual attraction energy may be maximized. As noted in Chapter 2, such precipitation happens when the probability of two particles being near to each other is substantially larger than the probability that they are far apart. As we saw, this amounts to saying that the interaction free energy of two particles is attractive by an amount of order T or greater. If the mutual attraction is strong enough to bring two particles together, it is ordinarily more than strong enough to bring pairs or larger groups together. Thus the process of coming together feeds on itself; its ultimate result is that the particles have assembled into a compact, macroscopic mass or "dense phase." The nature of the dense phase and the amount of attraction or colloid concentration required to produce it are described by the statistical mechanics of phase transitions [5]. This well-developed branch of theoretical physics gives powerful, specific, and well-verified predictions about how mutually attracting particles undergo phase separation. We encountered this phenomenon briefly in Chapter 4, in discussing the "collapse" of a polymer in a poor solvent. For the most part the theory of phase transitions applies as well to colloidal particles as it does to small molecules. Thus the nature of precipitation and creaming is virtually the same as, e.g., the precipitation of steam or the liberation of carbon dioxide from a freshly opened soft drink. But in some cases the distinctive attraction of colloidal particles results in qualitatively new phenomena.

Phase transition theory applies only when the particles have come to thermodynamic equilibrium, at least locally. As we saw in Chapter 2, equilibrium is the state attained after a sufficiently long time. For any change of configuration the system has made, it must also have time to make the reverse of that change. For example, any given pair of particles that are together should have had enough time to separate and recombine several times in order to attain equilibrium. With small molecules, phase separation is usually slow enough that each local region (containing a few molecules) is close to thermal equilibrium. But in colloids, one encounters a new regime of *irreversible* attraction, far from equilibrium.

As we have noted, most forms of colloidal attraction grow in strength with the particle size. For large particles, the attractive energy may far exceed the thermal energy T. Then the attraction becomes for practical purposes irreversible, as discussed in Chapter 2. Two particles that stick together have virtually no chance to unstick in the time the system is observed: the strength of the attraction prevents even local equilibration. Thus phase transition theories cannot describe the assembly process; a new approach is needed.

The assembly process in this irreversible regime is called **kinetic aggregation** (Fig. 5.12). Kinetic aggregation of a colloidal solution ultimately produces large aggregates like the one shown in Fig. 1.5. These are strikingly different from the compact droplets or grains of normal precipitation: the aggregates are open, tenuous assemblies. The nature of these assemblies has been much studied in the last two decades and has been treated in comprehensive reviews [29, 30].

The colloidal particles of Fig. 1.5 were initially dispersed in a very dilute state. Four-nanometer charge-stabilized silica particles were suspended in water at a volume fraction of roughly 10^{-6} [31]. Then a large amount of salt was added, screening out the repulsive Coulomb barrier between the particles nearly completely, and exposing the particles' strong van der Waals attraction. A few minutes after the salt was added, the aggregate had formed. When two particles encountered each other in the course of their ordinary Brownian motion, they stuck together permanently. The resulting pair continued its Brownian motion until it encountered another particle or cluster. These two clusters stuck together permanently in their contacting configuration and continued to move. In this way the clusters grew to the size shown. One may readily simulate this process on a computer [29], to

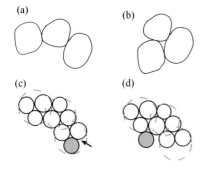

Fig. 5.12

Nonequilibrium aggregation. (a) Typical contacting configuration of three solid colloidal particles. The angle between the three is arbitrary. (b) Configuration of lower attraction energy. In thermal equilibrium this state is much more probable than (a), but in irreversible aggregation it is no more probable. (c) Typical contacting configuration of three triangular clusters of liquid droplets. (d) Compact configuration. Like configuration (b), this state is unreachable in the available time. Liquid drops can go from state (a) to state (b) with no energy barrier, but not from state (c) to state (d) [28]. To pass from (c) to (d) the shaded ball would have to break the contact indicated by the arrow and overcome the associated energy barrier.

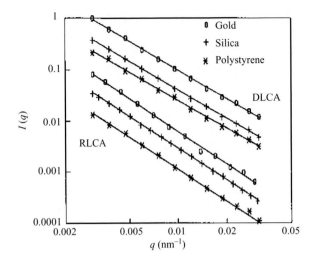

Fig. 5.13
Structure factors $S(q)$ from colloidal aggregates like that pictured in Chapter 1, obtained from light scattering, after [31]. The upper three data sets marked DLCA were from aggregates grown under diffusion-limited conditions using 7.5 nm gold particles, 3.5 nm silica particles, and 20 nm polystyrene spheres. Straight lines have the slopes expected for a fractal structure with $D \simeq 1.84$. The lower three data sets marked RLCA were from aggregates grown under reaction-limited conditions made from the same three dispersed colloids. Straight lines have slopes expected for a fractal structure with $D \simeq 2.1$.

produce an ensemble of representative clusters. These simulated clusters have the same treelike structure and the same fractal dimension seen in experiments [31] (Fig. 5.13).

Simplest aggregation model

The "Sutherland's Ghost" model of Ball [32] gives a simple way to understand why the structure is open. In this model we imagine that many single particles have combined to form dimers. Then all the dimers combine to form tetramers, then all these combine to form octomers, and so forth. When two n-clusters combine, they stick together as randomly as possible. We choose a particle on the first cluster and a particle on the second cluster at random. Then we choose a direction at random. Finally we combine the two clusters by connecting the two chosen particles in the chosen direction, as shown in Fig. 5.14. (There is no need to rotate the clusters, since this assembly process automatically produces all rotations of a given cluster.) In general, the combined cluster is not self-avoiding: its particles intersect. We freely allow such intersections: hence the name "ghost."

The average size of a large Sutherland's Ghost aggregate can be readily estimated. To define an average size, we select two particles in a large $2n$-cluster at random. These two are not in general connected directly: one must traverse some number b of connecting bonds to reach the second particle from the first. The two unshaded particles in Fig. 5.14 are separated by three bonds. (We do not count accidental intersections as connections.) The average of the bond distances b in an n-cluster will be called B_n.

Luckily, the average bond distance B_{2n} for a $2n$-cluster can be expressed simply in terms of the B_n of the n-clusters. Any $2n$-cluster is made from two subclusters of size n. If two points on the $2n$-cluster are chosen at random, there is one chance in four that the two belong to the first subcluster. Since the points on this subcluster are arbitrary, the average bond distance in this case is simply B_n. It is the same if both chosen points had been on the second subcluster. But there is one chance in two that the two chosen

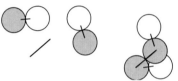

Fig. 5.14
One aggregation step in the Sutherland's Ghost model. Left: a (shaded) particle from two arbitrary dimers is selected at random, and the bond direction indicated by the line segment is chosen at random. Right: the selected particles are joined by the bond.

points belong to different subclusters. In that case the bond distance is a bit harder to find. Each subcluster has one particle that is joined directly to the other subcluster. To reach the second chosen particle from the first, we must first travel on the first subcluster to its joining particle, then bridge to the joining particle of the second subcluster, and finally travel to the second chosen particle. The bond distance is the sum of these three distances. Now, the average distance in the first subcluster is simply B_n, since the first particle and the joining particle were chosen at random. Likewise, the average distance in the second subcluster is also B_n. The distance between the joining particles is one bond. Thus the average bond distance between particles on different subclusters is just $2B_n + 1$.

We may now find the overall average bond distance B_{2n} by combining this distance with the single subcluster distance, each with its appropriate probability:

$$B_{2n} = \tfrac{1}{4} B_n + \tfrac{1}{4} B_n + \tfrac{1}{2}(2B_n + 1), \qquad (5.30)$$

or

$$B_{2n} = \tfrac{3}{2} B_n + \tfrac{1}{2}.$$

Since we know $B_2 = 1$, we can readily find any B_n by direct calculation. If n is large, we may ignore the $\tfrac{1}{2}$; then $B_n \simeq \tfrac{2}{3}\left(\tfrac{3}{2}\right)^{\log(n)/\log(2)}$, or

$$B_n \sim n^{\log\left(\frac{3}{2}\right)/\log(2)}.$$

The exponent is roughly 0.58.

From this B_n we may readily find the average *geometric* distance between two arbitrary points in terms of the particle diameter a. Clearly this geometric distance depends not only on the number of bonds traversed but also on their directions. In our model, these directions are completely random. Each bond along the connecting path was chosen randomly and independently of all the others. Thus the connecting path is a simple random walk, and the mean-squared distance r^2 between two points separated by b bonds is simply $a^2 b$. The mean-square distance R_n^2 between two arbitrary particles is evidently $a^2 B_n$. Thus $R_n \sim n^{\log(3/2)/(2\log(2))}$, or

$$n \sim R^D, \quad \text{where } D = 2\log(2)/\log(3/2) \simeq 3.4.$$

The mass scales with size like a fractal with dimension D.

This structure is very dense in three dimensional space, but in four or more dimensions, D is smaller than the spatial dimension and the Sutherland's Ghost structures are tenuous fractals[†]. Thus the model shows how the simplest features of kinetic aggregation can lead to an open fractal structure.

Effects of polydispersity and self-avoidance

While the Ghost model gives a qualitative account of the open structure of kinetic aggregates, it contains two gross simplifications of potentially great importance. The first of these is the assumption that all clusters are made from two equal-size subclusters. In reality, when a dimer encounters another colloidal particle, this particle may be a free particle, another dimer, or a

[†] We have only shown that the overall mass n scales with overall size according to a fractal law. The same fractal dimension D is believed to describe the average mass $n(R)$ within some local region of radius R around an aggregate particle. This is clearly true if one takes the sphere to enclose one of the subclusters making up the aggregate. (Incidental incursions into this sphere from other subclusters do not contribute qualitatively to the enclosed mass if $d > D$.) Since we do not expect the mass for a given size R to vary widely with the position of the sphere on the aggregate, we expect that $n(R)$ for an arbitrary sphere is of the same order as that for spheres covering subclusters. Such arguments, together with computer experiments, lead to the accepted belief that these aggregates are fractals at all scales between that of the particles and that of the aggregate [29].

larger cluster. The two joined clusters are typically unequal, and they may be greatly unequal. The second drastic simplification is that self-intersections are ignored. We may see the effect of both of these simplifications by simple modifications of the model.

It is easy to modify the Ghost model to account in a primitive way for unequal clusters. This modification is called the "fixed-ratio Sutherland's Ghost" model [32]. This model is like the original model, except that the reacting clusters are forced to have a particular mass ratio, such as $3:1$. One may readily modify the above reasoning to determine the average bond distance, B_{4n}, in terms of those of its constituents, B_n and B_{3n}. Now when two particles are chosen at random to compute the combined bond distance, B_{4n}, the probability that both particles are on the smaller cluster is now reduced (to $1/16$); the other probabilities are also altered. For a general ratio $r:1$, the probability that an arbitrarily picked point belongs to the small cluster is $1/(r+1)$. The probability for the larger cluster is $r/(r+1)$. Eq. (5.30) becomes

$$B_{(r+1)n} = \left(\frac{1}{r+1}\right)^2 B_n + \left(\frac{r}{r+1}\right)^2 B_{rn} + 2\frac{r}{(r+1)^2}(B_n + B_{rn} + 1).$$

This equation also allows power-law solutions of the form $B_n \sim n^x$. Substituting into the above equation, one finds an implicit equation for x.

$$(r+1)^x = \left(\frac{1}{r+1}\right)^2 + \left(\frac{r}{r+1}\right)^2 r^x + 2\frac{r}{(r+1)^2}(1 + r^x). \quad (5.31)$$

The exponent x decreases with increasing ratio r. For small r, Eq. (5.31) reduces to

$$x = 2(1 + r^x).$$

As in the simple Ghost model, the geometric path between any two sites on a cluster is still a random walk, so that $n \sim R^D$ with $D = 2/x$. Evidently the nominal fractal dimension D goes to infinity as the ratio $r \to 0$. It is possible to extend this treatment to include a complete distribution of aggregating sizes, not merely a fixed ratio [33]. For the expected mass distributions[†], D increases moderately from the 3.4 of the simple Sutherland's Ghost model. Clearly, to account for D of kinetic aggregates quantitatively requires a realistic treatment of the relative masses of the aggregating clusters.

The second glaring defect of the model is its neglect of self avoidance. We may evaluate the importance of this defect in the same way we have done for polymers in Chapter 3. We modify the process to assure self-avoidance. Pairs of clusters are joined as in the original model. Then the combined cluster is checked for self-intersections. If any are found, the combined cluster is discarded. If this discarding probability approaches unity for large clusters, the average properties of the remaining ones may be qualitatively affected. But if the discarding probability does not approach unity, the average properties cannot be qualitatively affected and scaling exponents such as D must be unchanged.

By this reasoning, we may see that self-avoidance has no impact on D in sufficiently high spatial dimensions d. We saw in Chapters 3 and 4, that

[†] The probability $P(n,t)$ that a given cluster has mass n at time t is called the mass distribution. To account for the change of this distribution as the aggregation proceeds, one must consider all processes which may increase or decrease the number of n-clusters. The joining of an i cluster with an $(n-i)$-cluster increases $P(n,t)$. The joining of an n-cluster with any other cluster decreases $P(n,t)$. The rate of joining of i-clusters with j-clusters is evidently proportional to their numbers, $P(i,t)$ and to $P(j,t)$. But two different mass pairs with the same cluster numbers P need not have the same joining rate. A given pair of clusters with masses i and j has a specific joining rate, denoted $K(i,j)$. Adding together all the processes that change the number of n-clusters, one obtains the so-called Smoluchowski equation [34]:

$$\frac{\partial P(n,t)}{\partial t} = \frac{1}{2}\sum_{i,j} K(i,j)P(i,t)P(j,t)$$

$$\times [\delta_{i+j,n} - \delta_{i,n} - \delta_{j,n}].$$

The Smoluchowski equation gives several different types of P distributions, depending on the specific joining rate K. The equation itself is an approximation, for it accounts for only the masses of the clusters and ignores other variables, such as the cluster shapes and their positions in space. Nevertheless, it is believed to be well-justified for three-dimensional colloidal solutions. These subjects are treated at length in [34].

when two fractals of size R are placed at random in the same volume, the average number of intersections between them within distance R goes as $R^{D_1+D_2-d}$. When this exponent is negative, the number of intersections goes to zero as $R \to \infty$. Thus our two aggregates with dimension D have no intersections in spatial dimension $d > 2D$. Even if the two aggregates are connected at the origin, the number of intersections is limited and does not grow indefinitely with R. Thus in these high spatial dimensions the probability of self-avoidance is finite and does not go to zero as the cluster sizes go to infinity. Then the aggregation process is not qualitatively affected by the self-avoidance constraint, and the D of the aggregates is unchanged. For Sutherland's Ghost aggregates with $D \simeq 3.4$, self-avoidance has no affect on D above $d = 6.8$. If the aggregation process were carried out on a computer in a virtual space of seven or more dimensions, the resulting D should be 3.4 even with self-avoidance imposed. We have encountered the analogous property for polymers: their self-avoidance does not affect their fractal dimension in more than four spatial dimensions.

In lower spatial dimensions, self-avoidance becomes relevant. In polymers the discarding process has the greatest impact on the more compact configurations. The remaining, self-avoiding, ones are generally increased in size, and their D is reduced. The same qualitative behavior is expected for aggregates. D should decrease progressively as the spatial dimension d is reduced.

Diffusion-limited and reaction-limited

The Ghost model has another unrealistic aspect. It gives no account of how the clusters must move in order to join together. At first sight, this omission does not appear serious. For polymers any questions of motion are irrelevant to their fractal dimension. This is true because the configurations of a polymer form an equilibrium ensemble: the probability of a given configuration is prescribed by the Boltzmann principle. This Boltzmann probability depends only on the energy of that configuration and not on any motion. But our aggregates are not an equilibrium system. The configurations and their probabilities are defined by an irreversible, kinetic process. Accordingly, we must consider how various types of motion may affect the configurations. We imagine a particular joining configuration of two clusters from the Ghost model and consider the probability that this configuration occurs.

In a real colloidal suspension, two clusters that encounter each other move by Brownian motion. Instead of simply being placed next to each other as in the Ghost model, the two clusters move together by a random walk. The walk influences the probability that a particular contact will be made. One limit of interest is when the two clusters stick together on first contact. Then only those random walks that happen to reach the joining configuration with no intervening contact between the clusters are possible. This constraint is a form of self-avoidance constraint. We can phrase it in geometric terms. For each cluster in our ensemble, we construct a typical random-walk history by moving the cluster in a random walk and including all the spatial points thus traversed, as shown in Fig. 5.15. (For simplicity we consider only translational motion in this random walk; rotational motion also occurs but

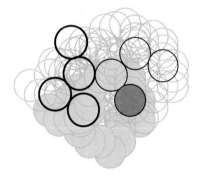

Fig. 5.15

A candidate joining configuration in diffusion-limited aggregation. The previous steps in the random walk of the cluster on the right are shaded. One particle of this cluster is filled to indicate its history. Since the history of this cluster intersects the other cluster many times, this configuration must be discarded.

does not affect the conclusions.) The resulting object is a fractal whose dimension is two greater than that of the cluster itself. We now choose a random point on this cluster and on another cluster. We then join these according to the Ghost model. However, for a valid joining configuration, the history must not intersect the other cluster. If it does, it must be discarded. The remaining configurations are valid joining configurations, in which the one cluster has avoided the history of the other[†].

This constraint is again that of the mutual avoidance of two fractals. The Brownian-motion history of a fractal of dimension D is a fractal of dimension $D + 2$. The history of each particle of the fractal lying in a sphere of radius R is a random walk, with roughly R^2 particles in the sphere. The number of particles leaving these random-walk trails is the number of particles of the fractal—roughly R^D. The total number of particles in the history within the sphere is thus roughly $R^2 R^D = R^{2+D}$. As before, it must be irrelevant in sufficiently high spatial dimensions d. If the clusters have dimension D, then the avoidance has no effect on D provided $(D + 2) + D - d < 0$. For our Ghost aggregates, this condition holds for $d \gtrsim 8.8$. Evidently this random-walk avoidance has a stronger effect than the simple avoidance discussed previously. D begins to decrease from its asymptotic value for higher d (8.8 versus 6.8), and at a given d we would expect it to decrease further. When clusters move by Brownian motion and stick on contact, the aggregation is called **diffusion-limited**. This form is very prevalent in colloidal solutions. The aggregate of Fig. 1.5 was produced in diffusion-limited conditions.

Sometimes diffusion is not the rate-limiting aspect of the aggregation process. It can happen that particles stick irreversibly with very low probability even when they are adjacent. Indeed this is the case for stabilized colloidal particles. If the stability is diminished somewhat, e.g. by adding a little salt, the sticking occurs fast enough to observe. Still the sticking rate may be much slower than the rate of encounters via Brownian motion. This regime is called **reaction-limited** aggregation. In the reaction-limited regime, two clusters which join at a particular moment have had ample opportunity to visit each other's neighborhood. Thus all configurations that do not intersect are equally likely, as in equilibrium. (Here we ignore the local repulsion that inhibits very close approaches; we implicitly count such approaches as intersections.) Now the joining proceeds as in the self-avoiding Sutherland model discussed above: the joined configurations are taken at random from all contacting configurations. As we have seen, this model is less constrained than the history-avoiding diffusion-controlled case. Thus we expect the resulting aggregates to have a fractal dimension D closer to the large $D = 3.4$ of the unconstrained model.

The diffusion-limited and reaction-limited regimes are the most important types of aggregation seen in colloidal suspensions. But other types of motion are also of interest. In ballistic aggregation, the joining clusters follow straight-line paths instead of random walks. One form of ballistic aggregation is sedimentation aggregation, in which the joining clusters are drifting downwards in a gravitational field. They join when a heavier cluster overtakes a lighter one [35]. Large, composite snowflakes are made by this form of aggregation.

[†] In making this construction we have treated one cluster as stationary and the other one as moving in a random walk. This amounts to working in the frame of reference of one cluster. This is equivalent to the actual situation in which both clusters execute random walks.

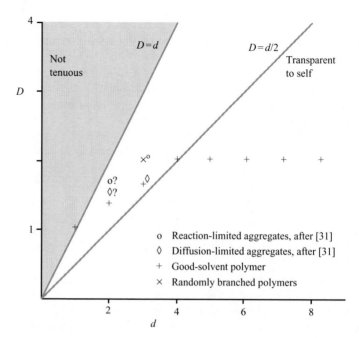

Fig. 5.16
Fractal dimension D of colloidal
aggregates and other fractals versus spatial
dimension d.

We have seen that a number of aspects of colloidal aggregation have a
profound effect on the structure and thence on the fractal dimension D.
To account for all these aspects—mass distribution, reaction rates, self-
avoidance—and thus obtain a realistic estimate of D is a complicated job.
This accounting may not yield great insight. But the qualitative nature of
these effects can be simply understood, as we have seen. In Fig. 5.16
we summarize the known information about D, based chiefly on simulations. The agreement between these simulations and experiment gives
confidence that the essential features responsible for the fractal structure are
recognized.

Properties of aggregates

The fractal structure of colloidal aggregates makes them broadly useful, as
noted in Chapter 1. The aggregates are used like polymers to thicken liquids
and solidify them by producing a gel network. They are used to reinforce
rubber and make it tougher. The origin of these special mechanical properties can be readily understood, adapting the reasoning used in Chapter 3
and 4.

A small volume fraction of colloidal aggregates in a liquid increases its
viscosity markedly. The reason is the same as in a polymer solution. The
aggregates, like polymers, are opaque to flow since they have $D > 1$: a
shear flow goes around rather than through them. Thus they impede flow as
though their pervaded volume were essentially filled in with solid material
instead of being nearly pure solvent.

Aggregates like polymers produce osmotic pressure in dilute solution.
Since these fractals are opaque to each other, their mutual excluded volume

is proportional to their pervaded volume, as with polymers. The osmotic compressibility produces detectable effects on the scattering from aggregate solutions. But this pressure is weak: since aggregates are generally larger than polymers, their concentration in dilute solution is small.

At higher concentrations the distinctive features of the aggregates begin to appear. Aggregates are rigid structures that hold their shape. Thus as solvent is removed the aggregates cannot readily interpenetrate as polymers can. Instead, they bend or break under the pressures from their neighbors. The macroscopic osmotic pressure opposing further removal of solvent is no longer due to thermal fluctuations. It is due to the microscopic elasticity of the structure. When the structure is compressed, the bonded-together colloidal particles are deformed. The deformation produces stress, as it would in a macroscopic chain of glued-together beads. The resulting pressure can be expressed in terms of an effective elastic modulus for the aggregates.

The elasticity of aggregates has been measured in several ways [36, 37], though its scaling has not been well established. Nevertheless, the measured elasticity appears to be consistent with a simple theoretical argument first published by Brown and Ball [36]. One way to measure this elasticity is to trap an aggregate between two parallel plates and then compress it slightly. This causes the few longest arms of the aggregate to bend. Only a small part of the structure stores energy in this process. By construction, a perfectly rigid, irreversible aggregate has no loops. Simulations allowing some flexibility and some rearrangement also have no loops except at the scale of a few particles. Thus an aggregate is a treelike structure with spanning arms and many side branches. Under deformation only the spanning arms bend and store energy. Since these grow more slowly than the aggregate as a whole, they can be made to constitute an arbitrarily small fraction simply by taking a larger aggregate.

A spanning arm displaced by a distance u exerts a restoring force proportional to u. The constant of proportionality $K(R)$ depends on the shape of the arm, just as it does for a macroscopic contorted wire. Simulations suggest that spanning arms of colloidal aggregates are themselves fractals, with a fractal dimension C somewhat lower than the overall fractal dimension. Thus the number of particles M_a in an arm of length R obeys $M_a(R) \simeq (R/a)^C$. The C of diffusion-limited aggregates is roughly 1.26 [38]. The C value of reaction-limited aggregates is consistent with unity. For both types of aggregate, the horizontal and vertical dimensions of the overall structure remain comparable. Given these facts, we may estimate the force constant $K(R)$ [39].

When the end of the arm is displaced, the dominant distortion is bending of the arm. The direction between successive beads is altered slightly, as shown in Fig. 5.17. The energy E_i stored at a given bead i is proportional to the square of the bending angle θ_i: $E_i = k\theta_i^2$. When the arm is bent, the θ_i's adjust so as to produce the required displacement u with the least energy $\sum_i E_i$. Each θ_i, acting on its own, would produce a displacement $u_i = \theta_i r_i$. Here r_i is the distance from the bead i to the displaced end. Clearly, beads further from the end produce the displacement more efficiently. The bulk of the bending will involve beads at distances of order R from the end. All such beads will have a comparable share of the bending and comparable angles

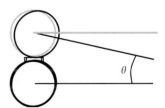

Fig. 5.17

Local bending of two particles in a spanning arm caused by displacement of one end. The undistorted configuration is shown in light shading. The two lines, drawn transverse to each bead, were parallel in the unbent configuration.

θ_i. This θ will be such that $u \simeq \sum_i \theta R$. The number of beads involved is roughly the number of beads M_a in the arm. Thus $\theta \simeq u/(M_a R)$. The resulting energy E is given by

$$E = \sum_i E_i \simeq k M_a \theta^2 \simeq k u^2/(M_a R^2).$$

This expression defines the desired spring constant $K(R)$, the coefficient of u^2. Using the fractal law for M_a, we find

$$K(R) \simeq k R^{-C-2}.$$

From this basic elastic constant $K(R)$, we may infer the bulk elasticity of a mass of aggregates at volume fraction ϕ. When an aggregate suspension is compressed to produce this elasticity, the individual aggregates have begun to press against each other. The aggregates are at their overlap concentration: $\phi = \phi_i \simeq (R/a)^{D-3}$. The longest spanning arms of a typical aggregate are deformed by these external contacts. If a small strain γ is now applied, then these longest arms are displaced by an amount $u \simeq \gamma R$. Each stores an energy

$$E = K u^2 \simeq k R^{-C-2} \gamma^2 R^2.$$

Each aggregate has only a few of these spanning arms: their number does not increase with R. Thus the energy stored per aggregate volume is of order E/R^3. Combining, we find that the overall strain γ produces an energy per unit volume of order $k R^{-C-3} \gamma^2$. The coefficient of γ^2 defines an elastic modulus G for the gel. This G, expressed in terms of volume fraction, obeys

$$G \simeq k \phi^{(C+3)/(3-D)} a^{-3}. \tag{5.32}$$

Figure 5.18 shows a comparison of the predicted power and experimental data.

Fig. 5.18
Modulus versus concentration in two different aggregate materials, after [40] (silica aerogel, filled dots) and [41], (fumed silica, open rectangles). These experiments approach the percolation threshold at low concentrations; this is expected to depress the moduli below the predicted power law. The straight lines, with slopes 3.2 and 3.55, indicate the power law predicted by Eq. (5.32), including the experimental uncertainties in D and C [38, 42].

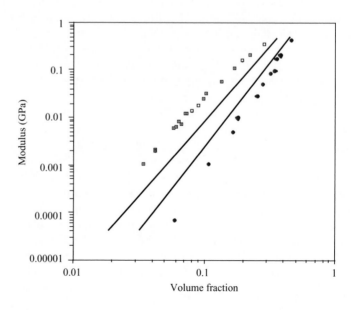

Aggregates, like polymer solutions, have elastic properties that scale in predictable ways in terms of the geometric scaling properties represented by D and C. Like polymer solutions, the aggregate gels have moduli for bulk, shear, or other types of strain that scale with the same power of ϕ. Our reasoning leading to this scaling did not depend on the particular type of distortion imposed.

References

1. P. Pincus, in *Lectures on Thermodynamics and Statistical Mechanics: XVII Winter Meeting on Statistical Physics*, ed. A. E. Gonzalez and C. Varea (Singapore: World Publishing Co., 1988).
2. D. H. Everett, *Basic Principles of Colloid Science* (London: Royal Society of Chemistry, 1988).
3. J. N. Israelachvili, *Intermolecular and surface forces*, 2nd ed. (London: San Diego, CA: Academic Press, 1991).
4. J. Mohanty and B. W. Ninham, *Dispersion Forces* (London: Academic Press, 1976).
5. H. E. Stanley, *Introduction to Phase Transitions and Critical Phenomena* (New York: Oxford University Press, 1971).
6. F. Brochard and P. G. de Gennes, *Ferroelectrics* **30** 33 (1980).
7. For an excellent review of polymer brushes see A. Halperin, M. Tirrell and T. Lodge, *Adv. Polym. Sci.* **100** (1991).
8. S. T. Milner, *Europhys. Letts.* **7**, 695 (1988).
9. H. J. Taunton, C. Toprakcioglu, L. J. Fetters, and J. Klein, *Nature* **332** 712 (1988).
10. A. N. Semenov, J.-F. Joanny, A. Johner, and J. Bonet-Avalos, *Macromolecules* **30** 1479 (1997).
11. P. G. de Gennes, *Scaling Concepts in Polymer Physics*, (Ithaca: Cornell University Press, 1979).
12. S. Bucci, C. Fagotti, V. Degiorgio, and R. Piazza, *Langmuir*, **7** 824, (1991).
13. S. H. Behrens and D. G. Grier, in *Electrostatic Effects in Soft Matter and Biophysics*, ed. C. Holm, P. Kekicheff, and R. Podgornik (Dordrecht: Kluwer, 2001).
14. M. Alexander, L. F. Rojas-Ochoa, M. Leser, and P. Schurtenberger, *J. Colloid Interface Sci.* **253** 35 (2002).
15. R. Kjellander, T. Akesson, B. Jonsson, and S. Marcelja, *J. Chem. Phys.* **97** 1424 (1992).
16. L. Guldbrand, B. Jonsson, H. Wennerstrom, and P. Linse, *J. Chem. Phys.* **80** 2221 (1984).
17. P. Pieranski, *Contemp. Phy.* **24** 25 (1983).
18. See e.g., M. Constantinos, Paleos ed., *Polymerization in Organized Media* (London: Taylor and Francis, 1992).
19. M. J. Stevens, and M. O. Robbins, *J. Chem. Phys.* **98** 2319 (1993).
20. L. Onsager, *Ann. N.Y. Acad. Sci.* **51** 627 (1949).
21. P.G. de Gennes, and J. Prost *The Physics of Liquid Crystals*, 2nd ed. (Oxford : Clarendon Press, 1993).
22. R. E. Rosensweig, M. Zahn, and R. Shumovich, *J. Magn. Magn. Mater.* **39** 127(1983); R. E. Rosensweig, *Ferrohydrodynamics* (New York: Cambridge Univ. Press, 1985).
23. A. P. Gast and C. F. Zukoski, *Adv. Colloid Interface Sci.* **30** 153 (1989).
24. E. J. Hinch and J. D. Sherwood, *J. Fluid Mech.* **132** 337 (1983).
25. D. Long and A. Ajdari, *Phys. Rev. Lett.* **81** 1529 (1998).
26. J. L. Anderson, *Ann. Rev. Fluid. Mech.* **21** 61 (1989).
27. This treatment extends P. G. de Gennes, *Macromolecules* **15** 492 (1982); **14** 1637 (1981).
28. P. Poulin, J. Bibette, and D. A. Weitz, *Eur. Phys. J. B* **7** 277 (1999).

29. P. Meakin, in *Phase Transitions and Critical Phenomena*, Vol. 12 C, eds. Domb, J. L. Lebowitz (New York: Academic, 1988) p. 335.
30. See e.g. R. Jullien and R. Botet, *Aggregation and Fractal Aggregates* (Singapore: World Scientific, 1987).
31. M. Y. Lin, H. M. Lindsay, D. A. Weitz, R. C. Ball, R. Klein, and P. Meakin, *Nature* **339** 360 (1989).
32. R. C. Ball and T. A. Witten, *J. Stat. Phys.* **36** 873 (1984).
33. R. Botet, *J. Phys. A: Math. Gen.* **18** 847 (1985).
34. P. G. J. VanDongen and M. H. Ernst, *J. Phys. A: Math. Gen.* **18** 2779 (1985).
35. P. Meakin, *Rev. Geophys.* **29** 317 (1991).
36. R. Buscall, P. D. A. Mills, J. W. Goodwin, and D. W. Lawson, *J. Chem. Soc., Faraday Trans. I* **84** 4249 (1988).
37. T. A. Witten, M. Rubinstein, and R. H. Colby, *J. Phys. (Paris) II* **3** 367 (1993).
38. P. Meakin, I. Majid, S. Havlin, and H. E. Stanley, *J. Phys. A* **17** L975 (1984).
39. Y. Kantor and I. Webman, *Phys. Rev. Lett.* **52** 1891 (1984).
40. T. Woignier, J. Phalippou, *Revue de Phys. Appliquee* **24** C4-179 (1989).
41. J. Forsman, J. P. Harrison, and A. Rutenberg, *Can. J. Phys.* **65** 767 (1987).
42. P. Dimon, S. K. Sinha, D. A. Weitz, C. R. Safinya, G. S. Smith, W. A. Varady, and H. M. Lindsay, *Phys. Rev. Lett.* **57** 595 (1986).

Interfaces

6

Most of our experience with liquids arises not from the bulk of the liquid but from its interfaces with other media. As we gaze onto the ocean or into our cup of coffee, it is the surface of the liquid that catches our eye. The property that distinguishes a liquid from a gas is the existence of this surface. This chapter deals with such interfaces—how they are structured, how much energy they have, and how fast they move and relax. Most importantly, it deals with how solute molecules, both simple and complex, affect these properties.

A great deal of the importance of structured molecules in liquids arises from their affect on interfaces. Molecules that migrate to an interface and affect its properties are called surfactants. Often such molecules are added to fluids like motor oil, cosmetics, cleaning products, and food, in order to achieve desired interfacial properties. In rubber and plastic composites like automobile tires and appliance cases, the strength of the material hinges on the properties of the interface between the components making up the composite. Discoveries involving the interface between water and fabric or solids and internal body tissues have had recent technological impact. But the most dramatic property of surfactants is their ability to cause spontaneous production of interfaces. Such surfactant-generated interfaces are the subject of the next chapter.

We begin this chapter by describing what the quantitative properties mean in experimental terms. We survey ways of measuring energy, structure, and motion associated with interfaces. Next we examine how interfacial energy comes about for simple liquids. We discuss wetting of an interface by a liquid, including the motion of the liquid in the process of wetting. Next we turn to the effect of added solute molecules on an interface. Such molecules can alter the energy of an interface and create structure there. Finally we turn to structured fluids like polymers. These create structures at the interface with distinctive scaling properties extending over distances as large as the polymers.

6.1 Probes of an interface

In this chapter we shall lay the groundwork for the next one by describing the general properties of liquid interfaces. Before we begin, let us survey the ways by which interfacial properties are measured. The most basic property of an interface is its **interfacial energy**—the work per unit area required to create the interface from a bulk material of the same composition.

(a)   (b)   (c)

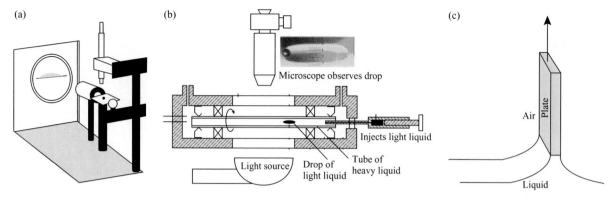

Fig. 6.1

Devices for measuring interfacial properties. (a) contact angle micrometer after [1]. Schematic light bulb in foreground illuminates the black droplet on the white sample plate from the side. Lens system behind the sample plate projects the shadow of the drop. Contact angle is measured from this shadow. (b) Spinning drop tensiometer, after [2], used for measuring small interfacial tensions. Horizontal tube at center rotates about horizontal axis. Centrifugal forces elongate the black drop of lower-density fluid. The microscope at the top views the distorted drop shown and measures its dimensions. (c) Wilhemeny plate, pulling upwards on a liquid surface. The support F is attached to an analytical balance to measure the force on the plate. A motor, not shown, raises or lowers the fluid so that the fluid film meets the plate vertically.

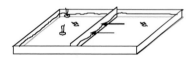

Fig. 6.2

Sketch of a Langmuir trough. Insoluble surfactant molecules are trapped on the left side of the barrier. They exert excess pressure against the slider bar. Arrows show the force needed to counter this pressure. A motor, not shown, compresses the slider bar, while a Wilhemeny plate, not shown, measures the surface tension inside.

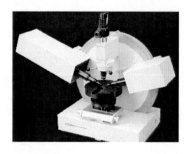

Fig. 6.3

Photograph of an ellipsometer from [3]. Laser at left illuminates a thin-film sample at the center of the black stage. Detector at right measures the intensity and polarization of the reflected light. Incident and reflected angles are controllable.

The simplest way to learn something about an interfacial energy is to measure the contact angle. Figure 6.1(a) shows an optical device for measuring contact angles. We will explain the connection between this angle and interfacial energies below. The spinning drop tensiometer of Fig. 6.1(b) measures this energy, as does the Wilhemeny plate of Fig. 6.1(c). These devices produce controlled increases in the interfacial area and measure the associated work. A related device of great importance is the Langmuir trough, Fig. 6.2. It is used to study surfaces with a fixed amount of surfactant trapped on them. The slider bar compresses the surfactants; a Wilhemeny plate monitors the associated decrease in interfacial energy. Langmuir troughs are the standard apparatus to prepare a controlled air–water interface containing surfactant. It forms the basis of many of the microscopic probes described below.

A more fine-scale view of interfacial forces can be gained using the surface forces apparatus of Fig. 2.5. It measures the tiny forces exerted by a $10\,\mu m^2$ surface on another such surface, one or more nanometers away. Often the forces arise from molecules bound to the surfaces. Thus the apparatus probes these interfacial molecules. It can sense how deeply they extend into the gap between the two surfaces, for example.

We have seen in Chapters 2, 3, and 4 that scattering is a powerful means of measuring microscopic structure. Such probes have been ingeniously adapted to the study of interfaces. Ingenuity is needed to assure that the signal comes from the molecules near the interface and not from the far more numerous molecules in the bulk of the liquid. The standard technique of ellipsometry Fig. 6.3, senses the amount of relative surfactant on a surface by measuring the phase shift of a light beam reflecting from it, relative to that from a bare surface. To sense the spatial distribution of the adsorbate, ellipsometry is not enough. If the distances involved are many tens of nanometers, evanescent-wave fluorescence can be used.

The idea is sketched in Fig. 6.4. The incident light is arranged to send an evanescent wave along the surface in question. The light intensity decays exponentially with distance from the interface, and the decay length may be controlled within limits. This light excites the adsorbate molecules, which have to be labeled=[†] with a fluorescent dye. By measuring the intensity of the fluorescence as the evanescent decay length is varied, one may obtain information about how the fluorescent molecules are arranged in space [4].

Often one wants to know about structure at the single nanometer scale. Here x-ray and neutron scattering are used. To infer the profile of adsorbate density with distance from the surface, one may measure reflectivity near conditions of total internal (or external) reflection. As the incident beam angle is increased, the fraction reflected begins to drop from 100% as transmission becomes possible. If the sample is a thin film, the reflectance variation with angle can be used to infer the composition as a function of depth. The inference is not straightforward, though. One must guess the profile and keep improving the guess until the calculated reflectivity versus angle matches the measurements.

Quite often molecules are not spread uniformly over an interface, but are heterogeneous or patterned in some way. This kind of spatial structure can make a big difference in the properties of the interface, yet this in-plane structure is not apparent from the depth profiles treated above. To find such information from scattering, one must probe scattered wave vectors with components parallel to the surface. One must detect not just the reflected waves, but waves scattered at an azimuthal angle to the incident beam Fig. 6.5. Such experiments can detect moving inhomogeneity as well as static patterns, using the methods of dynamic scattering sketched in Chapter 4. Naturally, since the scattering occurs from only a small number of molecules at the surface, the scattering is weak and hard to detect. Thus these measurements require large-scale neutron or x-ray facilities that are based on nuclear reactors and particle accelerators.

Another way to see irregular coverage of an interface is to use microscopy, as we previously discussed in Chapter 2.

6.2 Simple fluids

6.2.1 Interfacial energy

In a liquid the attraction between molecules is generally of order T or less, as explained in Chapter 2 in the discussion of viscosity. This suggests that

[†] That is, a fluorescent chemical group must be attached to the molecule to be probed.

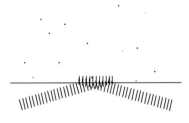

Fig. 6.4
Selective detection of molecules near a surface via evanescent wave fluorescence, after [4]. Beam of light hits the interface at a low angle and is totally internally reflected, entering the liquid only in the form of an evanescent wave near the interface. Any fluorescent molecules in the evanescent layer emit light of a lower frequency. The intensity of fluorescent light is measured by a photodetector, not shown.

Fig. 6.5
Schematic view of evanescent-wave scattering. The gray rectangle is the interface of interest, which is often an air–water interface in a Langmuir trough. The beam hits the interface at grazing incidence (from the left), so that practically all of it is reflected (darker beam). If the scatterers are nonuniformly spread over the interface, some of the incident beam is scattered, and may be detected at various azimuthal angles ϕ from the reflected beam (lighter beam).

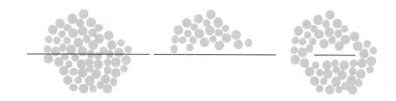

the interfacial energy per molecule in a liquid is also of order T or less. But interfacial energy is significant even when there is no attraction between the atoms. The boundaries of any fluid have a spatial structure and an energetic cost. If the fluid fills a box, then these fluid boundaries are at the walls of the box. We may account for the effect of these walls using the statistical mechanics principles introduced in Chapters 2, 3, and 4. We consider the simplest possible wall, a plane which adds an infinite energy cost to any particle intersecting it. We call this a hard wall. The effect of such a wall is to remove any configurations in which wall intersects particles. The result is something like the middle of Fig. 6.6. Clearly the wall disturbs the arrangement of particles in its vicinity. The density near the wall is reduced. The disturbance extends roughly a particle diameter into the liquid.

To add a hard wall to a liquid requires work. As we know from Chapter 2, the work can be found by considering properties of the pure liquid. In a pure liquid there is probability p_A that a given region A has no particles passing through it. As told in Chapter 2, the work required to remove the particles from such a region is $-T \log p_A$. If the region happens to be the slab shown on the right side of Fig. 6.6, then this is the work required to insert the slab. If two slabs were inserted at separate places in the fluid, the probability that each region is empty is p_A, independent of the other region. The probability that both regions are empty is p_A^2, and the work required is just double that for one of the slabs. That is, the work is proportional to area of the slabs. The same is true if two large slabs are joined to form a single slab of twice the area. Then almost all of each slab is far from the other slab, and independent of it. Thus the work to insert the slab has the form α(area), where (area) means the area (on both sides) of the slab. The coefficient α is called the interfacial energy or interfacial tension. If the liquid is a hard sphere liquid like the one pictured, the probability p_A is some number of order unity if the slab is no larger than a particle radius a. That means the interfacial tension $\alpha \simeq T/a^2$ for such a liquid.

Real liquids generally have interfacial energies of this order, as well. The fact that the molecules attract or repel each other somewhat does not change the picture above qualitatively. Likewise, if the interface has some attraction or repulsion for the molecules, or some mild curvature or roughness, this does not change the interfacial energy greatly. Even the free surface between a liquid and its vapor is qualitatively similar. Like a hard wall, a free surface is a surface where the density must be a small fraction of the liquid density. Thus the disturbance it causes is qualitatively as great as that of a hard wall.

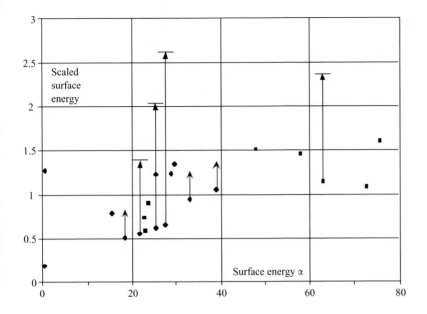

Fig. 6.7
Surface energies for the liquids tabulated in [5]. Horizontal scale is the surface energy, in millijoules per square meter (or dynes per centimeter). Molecules with NH or OH groups are denoted by squares; others are denoted by diamonds. Vertical axis is the scaled surface energy $\alpha/(T/\delta A)$. This δA is the area per flexible segment of the molecule. It is calculated by first determining the volume of a flexible segment from its mass m and the liquid density. Then δA is the projected area of a sphere containing that volume. For small, rigid molecules, the flexible segment is the whole molecule. For flexible molecules, it is the smallest freely jointed unit larger than a C_2H_2 unit. For cases where the flexible unit is taken as smaller than the whole molecule, an arrow extends upward to indicate the range of values resulting from choosing larger pieces of the molecule as the flexible unit. Upper limit of the range is indicated by a ceiling on its arrow. Arrows with no ceiling are polymers. The lowest α's shown are for liquid helium at $4°$ K, $m = 4$ and liquid helium at $1.6°$ K, $m = 4$. The remaining liquids, at room temperature, are, from left to right, n-pentane, C_5H_{12}, $m = 72.15$, polytetrafluoroethylene (Teflon), $(C_2F_4)_n$, $m = 100$, n-octane, C_8H_{18}, $m = 29$, ethanol, C_2H_5OH, $m = 46$, methanol, CH_3OH, $m = 32$, acetone, CH_3COCH_3, $m = 58$, n-dodecane, $CH_3(CH_2)_{10}CH_3$, $m = 28$, cyclohexane, $C_6H_{11}OH$, $m = 100$, n-hexadecane, $CH_3(CH_2)_{14}CH_3$, $m = 28$, benzene, C_6H_6, $m = 78$, carbon tetrachloride, CCl_4, $m = 153$, polystyrene, $(CH_2CH(C_6H_5))_n$, $m = 52$, polyvinyl chloride, C_2H_3Cl, $m = 62$, ethanediol, $HOCH_2CH_2OH$, $m = 62$, formamide, $HCONH_2$, $m = 45$, glycerol, $HOCH_2CH(OH)CH_2OH$, $m = 31$, water, H_2O, $m = 18$, hydrogen peroxide, H_2O_2, $m = 34$. For this wide range of liquids, the scaled surface energy remains near 1.

Figure 6.7 shows some representative interfacial energies. As argued above, liquid interfacial energies are of the order of T per surface degree of freedom. For small rigid molecules, this amounts to T per molecular area. For large flexible molecules, there is a characteristic area for each independently moving piece[†], which displaces a characteristic area at the surface. Then the interfacial energy is roughly T for each such area. Within this generalization, polar molecules such as water tend to have more energy than nonpolar ones. Solids as well as liquids have interfacial energy. To create new surface (e.g. by breaking a block of the solid in half) requires work. One must overcome the attraction that binds the molecules of the solid together. Thus the surface energy is of order of the binding energy of a molecule per area occupied by a molecule on the surface. The binding energy of a solid is generally much larger than the thermal energy T. Thermal fluctuations are minor and each atom is fixed at a position that minimizes its potential energy (*cf.* the argument in Chapter 2 for viscosity). Thus the surface energy of a solid is often much more than T per surface atom. Among solids, metals and ionic crystals have high surface energies (which moreover depends on the orientation of the surface with respect to the crystal axes). Hydrocarbons have low surface energies. Rare-gas solids have the lowest ones

The interfacial energy shows up concretely whenever a liquid can change its interfacial area. If a drop of liquid is suspended in space, it minimizes its interfacial energy by adopting a spherical shape. If an external agent alters this shape, the work it does is α times the increase in area. If the fluid is distorted and then released, it begins to move so as to decrease its interfacial area, converting interfacial energy into kinetic energy. The kinetic energy is equal to α times the reduction in area (less any loss from viscous dissipation). If a spherical drop of radius R could reduce its volume by an amount

[†] For a polymer this would be a few backbone bonds and their attached atoms.

$\delta V = 4\pi R^2 \delta R$, the interfacial area would decrease by $8\pi R \delta R = 2\delta V/R$. Thus the interfacial energy would decrease by $(2\alpha/R)\delta V$. Since energy is gained by decreasing the volume, the interior of the drop experiences a pressure $2\alpha/R$. This pressure, called the **Laplace pressure**, is present generally inside a liquid with a curved surface.

When electric or magnetic fields are present, the energy cost of deforming a surface can be greatly altered. A familiar example occurs when one draws a dielectric fluid up into the gap of a parallel-plate capacitor by applying a voltage to it. The force holding up the fluid acts at the interface. The strongly paramagnetic fluids called ferrofluids experience analogous forces in a magnetic field, as noted in Chapter 1. These interfacial forces are important for describing the behavior of some fluids in external fields. To characterize what these forces do is difficult in general, since one must determine the field profile and the interfacial profile simultaneously. The energy associated with the interface depends in a nonlocal way on the shape of the interface, in contrast to the surface energy, which is simply proportional to the area.

6.1. *Characteristic sizes of droplets.*

(a) A drop of water of radius R has an interior Laplace pressure of one atmosphere, 10^5 ergs/cm^3. Water has an interfacial energy of α of 70 ergs/cm^2. How big is R?

(b) Another drop of water lies on a neutrally wetting surface and thus has a hemispherical shape in the absence of gravity. Find the change of gravitational energy if this hemispherical drop is flattened onto the surface. For what radius R is this gravitational energy equal to the original surface energy?

(c) (**Harder**) What is the actual shape of a drop with this radius, accounting consistently for gravitational distortion. (To find the shape of least energy requires knowledge of Euler–Lagrange and Lagrange multiplier methods.)

6.2.2   Contact angle

The forces associated with interfacial energy are readily seen when two interfaces meet. The most common example is a drop of liquid sitting on a solid surface. If the drop is smaller than a few millimeters in diameter, the effects of gravity are not important. The interface between the solid and the liquid generally has a different interfacial energy α_s from the air interfacial energy α. Moreover, the solid–air interface has its own interfacial energy α_{s0}. The body of liquid must adopt a shape which minimizes the total interfacial energy: this is no longer a spherical shape with no contact with the surface. Instead, the liquid may spread into a film that covers the solid surface. This occurs when the interfacial energy of each area δA of the liquid-covered surface is smaller than that of the same area uncovered. The uncovered surface has energy $\delta A\alpha_{s0}$. The covered surface has two interfaces: the liquid–solid interface with energy $\delta A\alpha_s$ and the liquid–air interface with energy $\delta A\alpha$. Thus the spreading occurs when $(\alpha_s + \alpha) - \alpha_{s0} < 0^\dagger$. From this condition we note two things. First since the α's must be positive, α_s must be smaller than α_{s0}. The liquid molecules must be attracted to those of the solid. Second this attraction must be sufficiently strong to pay for the cost of extra free surface, with its energy α. This situation is

---

† In this comparison we have not considered changes in the liquid interface area away from the δA region being considered. In fact there is no need to do so. When the liquid spreads to cover an additional area δA of bare surface, the necessary fluid can be supplied by an overall thinning of the film. This thinning does not change the interfacial area elsewhere.

known as **complete wetting**. The liquid spreads to cover all the exposed area of the solid surface (as much as the supply of liquid will allow). Complete wetting tends to occur on high-energy solid surfaces, where α_{s0} is large. For example, hydrocarbon oils typically wet metal surfaces completely. Often a liquid is attracted to such a high-energy surface by enough to justify complete wetting [6].

At the other extreme, it can happen that any contact between the solid and the liquid costs energy. This occurs when the energy of an air–liquid plus an air–solid interface is lower than that the of the solid–liquid interface: $\alpha + \alpha_{s0} < \alpha_s$. Clearly this can only happen when $\alpha_s > \alpha_{s0}$, so that the surface repels the liquid. In this situation, a spherical drop in grazing contact with the surface would separate from the surface (in the absence of gravity). Such a surface is said to be **completely nonwetting** for the liquid.

For both complete wetting and complete nonwetting, α is smaller than $|\alpha_s - \alpha_{s0}|$. If instead α is larger than this difference, **partial wetting** occurs. The liquid surface meets the solid surface at **contact angle** θ. Figure 6.8 shows how interfacial energy is related to forces in this familiar case. If a section of the liquid of length L advances, a work $-L\delta x \alpha_{s0}$ is done in decreasing the bare surface. This is just the work that would be done if the solid exerted a tensile force $L\alpha_{s0}$ on the liquid trying to advance it. The interfacial energy thus can be viewed as a force per unit length. This is why interfacial energies are often called interfacial *tensions*. The liquid-covered surface exerts an opposing tension $L\alpha_s$. Finally, the advance would create free liquid surface of amount $L\delta x \cos(\theta)$, so that the free liquid exerts a force $L\alpha \cos(\theta)$ opposing the advance. The sum of these forces must be zero; otherwise the fluid would advance or retreat spontaneously. Thus

$$\alpha \cos(\theta) = \alpha_{s0} - \alpha_s. \qquad (6.1)$$

This is **Young's law** of partial wetting [6]. Young's law determines the contact angle; for a circular drop the contact angle in turn determines the radius R. The free surface of a drop must have a constant spherical curvature everywhere; otherwise, the Laplace pressure would be different in different places. When the solid neither attracts nor repels the liquid, $\alpha_s = \alpha_{s0}$. This is the case of **neutral wetting**. Here the contact angle θ is 90° for any value of α. In general α_s is different from α_{s0}. Then if α is made smaller, the contact angle decreases to zero to produce complete wetting or increases to 180° to produce complete nonwetting.

The surface tension force of a liquid against its supporting structures can be mechanically significant. Such **capillary forces** can make fluid rise to macroscopic heights in a thin tube. Fluid transport in plants is accomplished largely by capillary forces. In fine granular materials with large surface to

Fig. 6.8
Left: a partially wetting droplet on a horizontal surface, viewed from the side. The contact angle θ is indicated. The liquid surface meets the solid surface at the contact line. As θ approaches 0 the droplet approaches the state of complete wetting. As θ approaches π, the droplet approaches the state of complete nonwetting (center). Right: motion of the contact line by an amount dx along the surface stretches the liquid surface by an amount $dx \cos \theta$. The vectors show the forces on a unit length of the contact line, including the bare solid surface tension α_{s0}, the liquid-covered solid surface tension α_s, and the liquid–air surface tension α.

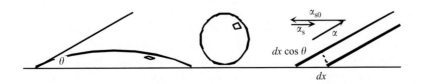

volume ratios, these capillary forces can distort their supporting structures or bind them together. Thus small amounts of water in the air can readily form wetting layers on fine particles, turning a free-flowing powder into a solid cake.

6.2. *Capillary solid*   A person standing on a cube of wet sand exerts a stress of 10^4 N/m^2. This stress must be supported by sand grains whose size is a.

(a) Roughly how much force does a given grain support for a given size a? On some of the grains this force is a tensile force trying to separate two adjacent grains. Treat these two grains as cubes of length a and suppose that the cube faces are separated by a thin water film. Separating the cubes increases the thickness of this film and reduces its cross-sectional area to maintain constant volume.

(b) For what size a is the interfacial tension equal to the applied force? Grains smaller than this ought to support the weight of the person.

6.2.3 Wetting dynamics

If a drop of completely wetting fluid is placed on a surface, the interfacial tensions discussed above are not balanced and the droplet begins to spread. An extended treatment of this spreading dynamics is given in a big review article by de Gennes [7]. The contact angle θ decreases with time. The speed of advance $v(t)$ must be such that the rate of decrease of interfacial energy is equal to the rate of viscous dissipation. When a section of the drop moves outward by an amount δx, the energy decreases by $L\delta x[\alpha_s - \alpha_{s0} + \alpha \cos(\theta)]$. The quantity in [...] is called the **spreading pressure** S. For small θ it takes the form $S = S_0 + \frac{1}{2}\alpha\theta^2$. Here S_0 is the spreading pressure at $\theta = 0$: $S_0 \equiv \alpha_s - \alpha_{s0} + \alpha$. The nonzero wetting angle increases the spreading pressure. Energy is lost at a rate LvS. The driving force for spreading is that which would occur in a flat film plus an additional force due to the nonzero contact angle. The dissipation opposing this driving force can involve thin film flow of the liquid near the surface, which should not depend on θ, and also the flow in the wedge shaped region beyond the contact line, which does depend on θ. In order to sidestep the thin film effects we focus on the case where the flat spreading pressure is nearly zero, so that $S \simeq \frac{1}{2}\alpha\theta^2$.

The dissipation rate depends on the velocity gradient inside the drop. At a distance x from the contact line, the drop has thickness $h = x \sin\theta$. The average velocity gradient $\dot{\gamma}$ is of order $v/h = v/(x \sin\theta)$ (since the velocity goes to zero at the solid surface). The dissipation rate per unit volume w for a fluid with viscosity η is $\eta\dot{\gamma}^2$ or $w \simeq \eta v^2/(x \sin\theta)^2$. To obtain the total dissipation rate W, we must integrate over position dx and multiply by the thickness h at each x: $W \simeq L\eta v^2 \int_b^R dx/(x \sin\theta)$. Here we have taken a minimum value of x to be that where the continuum description of the fluid breaks down; b is roughly the size of a fluid molecule. The maximum distance is of the order of the droplet radius R. Evidently, the integral is a logarithm. For small θ, $W \simeq L\eta v^2 \log(R/b)/\theta$. Equating the energy loss rate with the dissipation rate, we find $\frac{1}{2}\alpha\theta^2 = \eta v \log(R/b)/\theta$, so that $v \log(R/b) \simeq (\alpha/\eta)\theta^3$. This speed v is simply the time derivative dR/dt.

The geometric properties of the drop give a further relation between v and θ. The drop is a spherical cap of constant volume V. For small θ,

$V \simeq R^3\theta$. We can now combine this with the spreading-pressure relation above for small θ to obtain

$$\frac{dR}{dt} \log\left(\frac{R}{b}\right) \simeq \frac{\alpha}{\eta} \left(\frac{V}{R^3}\right)^3$$

This equation readily tells how R varies with time t, if we neglect the slowly varying $\log(R/b)$:

$$R \simeq [(V^3\alpha/\eta)t]^{1/10}. \tag{6.2}$$

This is the well-known **Tanner law of wetting** [8]. It says that the spreading slows in a universal way with time in the marginal case where $S_0 \simeq 0$.

Remarkably, Tanner's law is observed even when the flat spreading pressure S_0 is large and should dominate the full spreading pressure. Instead of advancing rapidly in response to the full spreading pressure, the macroscopic contact line advances gradually as though there were no flat spreading pressure. As anticipated above, the extra spreading pressure is opposed by an extra dissipation mechanism, not considered above, that is independent of the angle θ. As explained by de Gennes and Hervet [9], the fluid spreads far ahead of the macroscopic contact line in a thin **precursor film** or **foot**. The larger the flat spreading pressure is, the faster this film advances. The flat spreading pressure is balanced by dissipation in the precursor film. Only the residual spreading pressure $\frac{1}{2}\alpha\theta^2$ remains to be balanced by the dissipation in the macroscopic drop.

6.2.4 Surface heterogeneity

Real solid surfaces are not completely uniform. Impurities and irregularities cover the surface in a way that varies from place to place on the atomic scale. Thus at some places the $\alpha_{s0} - \alpha_s$ is bigger than average, so the local contact angle is smaller. At other places the local angle is larger than average. If the contact line is to advance for a macroscopic distance, it must be able to move over even those places with the largest contact angle. If it is to recede, it must detach from even those places with the smallest contact angle. If the contact angle lies between these two angles it can neither advance nor recede; it can only make local readjustments. The contact line is **pinned**. This pinning means that the contact angle is not uniquely determined but depends on how the fluid was placed on the surface. The wetting is **hysteretic**.

Wetting hysteresis is readily observable on most surfaces. Even nominally smooth and uniform surfaces show a several-percent spread between advancing and receding contact angles. Deliberately nonuniform surfaces can pin a contact line over a wide range angles, from nearly vertical to nearly zero [10].

An important form of surface heterogeneity is surface roughness. Microscopic roughness increases the surface area relative to the projected area by some factor β larger than 1. Thus the energy per unit projected area α_{s0} is increased by a factor β relative to a flat surface. So is the liquid–solid energy α_s. The contact angle θ is now given by $\alpha \cos\theta = \beta(\alpha_{s0} - \alpha_s)$. We see that roughness amplifies the wetting preference of the surface: it increases the departure from the neutral conditions where θ is 90°. In addition to this effect roughness causes wetting hysteresis.

6.2.5 Other interfacial flows

The simple flow represented by a spreading drop is just one of a rich variety of interfacial flow phenomena. As we saw, the driving force for this spreading was the surface interfacial energy; such flows are termed **capillary flows**. This same force can drive the reverse phenomenon of **dewetting**. Dewetting occurs when a surface is coated with a film of liquid that does not wet that surface in equilibrium [11]. It may be prepared by letting a thick layer of liquid evaporate. Such a film is metastable. If a small dry spot forms, it spontaneously spreads to form a growing circular dry patch. The liquid from the dry region collects near the border faster than it can spread through the film, thus forming a thick ring of spreading liquid. Since the interfacial energy gained is dissipated in the adjacent ring, the spreading velocity becomes independent of the ring's radius. Both spreading and dewetting flows are qualitatively different and faster if the solid substrate is replaced by a viscous fluid.

Even without dewetting, evaporation causes flow when the thickness of a fluid layer is not uniform. For example, a thin circular droplet with a pinned contact line must thin further without contracting as it evaporates. The thickness at a given place must thus decrease at a rate proportional to its local thickness. But the local evaporation rate does not depend on thickness in this way. Thus to maintain the proper droplet shape, lateral flows are necessary. These flows are strong enough in practice to bring material from the center to the contact line during the drying time [12].

Another important driving force occurs when a liquid is nonuniform in composition or temperature. In such cases the interfacial energy is in general nonuniform, as well: part of the surface could have a lower energy than an adjacent part. For example, one part of the surface could have a higher concentration of some solute molecule, leading to a lower interfacial energy there. We shall explore this solute effect in the next section. Whatever the cause of the nonuniform interfacial energy, the nonuniformity results in a driving force to expand the region of lower interfacial energy, just like that which causes spreading of a wetting liquid on a surface. The resulting flow is called a **Marangoni flow** [13]. The classic example of Marangoni flow is readily seen in a freshly poured glass of wine. "Tears of wine" climb up the sides of the glass. The upward flow that forms the "tears" results from a surface tension gradient. The top of a tear has a higher surface tension because more alcohol has evaporated there than elsewhere. The wine does not climb the glass in a uniform ring, but rather climbs higher in some regions than in others, thus forming the tears. The reason is that a uniform ring is unstable. The higher a climbing region rises, the further it gets from the rest of the wine. This accentuates the evaporation and allows that region to climb still higher. Such instabilities are common features of Marangoni flows.

Gradients of temperature or composition on a solid surface can also give distinctive features to liquids spreading over them. The flow can even advance and retreat many times before coming to rest [13]!

6.3 Solutes and interfacial tension

As we noted above, molecules dissolved in a liquid generally influence its interfacial tension. In this section we investigate how large this influence is. A very simplified approximation gives the right order of magnitude. We imagine a spherical droplet of liquid with N_s solute molecules. We further imagine that these molecules have a strong attraction for the surface—so strong that they are all at the surface. We suppose that N_s is small enough so that the molecules are far apart on the surface and have negligible interactions. We now distort the shape of the interface, thus increasing its surface area by an amount δA. Without the solute, this distortion requires work. If the liquid's surface tension is α_0, then the work is $\alpha_0 \delta A$. The N_s solute molecules alter this work. These molecules are confined to an area of A/N_s apiece. Increasing the area allows each molecule to take on more configurations c and thus decreases its free energy; for all the molecules this free energy $\mathcal{F}$ is a constant minus $N_s T \log(A/N_s)$. This form and the rationale for it were discussed in a three-dimensional context in Problem 2.3). The work δW is $\delta A \partial \mathcal{F}/\partial A = -\delta A T (N_s/A)$. Not surprisingly, this work is the same as we would encounter in an ideal, two-dimensional gas: the surface pressure is T times the number density N_s/A.

As we see, the solute acts to reduce the surface tension by means of an opposing surface pressure. Generally, this opposition is weak. We have argued above that most simple liquids have a surface energy of the order of T per molecular area. If the solute molecules are dilute, the area per solute molecule is necessarily much bigger than that of the base liquid. Thus these solute molecules can only reduce the surface tension by a small fraction. To attain a big change of surface tension, the adsorbed molecules must be strongly concentrated at the surface and must interact strongly. We shall explore such strongly interacting adsorbates in the next chapter.

We may improve the primitive picture above by relaxing our assumption that the solute molecules are confined to the interface. The effect of the solute molecules can be treated systematically by first looking at how one solute molecule affects the free energy of the droplet. The main part of the free energy has the form $\mathcal{F}_1 - T \log \Omega$, where Ω is the volume of the drop and the constant $\mathcal{F}_1$ gives the work required to insert the molecule at some given place (removing solvent molecules so that the volume stays fixed). Now we consider the effect of the surface. In general our molecule interacts with the surface and thus it has a nonuniform probability density. We may represent the local probability density as $\rho_0 f(z)$, where $\rho_0 = 1/\Omega$ is the density of solute molecules, z is the distance to the interface and $f(z)$ is the relative probability that the molecule is at a particular position at distance z from the interface. Thus $f(z)$ is an un-normalized configuration probability, as discussed in Chapter 2. In that chapter, we showed how the free energy of such a system could be expressed in terms of f:

$$\mathcal{F}_2 = \mathcal{F}_1 - T \log\left(\int d^3 r f(z(r))\right)$$

We notice that if the interface has no effect on the solute, $f(z)$ is a constant and the expression for $\mathcal{F}_2$ reduces to our initial expression $\mathcal{F}_1 - T \log(\Omega)$.

Our interest is in a macroscopic interface with little or no curvature. By comparison, any nonuniformity in $f(z)$ is a short-ranged enhancement or suppression. In this situation we can simplify $\mathcal{F}_2$. We shall first add and subtract 1 from $f(z)$ to obtain $\mathcal{F}_2 = \mathcal{F}_1 - T\log(\Omega + \int d^3r[f(z(r)) - 1])$. The last integral vanishes except near the surface and becomes in the limit of small curvature $A \int dz[f(z) - 1]$. For sufficiently large Ω, the first term of the log must always be much larger than the second. Thus the log can be expanded: $\mathcal{F}_2 \rightarrow \mathcal{F}_1 - T\log\Omega - A\int[f(z) - 1]/\Omega$. The nonuniform $f(z)$ creates a contribution to $\mathcal{F}$ proportional to the interfacial area. Now we can readily write the overall free energy $\mathcal{F}$. It has a contribution $\mathcal{F}_0(A)$ in the absence of solute molecules, plus a second contribution $N\mathcal{F}$ if there are N noninteracting solute molecules:

$$\mathcal{F} = \mathcal{F}_0(A) + N\mathcal{F}_2 = \mathcal{F}_0(A) + N\mathcal{F}_1 - TN\log\Omega - TNA \int \frac{f(z) - 1}{\Omega}$$

As we have explained above, the liquid without solute has its own interfacial energy α_0, so that $\mathcal{F}_0(A) = \mathcal{F}_{00} + A\alpha_0$. We notice that N/Ω is simply the average density of solute molecules ρ_0. We thus see that $\mathcal{F}$ has the form $constant + A\alpha$, where

$$\alpha = \alpha_0 - T\rho_0 \int [f(z) - 1]\,dz$$

The $\rho_0 \int [f(z) - 1]\,dz$ has a simple physical interpretation. It is the average number of solute molecules per unit area near the surface in excess of those present with no surface interaction. It is called the **surface excess** Γ[†]. We see that whenever the surface excess is positive, the surface tension is reduced. The reduction is just the pressure that would be exerted by an ideal surface gas of areal (mass) density Γ, as in our initial example above. Conversely, if the solute is repelled from the surface, the interfacial energy is increased.

6.3.1 Fluid mixtures

The above picture remains qualitatively true when the solute concentration becomes substantial, so that our liquid is a two-component mixture. In general, one of the two components will be attracted to the interface. We denote this attracted component as A, and denote the other as component B. From the above discussion adding more A to the mixture should lower the interfacial tension. A further phenomenon can happen if A and B are immiscible. Then when enough A is added, the liquid partitions into two phases, as sketched in Fig. 6.9. The phase richer in A is then attracted to the interface and sits next to it. If the two phases have a high interfacial tension between them, then the A-rich phase forms partially wetting droplets on the external interface. One important limiting case is a **critical mixture**, in which the two phases become nearly identical in composition. It has been shown [14] that such a critical mixture always completely wets any external interface: the A-rich phase coats the interface.

It appears that adding immiscible solute molecules should be a good strategy for lowering interfacial energy: the immiscible molecules are

[†] In normal usage Γ is defined as the excess *mass* of solute per unit area. Thus if the solute molecules have mass m, the usual surface excess is m times our Γ.

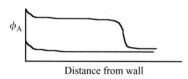

ϕ_A

Distance from wall

Fig. 6.9
Concentration profile of a fluid mixture near an attracting wall. ϕ_A is the volume fraction of species A. Lower curve shows profile for when the two constituents mix; the attraction increases the interaction, which increases the concentration of A near the wall. Upper curve shows the profile after more A is added to the mixture so that it demixes into an A-rich and an A-poor phase. Now the A-rich phase condenses onto the wall and extends a macroscopic distance from it. If the A–B interfacial energy is large, the A-rich phase may form droplets on the wall.

driven to the surface, thus lowering the surface energy. Making the solute more immiscible in the liquid, amounts to increasing the free energy cost $\mathcal{F}_1$ defined above. But in fact, making the solute immiscible is generally *not* a good strategy for lowering surface tension. Immiscible solute molecules generally experience an effective attraction for each other owing to their mutual repulsion from the solvent. Such attractions are present even when the solute molecules are driven to the surface. Thus instead of spreading over the surface and producing surface pressure, the solute molecules may phase separate on the surface, thus cutting the desired surface pressure. A way around this dilemma is to use **amphiphilic** solute molecules. Such molecules have a part that by itself would be insoluble and another part that by itself would be highly soluble. As we shall see in the next chapter, such amphiphilic molecules can segregate strongly to the interface without phase separating there.

6.4 Polyatomic solutes

Up to now we have dealt with small-molecule solutes. What new interfacial phenomena appear when the solutes are large, polyatomic objects: colloidal particles, colloidal aggregates, or flexible polymers? For colloidal particles, the situation is much the same as for small solute molecules. The main difference is that interactions are strong. A colloidal particle at a free surface can deform it, thereby inducing attractive interaction with neighboring particles, like Cheerios on the surface of milk, or like the mattress effect of Fig. 5.1. The colloidal aggregates treated in the Appendix to the Chapter 5 experience similar capillary forces; these are often strong enough to crush an aggregate.

6.4.1 Polymer adsorption

In comparison to the cases above, the adsorption of polymers is rich and subtle. Chapter 5 gave us a hint of this subtlety. The subtlety arises because polymers are deformable, so that small increments in adsorption strength can lead to significant changes in the polymer's shape. Polymers adsorb because their monomers have a short-ranged affinity for a fluid boundary. We may represent this affinity by a monomer interaction potential $u(z)$, where z is the distance from the surface. Since this affinity arises from the local liquid structure, its range a is no larger than a monomer and is much shorter than the polymer coil size R. The attractive potential $u(z)$ is sometimes called a **phantom attraction**. Real interfaces have additional features, to be revealed below.

How big must u be in order to cause significant adsorption? We wish to enhance significantly the equilibrium probability that the polymer will be found next to the adsorbing surface. This means that the total free energy of the polymer at the surface must be smaller than its free energy far away by an amount T or more. To estimate the u necessary, we consider a penetrable or phantom surface, rather than a hard wall. Further, we suppose initially that u is so weak that the polymer does not get distorted by the adsorption. Then the adsorption energy U is simply u times the number of monomers within a distance a of the surface. To find this number, we note that it is the number of intersections M_{AB} of two fractals, as often encountered in the Chapters 3 and 4. One of the fractals is the surface, with dimension $D_A = 2$; the other is the polymer, with $D_B \simeq \frac{5}{3}$. Accordingly for a polymer of radius R having $n \sim R^{D_B}$ monomers the number of intersections is given by $(const) R^{D_A + D_B - 3} \sim R^{2/3} \sim n^{2/5}$. Evidently in order to have an adsorption energy of T, one needs $u \sim n^{-2/5}$. The energy u_{min} needed to adsorb a polymer is far smaller than that needed to adsorb a monomer.

We now consider the effect of increasing u above the value u_{min} needed for adsorption. It is easily possible for u to be much larger than u_{min} and still much smaller than T. This means that short sections of the polymer are affected hardly at all by the surface, but long sections have large energy and are affected strongly. We have seen that the energy needed to make a significant distortion in a polymer coil is only of order T. Thus we must anticipate that adsorption energies much larger than T should cause large distortions.

We anticipate that the polymer will be compressed towards the attracting surface, so that its typical thickness is some $\xi_u \ll R$. We may find an optimal value of ξ_u by considering the two forms of energy that depend on ξ_u. To find the absorption energy U, we note that the compression multiplies the density of monomers at the surface by R/ξ_u. Thus the energy is $u M_{AB} R / \xi_u \simeq un(a/\xi_u)$. By itself, this adsorption energy U favors ξ_u being as small as possible. But opposing U is the energy of confinement; let us call it C. We touched on this confinement energy in Chapter 5 in the Section on repulsive forces. We considered another form of confinement in Problem 3.15. Paraphrasing that discussion, we imagine confining polymer to a slab of width ξ_u. To achieve this confinement we divide the initial unconfined chain configuration into k blobs of size ξ_u. Finally, by manipulating one degree of freedom in each blob, we twist the chain so that it fits into the slab. We have thereby constrained k degrees of freedom and thus increased the free energy C by roughly Tk. Since $k(\xi_u/a)^{5/3} \simeq n$, we conclude $C \simeq Tn(a/\xi_u)^{5/3}$. Naturally, C resists confinement; it increases as ξ_u decreases.

The balance of the opposing effects of C and U gives an optimal ξ_u. For the optimal ξ_u, $C \simeq U$, as with all such optimizations (*cf.* the Flory argument in Chapter 3). We note that both C and U are proportional to the number of monomers n: doubling n (at fixed ξ_u) doubles the number of attracting monomers; it also doubles the number of confined blobs. Thus n is a common factor in C and U, and the optimal ξ_u cannot depend on n (or R). Comparing C and U, we infer $T(a/\xi_u)^{5/3} \simeq ua/\xi_u$, or $\xi_u \simeq a(u/T)^{-3/2}$. The balance of energy occurs blob by blob, with each blob having an attracting energy and a confinement energy of order T.

Having found the width ξ_u, we can now ask what is the lateral dimension R of the adsorbed chain. The adsorbed blobs repel one another just as unadsorbed blobs do. Thus they form a self-repelling random walk on the surface. The self-repulsion causes lateral expansion as in three dimensions. As we saw in Chapter 3, the scaling law for this expansion is $R \sim n^{3/4}$

Since this power is larger than its three-dimensional counterpart $\frac{3}{5}$, R must become larger on adsorption due to increased self-repulsion effects.

If there are other chains in the solution, there is plenty of room for these to adsorb on the surface along with the single chain discussed above. The entire surface can be covered densely with adsorbed blobs with little interference from other chains. The surface is then a two-dimensional semidilute solution of these adsorbed blobs. It has an osmotic pressure owing to the mutual repulsion of the blobs of order T per blob—the same magnitude as the adsorption energy U and the confinement energy C. This osmotic pressure reduces the net binding energy, but only by a factor of order unity. The result is to increase the size of the adsorbed blobs by a finite factor. Thus we arrive at a picture of how polymers adsorb from dilute solution in the presence of weak attraction. There is a dense layer of the thickness of one adsorption blob of size ξ_u. This layer contains $(\xi_u/a)^D/\xi_u^2 \sim \xi_u^{-1/3}$ monomers per unit area. This surface excess produces a surface osmotic pressure T/ξ_u^2, opposing the surface tension of the bare surface. We have seen that bare surfaces have energies of order T/a^2 or more; thus, the polymers in the weak-adsorption regime have little impact on the overall surface energy.

6.4.2 Concentration profile

This picture describes the dominant energies and length scales in an adsorbed polymer layer. In this simple picture, the monomer concentration falls rapidly to zero beyond distances of order ξ_u. For some purposes it is good to know just how rapid this falloff is. The problem is to find the optimal concentration profile, given that the surface concentration is $(\xi_u/a)^D/\xi_u^3$. In our simple picture, there are k blobs per chain, and thus $1/k\xi_u^2$ chains per unit area. One possible way to optimize the concentration profile is to adsorb more chains but with fewer adsorbed blobs per chain. Attracting additional chains costs free energy. On the other hand, all the adsorbed chains can be less confined in this situation, thus reducing the confinement energy. Our problem is to find the best tradeoff between these two effects.

We may analyze this tradeoff most simply by supposing that our polymers have an enormous molecular weight, so that the bulk polymers are much larger than ξ_u. From the reasoning above, we expect the surface to be saturated with blobs of size ξ_u. We now consider a chain that passes through a point at some height $z \gg \xi_u$. In the absence of the adsorption, this chain would intersect the wall at a distance of order z from the starting point. However, with the attraction and its associated layer of blobs, our chain is perturbed. We now consider a sequence of a few blobs on our chain near the surface, as shown in Fig. 6.10, *top*. Since the surface is assumed to be saturated with blobs, our chain cannot crowd into the layer without removing some other blobs. However, if blobs from our chain simply exchange with those of the adsorbed chains, as in Fig. 6.10, *bottom*, there is no change in the adsorption or confinement energy. Thus the new configuration is roughly as probable as the old one. Thus any segment of our chain that

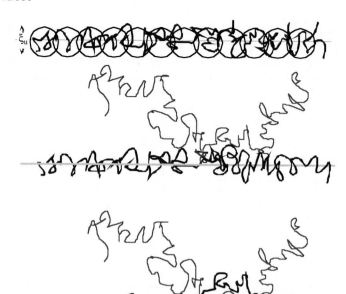

Fig. 6.10
Chains adsorbed on a phantom surface.
Top: a single chain adsorbs in a thickness
ξ_u as described in the text. Middle: a chain
from the bulk solution approaches a
saturated adsorbed layer. Bottom: blobs of
the bulk chain exchange with those of the
surface, thereby binding the bulk chain and
weakening the attachment of the
pre-adsorbed chain.

approaches the wall has a finite probability of being adsorbed, displacing some of the previously adsorbed chains. The displacement process has little effect on the density near the wall. Still, the tightly adsorbed chains of our initial state evolves into a greater number of chains that are more loosely adsorbed and that extend further from the surface. We now ask how many such chains are present at a height $z \gg \xi_u$.

When the adsorbed layer has reached equilibrium, chains at any height z feel no net force. We consider a segment of chain of size z that is at height z above the surface. This segment is part of a much longer chain, thus it is typically attached to the wall. There is some tensile force in the attaching ends tending to pull the segment towards the wall. Our arguments above lead us to suspect that these chains are weakly bound, with an energy of order T. This would mean that there is not enough energy to distort the segments substantially. Accordingly we shall suppose that the adsorbed segment keeps its unperturbed size. Then we will argue that this state is self-consistent. The tensile force is then of order T/z. If one chain can be marginally bound at this height, others can as well. Chains congregate at height z until their mutual repulsion prevents them from congregating further. At this point their repulsive interaction energy is equal to their adsorption energy, which in our picture is of order T. These loops of height z are thus in the same situation as the original ξ_u blobs. They form a layer of loops at a separation of order z. The concentration at this height z is thus roughly the average concentration within our segment, *viz* $\langle \phi \rangle_0(z) \sim z^{D-3}$.

Our assumption of weakly distorted segments is self-consistent. Now we further define our segment to be that part of a chain which extends from the original height z to $z/2$. This segment experiences a gradient of osmotic pressure, tending to push it away from the surface. Since the diameter of our segment is roughly z, the force that it feels is roughly $(\Pi(z/2) - \Pi(z))z^2$. Now, the blobs at height z have a size that is also roughly z according to our picture. Thus $\Pi(z) \simeq T/z^3$. The force is thus of order $(T/z^3)z^2$ or T/z. This is of the same order as the tensile force we supposed above. Thus under our assumption, the forces acting on any typical segment are of the same order. Any adjustment required in order to bring the segments to an equilibrium state of balanced forces is a minor adjustment that does not alter the scaling behavior we have deduced.

This picture of the height concentration profile, originally postulated by de Gennes [15], has been supported by several calculational approaches. It has also been confirmed by realistic computer simulations [16]. Direct experimental confirmation is difficult because the profile involves only a tiny fraction of the adsorbed monomers. Indeed, the above argument says that $\phi(z) \sim z^{-4/3}$, so that the total surface excess $\Gamma \sim \int dz\phi(z)$ is dominated by the smallest z's—i.e., $z \simeq \xi_u$. Thus $\Gamma \simeq \xi_u \phi(\xi_u)$, as argued above. Still, scattering investigations of the profile show behavior consistent with the de Gennes picture [17].

6.4.3 Hard wall

This picture describes the main features of adsorbed polymer structure. But these features are modified subtly by additional effects. Up to now we have considered adsorption by a penetrable, phantom surface. Real adsorbing surfaces are normally impenetrable. Impenetrable surfaces without adsorption are not neutral; they actively suppress close approaches of the polymer coils. Any monomer that is adjacent to the surface suffers extra constraints: the subsequent monomer may not pass through the surface. Thus, its angular freedom is restricted. This amounts to a free energy cost of order T for each such monomer. A weak attractive energy u is not sufficient to overcome this repulsion. Instead, u must be of the order of T if its range is a statistical segment of length a. Adsorption begins to occur only if u exceeds a threshold of order T. Weak adsorption with small surface excess then occurs when u exceeds this threshold by a small fraction of T. In this regime, one may define the adsorbed blob size ξ_u and discuss the concentration profile for distances larger than ξ_u as above.

Over heights z smaller than ξ_u, the impenetrability has an interesting effect. For a phantom surface, the concentration is virtually independent of z in this **proximal region**. For segments much smaller than the blob, the potential energy of adsorption U of the segment is much smaller than T. Thus it can be little affected by the surface. But for real surfaces there is an interplay between the impenetrability constraint and the self-repulsion of the chain. This leads to new effects [18]. Even a single chain constrained to touch an impenetrable surface with a large number of its monomers has a power-law falloff of concentration near the surface varying z^{-p}, where $p \simeq \frac{1}{3}$. This extra concentration at the surface arises because the adsorbed monomers occur not independently but in correlated *trains*. These correlations can arise because the adsorption energy per monomer for an impenetrable wall must be significant on the scale of T. When adsorption is very weak, this proximal power law can lead to measurable effects on the surface excess and the concentration profile. The reasons for the proximal power are difficult to account for in simple terms, and direct evidence for it has proved difficult to establish experimentally [19]. There is in particular no simple connection between the proximal power and the fractal dimension D of a self-avoiding polymer.

Our discussion of the density profile assumed that the chains had indefinitely long length. When their length is finite, the power-law regime is limited to distances z smaller than the unperturbed size of the chains. In addition, the finite-length chains appear to show a second power-law regime that implies a falloff of concentration slightly weaker than the $z^{-4/3}$ found above for sufficiently large z, owing to tails of chains adsorbed at only one end [20].

6.4.4   Kinetics of adsorption

As we have seen above, the attainment of an equilibrium adsorbed layer requires the cooperative motions of many chains as the initial tightly bound chains are gradually displaced by additional loosely bound chains. A new unabsorbed chain must overcome an activation barrier (discussed in Chapter 2) in order to be adsorbed. It must penetrate into regions of increasingly high concentration against an osmotic pressure gradient before any of its monomers can experience the attracting surface. If the adsorption is strong, the region near the surface is often so crowded with monomers that they are virtually immobilized as in a glass. In such cases the relaxation to equilibrium can take hours [21].

6.4.5   Surface interaction

As we saw in Chapter 5, when surfaces perturb the liquid around them, there is an induced attraction or repulsion between those surfaces. The adsorption of polymers induces interactions between two surfaces in this way. If the polymers are irreversibly adsorbed, the osmotic pressure at the midpoint increases as the two surfaces approach. Thus external work is required to decrease the separation and the interaction is repulsive. Even though most of the monomers live at distances of order ξ_u from the surfaces, the range of the significant repulsion may be much greater than ξ_u. If two colloidal particles have radii much larger than the polymer size R, the repulsive energy at separation R is many times T. If the adsorption is reversible and reaches equilibrium, two surfaces with overlapping adsorbed layers may release polymers, since one surface may now use chains from the other surface in order to attain its optimal coverage. Now the forces crossing the midplane come not only from the repulsive osmotic pressure, but also from the tension in chains that bridge between the two surfaces. The net effect is attractive for sufficiently small separation[†].

6.4.6   Flow

An adsorbed polymer layer has an important effect on the flow past a surface. Near a bare, solid surface, the velocity goes to zero linearly as the surface is

[†] The recent theory of Semenov and Joanny [20] predicts a net repulsion at larger separations. Such a repulsion need not violate the attraction theorem discussed in Chapter 5. That theorem concerns of perturbed fields whose energy senses nonuniformity only through local gradients. However, polymer solutions can in principle have energies that depend nonlocally on the concentration.

approached, yielding a uniform shear rate and a uniform viscous stress. But flow through an adsorbed layer need not have a uniform viscous stress. Instead, force can be transmitted from the fluid to the polymers, and thence directly to the surface. We saw in Chapter 4 that fluid velocity decays exponentially in a semidilute solution of blob size ξ, with a decay length of the order of ξ. Similar decay occurs in an adsorbing layer. We may picture the layer as having an outermost sublayer of thickness $R/2$, extending from height $R/2$ to height R and having blobs of size $R/2$. Then the adjacent denser sublayer has thickness $R/4$ and blob size $R/4$, followed by successively denser sublayers. We suppose that the flow velocity at the outer boundary of the layer is v_0. Then on traversing the outer sublayer, the velocity must decay to a finite fraction of v_0, say v_0/χ. If this χ is larger than 2, the velocity decreases faster than the height z. After k sublayers, $z = R \times 2^{-k}$ and $v = v_0 \times \chi^{-k}$. Solving for $-k$ yields

$$\frac{\log z/R}{\log 2} = k = \frac{\log v/v_0}{\log \chi},$$

so that $v/v_0 = (z/R)^{\log \chi / \log 2}$.

This χ must indeed be larger than 2. If χ were 2, the velocity would interpolate linearly between 0 and v_0 as one traverses the layer. This is exactly the velocity profile near the bare layer, with only viscous drag. Now, the polymer layer certainly has substantial drag in addition to the viscous drag, and it must decrease the velocity within the layer by at least a finite factor. The only way to attain this decrease is for χ to be bigger than 2. We conclude that the flow velocity extrapolates to zero at some height of order R, greater than zero. Thus as far as the flow is concerned, the layer behaves as though it had a thickness t of order R rather than ξ_u, as shown in Fig. 6.11. Our reasoning depends on the flow being weak, so that it does not disturb the adsorbed layer appreciably. The effect of adsorbed layers on flow has been tested using the surface-forces apparatus [22]. These experiments confirm our conclusion that the hydrodynamic thickness of order R, though they have not demonstrated the form of the velocity profile. The factor χ depends on the screening properties of polymers in a semidilute environment. We have seen that these are characterized by a screening length of order ξ, independent of the chemical structure of the polymer or the solvent. We argued that the screening length is a *universal* multiple of the structural correlation length. Similar reasoning led to our conclusion that the χ factor above is a finite number for any adsorbed layer in a good solvent, independent of such structural factors. Thus we expect χ to be universal, and accordingly the exponent $\log \chi / \log 2$ for the velocity profile should be universal, as well. To our knowledge this universal exponent has not been calculated or measured.

Polymers influence flow past a surface in an important non-adsorbing situation. Here we consider an entangled polymer solution or melt flowing past a smooth wall. Now the stress reaches the wall via two different processes in series. Within the fluid, the stress is transmitted through entanglements and is governed by the large viscosity of the polymer solution. But there are no entanglements connecting the wall to the adjacent fluid.

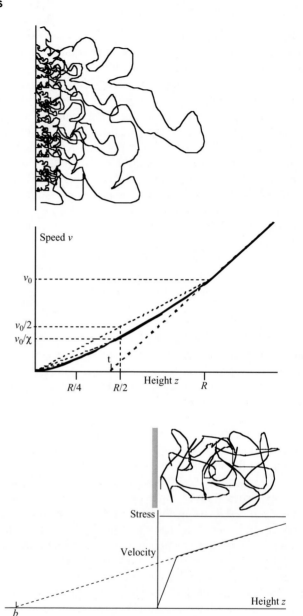

Fig. 6.11
Top: sketch of a saturated adsorbed
polymer layer in equilibrium with a
solution. Layer consists of loops at various
heights such that the distance between
loops at height z is comparable to z.
Bottom: illustration of the procedure in the
text for accounting for the velocity profile.
Beyond the adsorbed layer, the velocity
grows linearly with distance from the
surface. Within the adsorbed layer, the
velocity decreases by more than a factor 2
for each halving of the distance, resulting
in a power-law increase of velocity whose
exponent is larger than 1. The velocity at
large distances extrapolates to a height t,
the effective hydrodynamic thickness. It is
comparable to the layer thickness R.

Fig. 6.12
Flow of an entangled polymer solution past
a wall, sketched at top. Upper curve shows
the (uniform) shear stress. Lower broken
curve shows the velocity profile required to
attain the uniform stress. The extrapolation
length b is shown on the horizontal axis.

Here the stress is conventional fluid stress, caused by fluid and polymer
atoms hitting wall atoms. It can be characterized by the viscosity of the
solvent or of disconnected monomers. Though the means of transmitting
stress in these two regions are very different, the amount of stress must be
the same. The stress must pass from the bulk of the fluid to the container
via the wall layer. To attain the same stress as the bulk, the wall layer must
have a much larger velocity gradient, as sketched in Fig. 6.12. The velocity
extrapolates to zero beyond the wall, at a distance b called the **slip length**.
For simple fluids the slip length is of the order of an atomic size a. But if the

polymer viscosity is larger than the solvent viscosity by a large factor, the slip length exceeds a by an equally large factor. Such large slip lengths are readily observed, and they may reach the scale of microns [23]. Sometimes one wants to avoid this slipping behavior. One way to achieve this is to fasten polymers to the wall. Then the fluid polymers must disentangle from the wall polymers in order to flow. One may envisage all kinds of exotic behavior when these flows are fast enough to perturb the fastened polymers or to inhibit re-entanglement. The engineering literature has no lack of examples of such exotic behavior. Perhaps the best known are the "sharkskin" patterns caused by stick-slip flow past a wall under these nonlinear conditions [24].

6.5 Conclusion

In this chapter we have accounted for the energy scales associated with liquid surfaces and interfaces. We have seen how these energies create geometric and kinetic effects e.g. in wetting phenomena. Polymers and colloidal suspensions interact strongly with surfaces and this interaction can alter the interaction between two surfaces. It can also alter flow past a surface. With these interfacial phenomena in mind, we are ready to explore the distinctive phenomena caused by surface-loving molecules, or surfactants.

References

1. Tantec contact angle micrometer Tantek Co. 630 Estes Avenue, Schaumburg IL 60193 http://www.tantecusa.com/camplus2.html (2002).
2. Pamphlet "Spinning Drop Tensiometer Site 04" KRUESS USA, Instruments for Surface Chemistry 9305 Monroe Road, Suite B, Charlotte, NC 28270 - 1488 http://www.kruss-usa.com/public_pdf/Site04e.pdf
3. SE400: Sentech Discrete Wavelength Ellipsometer http://www.sentech.com/ellipsometer.htm SENTECH Instruments GmbH Carl-Scheele-Strae 16, 12489 Berlin, Germany.
4. F. Rondelez, D. Ausserre, and H. Hervet, *Ann. Rev. of Physical Chemistry* **38** 317 (1987).
5. J. Israelachvili, *Intermolecular and Surface Forces* (New York: Academic, 1985) Table XVII, p. 158.
6. J. Israelachvili, *op. cit.* Chapter 14.
7. P. G. de Gennes, *Rev. Mod. Phys.* **57** 827 (1985).
8. L. Tanner, *J. Phys. D.* **12** 1473 (1979).
9. H. Hervet and P. G. de Gennes, *Comptes. Rendus Acad. Sci.* **299II** 499 (1984).
10. L. Leger and J. F. Joanny, *Rep. Prog. Phys.* **55** 431 (1992).
11. C. Redon, F. Brochard-Wyart, and F. Rondelez, *Phys. Rev. Lett.* **66** 715 (1991).
12. R. D. Deegan, O. Bakajin, T. F. Dupont, G. Huber, S. R. Nagel, and T. A. Witten, *Nature* **389** 827 (1997); *Phys. Rev. E.* **62**, 756–765 (2000).
13. S. H. Davis, *Ann. Rev. Fluid Mech.* **19** 403 (1987).
14. D. Jasnow, *Rep. Prog. Phys.* **47** 1059 (1984).
15. P. G. de Gennes, *Adv. Colloid Interface Sci.* **27** (1987).
16. J. de Joannis, R. K. Ballamudi, C. W. Park, J. Thomatos, and I. A. Bitsantis, *Europhysics Lett.* **56** 200 (2001).
17. L. Auvray and J. P. Cotton, *Macromolecules* **20** 202 (1987).
18. E. Eisenriegler, *J. Chem. Phys.* **79** 1052 (1983).
19. J. M. DiMeglio and C. Taupin, *Macromolecules* **22** 2388 (1989).

20. A. N. Semenov, J. F. Joanny, A. Johner, and J. BonetAvalos, *Macromolecules* **30** 1479 (1997).
21. K. Kremer, *J. Phys-Paris* **47** 1269 (1986).
22. J. Klein, *Ann. Rev. Materials Sci.* **26** 581 (1996).
23. F. Brochard and P. G. de Gennes, *Langmuir* **8** 3033 (1992).
24. R. G. Larson, *Rheol. Acta* **31** 213 (1992).

Surfactants

7

7.1 Introduction

In the previous chapter we saw that the boundary of a liquid is responsible for significant, controllable forces—capillary forces. In this chapter we encounter an enormous expansion of that control through the use of "amphiphilic" molecules. An amphiphilic molecule contains a part that by itself would be soluble in the liquid and another part that would be insoluble in it. The insoluble parts have a more favorable free energy when they are away from the liquid; thus these molecules tend to concentrate at the liquid boundary. Because they influence surfaces and other boundaries, these molecules are called surfactants.

In a sense a surfactant incorporates a bit of interface within itself. The two dissimilar parts of the molecule, having opposite affinities for the liquid, generally have a repulsive interaction with each other. This repulsion amounts to a free energy that was stored in each surfactant molecule when it was made. The boundary between the two parts of the molecule often plays the role of an interface. To minimize the repulsive energy inherent in the amphiphilic molecules, they often organize to create real interfaces throughout the liquid, as shown in Fig. 1.2. The resulting structures can dramatically alter the forces within a liquid. In this way capillary forces come to control the bulk properties of the liquid, not just its surface properties.

A second feature of surfactants is that they can cause immiscible liquids to mix. Here one considers surfactant amphiphiles one of whose parts is soluble only in the first liquid and the other of whose parts is soluble only in the second liquid. Such surfactants are at equilibrium at the interface between the two liquids. Adding more surfactants creates more interface. With a sufficient amount of surfactant, this interface may grow so much that any point in either fluid is only a microscopic distance from the interface. Then the immiscible fluids are in effect mixed.

This chapter explores the special properties that amphiphilic molecules impart to a liquid. As in previous chapters, we focus on the simplest level of quantitative, predictive analysis. We identify the characteristic spatial scales, energy scales, and time scales that make these fluids act the way they do. We begin with a section on mixing; we recall the empirical rules that describe the mixing of different molecules. The next section describes the most common surfactant molecules, noting their resemblances and differences. The following section examines how strong amphiphiles should interact in a liquid, and why they should aggregate into micelles when the concentration is sufficiently large. Then comes a section discussing how the fluctuations of micelles lead to interaction and generate forces. The following section consider how surfactants alter the properties of a fluid interface. We will discuss how it is possible for surfactants to lower an interfacial tension by orders of magnitude, and to induce spontaneous mixing of two immiscible liquids, thus forming a **microemulsion**. A section on amphiphilic polymers shows how the large size of polymers gives rise to scaling laws that predict their micelle structure. Finally we touch on the way micellar structures move and relax mechanically, and we mention the various nonequilibrium structures that can result when surfactants are aggregated in a fluid.

7.2 Mixing principles

The behavior of surfactants in solution hinges on the principles controlling miscibility in liquids. In the previous chapter we saw that the free energy of a fluid A changes when one adds a foreign molecule B at a specific place. To be specific, we consider the work W required to bring molecule B from a reference liquid of pure B to the chosen point in liquid A. Work is required because the atomic environment of our B molecule changes when it is transferred from a liquid of identical B molecules to a liquid of foreign molecules A. If the work is negative, then liquids A and B must spontaneously mix. Even if the work W is positive, the liquids still mix to some degree. If the A and B liquids are in contact, B molecules still are present in liquid A at some nonzero concentration. The probability that our B molecule resides at a specific point in the A liquid is $\exp(-W/T)$ relative to the corresponding probability in the native B liquid. If W is much greater than T, then the B concentration in A is small, the Bs' in the A solvent have little effect on each other, and the work to bring a B into the A solvent in equilibrium with B is nearly equal to W. Then the equilibrium volume fraction of A in B is $\exp(-W/T)$ if A and B displace the same volume. On the other hand, if the work W is no more than about T, then equilibrium the volume fraction of B in A is substantial and thus B mixes well with A.

Various aspects of the fluid give rise to the work W. One such effect is steric: the A molecules must arrange themselves differently around our B molecule than did the fellow B molecules in the reference liquid (*cf.* the discussion of Fig. 2.1 and Fig. 6.6). A second difference arises from electrical interactions. The molecules generally have a distribution of charge and each atom has an electric polarizability, as discussed in Chapter 5 under "induced dipole interactions." To treat these interactions accurately is a complicated enterprise that requires knowledge of many specific properties of the A and B molecules. Still, the qualitative mixing properties of two liquids can be understood in simple terms, by presuming that the interactions, both steric and electrical, are local in range, affecting mainly adjacent atoms.

The limiting case of this local point of view is to ignore difference in charge at different parts of the surrounding molecules, and to consider only the polarizability of the surrounding environment. This amounts to replacing the actual A solvent by a fictitious atomic solvent, whose electric polarizability matches that of the real A solvent. We treat the B solvent in the same way. By this approach one may justify five principles that are generally borne out empirically. (1) Positivity: the work required to transfer our B molecule is generally positive and increases with the molecular size. (2) Additivity: the work W may be found by adding contributions for each atomic constituent of the B molecule being mixed. (3) Ordering: the work is greater when the two liquids differ more in their static dielectric constant and their optical index of refraction. (4) Reciprocity: if A and B are of comparable size and if A is miscible in B, then B is miscible in A. (5) Transitivity: if A, B, and C are of comparable size and A and B are miscible in C, then A is miscible in B. We discuss each of these principles briefly below. These important aspects of physical chemistry are discussed much more carefully in Israelachvili [1] and Denbigh [2].

7.2.1   Positivity

In the discussion of van der Waals forces in Chapter 5 we found that two identical polarizable B atoms have an attractive interaction in vacuum or in any polarizable medium. Now, the van der Waals interaction perturbs each atom only slightly; thus the net interaction of many atoms is nearly the sum of pairwise interactions. This means that a B atom near a boundary between B and A liquids feels an attraction towards the B liquid—i.e., the work W is positive. Now if a few atoms are joined to form A and B molecules, the atomic polarizability of each atom is altered by its bonding environment. Still, each atom of B feels a net attraction for its counterpart on other B atoms even in an environment of A. Thus two B molecules feel a net attraction, and again the work W is positive. The removal of our B molecule to an A environment also affects the packing of the neighboring molecules around our B molecule. But for the small organic molecules we encounter in practice, these differences in packing are usually not dominant.

Furthermore, the group of atoms is itself a polarizable entity that must feel a net attraction towards similar entities in an environment of different polarizability. Our small B molecule will in general have a nonspherical distribution of charge, though it has no net charge. At large distances, this nonspherical distribution leads to a dipole field. This dipole may be oriented by electric fields; in other words it leads to a molecular polarizability. This polarizability creates van der Waals attraction just as atomic polarizability does. Thus it adds another positive contribution to the work W.

7.2.2   Additivity

Many molecules, such as hydrocarbon chains, are too large to be considered as small, fixed clusters of atoms. How can we understand the work W for carrying a hydrocarbon chain B from a liquid of similar chains into some other liquid A? As we have seen, this work results from the altered interactions between each hydrocarbon group and its immediate surroundings in the A medium compared with its original B surroundings. The interactions are local. Thus it is reasonable to approximate W as the *sum* of contributions W_0 that each hydrocarbon group would experience on being carried from the B liquid to the A liquid. Adopting this point of view leads us to an important conclusion, already encountered in Chapters 3–5. We have noted that W_0 is in general positive. Thus the work W required to transfer a large B molecule grows with its size. Ultimately the work becomes substantially larger than T and mixing becomes virtually nil: large molecules resist mixing. This conclusion holds even when the A solvent is nearly identical to B, so that W_0 is very small. Still, for sufficiently large B molecules, the total work W must become large. For example, polymers with a molecular weight of several million can demix strongly even when the only difference between them is the replacement of the hydrogen atoms with their isotope deuterium [3].

Sometimes the shape of the B molecules prevents additivity from working. Additivity requires that the environment of all small segments of the B molecule be the same, independent of the number of segments in the molecule. When B is transferred to the A medium, each segment is presumed to be surrounded by A. If B has a globular shape so that most of the atoms of B are surrounded by other atoms of the same B molecule, additivity clearly is not valid. Indeed, the B molecule may adopt a globular shape when inserted in the A medium because this globular shape reduces W. Such changes of shape generally require work of at least T in themselves, as we saw in Problem 3.15. Thus this work alone is enough to inhibit mixing substantially.

7.2.3   Ordering: like dissolves like

Returning to our atomic A and B liquids, we see that the van der Waals attraction must vanish when A and B become identical in atomic polarizability. Thus to mix well, molecules should have similar polarizability. Thus B molecules with large dipole moments mix well in A solvents with large dipole moments, such as water. Likewise hydrocarbon molecules with virtually no dipole moment, mix well with hydrocarbon solvents. These dipole moments control the static polarizability that produces the static dielectric constant of the liquid ϵ. However, the molecular dipole moment is not the only source of polarizability. Polarizability is the induced dipole produced by an applied electric field of arbitrary temporal frequency. Polarizability at any frequency gives rise to van der Waals attraction. Thus liquids of identical static dielectric constant ϵ may nevertheless be immiscible, because they differ in polarizability at nonzero frequencies. Another measure of polarizability at the optical frequencies that characterize electronic motion is the optical index of refraction. Thus two small-molecule liquids with similar static dielectric constants and similar indices of refraction are typically readily miscible.

7.2.4   Reciprocity

For *atomic* liquids of the same atomic size, only the contrast in polarizability leads to attraction and immiscibility. If A and B atoms differ in polarizability, this limits the miscibility of A atoms in B and B atoms in A to a comparable degree. The same is expected when A and B are small molecules. This reciprocity becomes less valid as the contrast between A and B increases.

Reciprocity breaks down completely if for example, B is a large molecule and A is a small one, for then B is generally immiscible in A even though A may be miscible in B.

7.2.5 Transitivity

The principle of transitivity says that two molecules that are miscible in a third should be miscible in each other. This is another consequence of the like-dissolves-like principle. We have argued that small-molecule solvents that are similar in dielectric constant and in index of refraction should be miscible in each other. If these properties are similar for solvents A and C, and for B and C, they must be so for A and B. Thus A and B should be miscible. If the molecules differ greatly in size, the reasoning breaks down, and the transitivity property need not hold. The most obvious counter example is the case where A and B are large hydrocarbon polymers and C is a small hydrocarbon molecule.

7.2.6 Effect of permanent dipoles: water

The principles stated above treated the solvent as a weakly polarizable A medium like a monatomic liquid. Then the effect of two different B molecules is approximately the sum of their individual effects. This view of the medium works badly in the case of a strongly polarizable solvent like water. For water, the dielectric constant ϵ is about 80. Thus the dipole moments of the water molecules respond so strongly to an external field that they reduce it by a factor of 80. Each water dipole lowers its energy by favorable interactions with nearby dipoles; it loses part of this energy if adjacent water molecules are replaced by nonpolarizable B molecules. This mutual alignment means that water molecules orient in concert much more than most solvents do. This alignment gives rise to longer range interactions between B molecules than in simpler solvents. The unusual properties of water as a solvent are discussed at length in Israelachvili's book [1].

7.2.7 Effect of charges: ionic separation

Our discussion up to now has considered only neutral A and B molecules. But sometimes B molecules can dissociate in the A medium to form ions. This ionization decreases the work W and thus promotes mixing. Now when the B molecule is brought into the A medium, with one atom at a fixed position, an ion is liberated. To find the Boltzmann weight $\exp(-W/T)$ we must now add a contribution for each possible position of this ion, using the definition of free energy from Chapter 2. Accounting for these possible positions multiplies the Boltzmann weight by a factor proportional to the volume per ion V. This amounts to a decrease in W by an amount $T \log V/V_0$, where V_0 is the volume per ion in the original B medium. (The $\log V/V_0$ is conventionally called the translational entropy.) The accessible volume of the ion is evidently much larger in the ionized state of the A medium, as compared with the un-ionized state of the native B medium. Thus the ionization gives a negative contribution to W and promotes mixing.

These benefits of ionization do not come without a cost: namely, the electrostatic energy of separating each ion from its oppositely charged partner(s) in the native B medium. The electrostatic cost must be smaller than about T. Otherwise, there would be a net positive cost to ionization even considering the entropy effect above. To avoid this cost, the mutual electrostatic energy of the two opposite ions must be made small. This energy U is reduced in proportion to the reduction of the electric field of, say, the negative B ion; it is smaller than in vacuum by a factor ϵ. Thus large ϵ means small U and small electrostatic cost of separating. A useful measure of this electrostatic energy is the Bjerrum length ℓ, defined in Chapter 3. It is defined so that the mutual energy of two singly charged point ions in the liquid at large separation r is given by $U(r) = T(\ell/r)$. Evidently ℓ varies as $1/\epsilon$. For water at room temperature $\ell \simeq 0.7$ nm. Accordingly, the energy cost to separate two ions from $r = 0.7$ nm to a large separation is just the thermal energy T.

In water most small singly charged ions dissociate and mix readily. The energy gained by bringing them to a distance of, say, 1 nm from infinity is only a fraction of T, in accordance with the formula above. At this distance, the organized layers of polarized water molecules around each ion begin to interfere with each other. Bringing the ions closer disrupts these "solvation shells" and costs free energy. The result is that little further net energy would be gained by bringing the ions into contact. The total energy of separation is of the order of T, and the ions may readily dissociate. For example, sodium and chloride ions dissolve in water at volume fractions up to about 10%. For volume fractions smaller than this, the gain in translational entropy is more than enough to compensate for any loss in electrostatic and solvation energy.

Evidently, the special properties of water are important in allowing ions to dissociate. In most solvents dissociation is a minor effect. Even in water dissociation is difficult for ions bearing more than one charge. Two divalent ions must be further than 2.8 nm apart in water in order to have an electrostatic energy smaller than T.

7.3   Surfactant molecules

As stated above, surfactant molecules have one part that is miscible in the solvent and one that is immiscible in it. For water solutions, the most common surfactants have ion pairs, like sodium sulfate, as their miscible parts. In water these singly charged sodium ions readily dissociate, so that such ions are strongly miscible. The immiscible part is a common molecule with small polarizability: a hydrocarbon chain. This chain increases the free energy of mixing W. Still, if the chain is short, the combined molecule is still readily miscible in water. Adding more CH_2 groups to the chain decreases the miscibility (according to the additivity principle above). Thus one finds that chains with as many as 12 carbons attached to a sodium sulfate group mix only at volume fractions below about $\frac{1}{4}$%. Thus this molecule, called

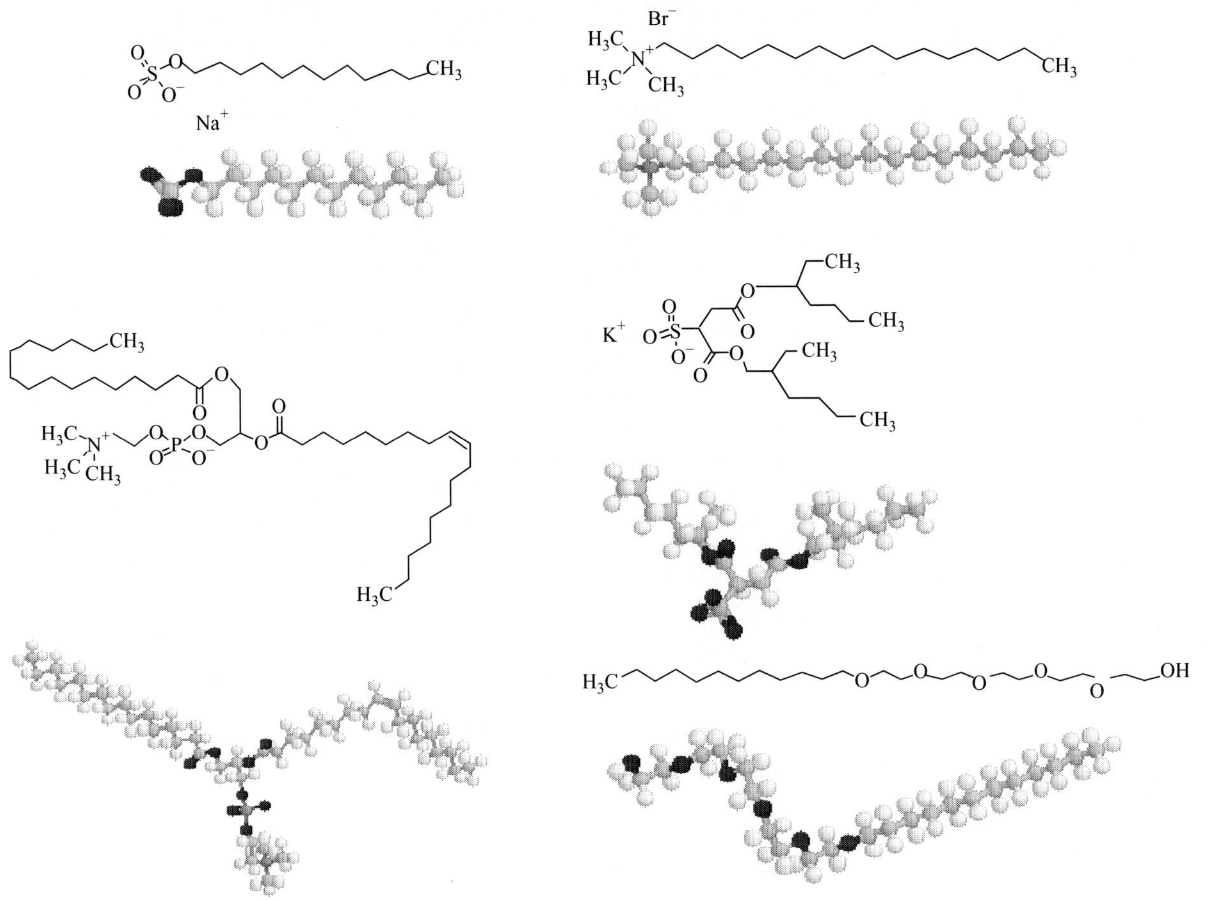

Fig. 7.1

Five common surfactant molecules from [4]. Top row: SDS, the cationic surfactant cetyl trimethyl ammonium bromide (CTAB). Bottom, the phospholipid 1-palmitoyl-2-oleoylphosphatidylcholine (POPC), sodium bis(2-ethylhexyl)sulfosuccinate (AOT), pentaethylene glycol monododecyl ether (C12 E5).

sodium dodecyl sulfate or SDS, is much less miscible than its miscible part. Figure 7.1 shows a diagram of it.

SDS, also called sodium lauryl sulfate, is the leading ingredient in household cleaning products, like soap, detergent, and shampoo. Many other surfactants are variations on this theme. The hydrocarbon part, called the tail, may be lengthened by four or more carbons. The polar sulfate head group SO_4^- can be replaced by carboxylate, CO_3^-, or by phosphate, PO_4^-. Finally, some head groups may have two hydrocarbon tails attached. As we shall see below, the significant differences in these molecules lie not only in the solubility of the different parts but also in the relative bulk of the head and tails and in the extensibility of the tails.

The polar head is most commonly a negative ion like the ones above. Then the molecule is called an anionic surfactant. Sometimes one chooses a cationic group such as ammonium, NH_4^+ as the polar head as in the CTAB molecule shown in Fig. 7.1. Naturally, cationic and anionic surfactants have a strong mutual electrostatic attraction. The dissociating ion is most commonly a singly charged metal ion like sodium. But this ion can be an organic group as well. Likewise, the ions may be weakly dissociating. An important example is an OH^- H^+ pair whose dissociation may be controlled by the ambient H^+ concentration or pH. Amine groups NH^- are a second important weakly dissociating group.

Some surfactants, called nonionic, have polar heads without dissociating ions. One example is a zwitter ion, in which an organic cation is attached to an organic anion by e.g., a short hydrocarbon chain. A second important example is the ethylene glycol group, whose polar oxygens associate strongly with surrounding water molecules and lead to a negative contribution to the work of mixing W.

To summarize, surfactants vary in strength from weak to strong. Weak surfactants have parts with little difference in miscibility. Strong surfactants have parts that differ greatly. Beyond this difference, surfactants differ in the relative bulk and deformability of their polar and nonpolar parts. Further, the polar parts differ in being anionic, cationic, or nonionic.

7.4 Surfactants in solution: micelles

Surfactants, like any molecules, must mix in any host liquid A at some nonzero concentration. If one mixes a small enough amount of surfactant in the liquid, the surfactant molecules must disperse through the liquid. But if one increases the amount, a point comes where this molecular dispersion no longer works. Beyond this point some of the B molecules typically demix into a separate B-rich phase. However, surfactant B molecules can find another way to reduce the work W of mixing. This work W has two large and opposite contributions. For definiteness we suppose that the solvent is a polar one. Then the polar part of the surfactant contributes a negative W, but the nonpolar tails contribute a larger positive part. In a native surfactant B phase, the mutually miscible nonpolar tails are adjacent to each other. The positive part of W arises from the cost of replacing this miscible environment with polar A solvent.

This cost can be avoided if the surfactants can be placed in the A solvent without the tails leaving their environment of other tails. Thus a number of surfactants tend to enter the A liquid together, with their tails adjacent to one another. The resulting aggregated structure or **micelle** can take forms like those shown in Fig. 7.2. If the surfactants enter the A solvent in the form of micelles, the net cost W_m can well be negative. For example, SDS forms spherical micelles which then mix with water at volume fractions that can exceed 10%. At higher concentrations the micelles' repulsive interaction forces other forms of aggregation. This is much higher than the 10^{-4} volume fraction that dispersed SDS achieves.

Any micelles present in the A solvent must be in equilibrium with dispersed surfactant molecules. For the case of spherical micelles with

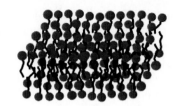

Fig. 7.2
Three types of micelles. Left to right, spherical, cylindrical, and bilayer.

K surfactants B, this equilibrium is expressed by the chemical reaction

$$KB \longleftrightarrow B_K.$$

This equilibrium dictates a condition on how the concentration of individual surfactants, $[B]$ may depend on that of micelles $[B_K]$. (Here we used the convenient notation where $[X]$ means "number of species X per unit volume.") This **law of mass action**, derived in all physical chemistry texts [2] is $[B]^K / [B_K] = $ constant. The constant is independent of concentration. The well known outcome of this constraint is that at low concentration the dispersed form dominates while at high concentration the aggregated form dominates, as explained below.

In practice, the aggregation number K of micelles is sizeable: for example, in SDS it is about 64 [5]. When K is large, one can understand the equilibrium in a simpler way, using the work W defined above. If micelles are present, one may define the work W to carry a surfactant from a micelle to a given place in the A solvent. Then by our reasoning above, the volume fraction of dispersed surfactant should be given by $\exp(-W/T)$. We are led to the conclusion that the micelles are in equilibrium with a fixed volume fraction of dispersed surfactant. Since adding surfactants cannot increase the dispersed concentration above this amount, all the added surfactants must be in the form of micelles. Likewise, removing surfactants reduces the number of micelles without changing the dispersed number. This continues until there is no excess surfactant available to form micelles. This threshold concentration of surfactants is known as the **critical micelle concentration**, or **CMC** for short.

This fixed CMC is consistent with the conventional chemical equilibrium picture mentioned above for large aggregation number K. If most of the surfactant is in the form of micelles so that the total concentration $C \simeq K[B_K]$, then $[B] = C - K[B_K] \ll C$ and we can write the equilibrium formula as

$$[B]/(C - [B])^{1/K} = \text{constant}$$

Neglecting the small $[B]$ part of the denominator, we conclude that

$$[B] \simeq \text{constant} \times C^{1/K}$$

For large aggregation number K, $[B]$ grows very slowly with C—$[B]$ is practically constant.

What determines the aggregation number K? Evidently it has something to do with the shape and deformability of the surfactant molecules. For example, an aggregate of SDS with $K = 2$ would not be very favorable, since the two tails could not avoid contact with the solvent. Thus nearly as much work would be required to put the aggregate into the solvent as to put the dispersed molecules there. But a spherical micelle with tens of molecules can readily shield the tails from the solvent. Beyond this point there is little to be gained by increasing the aggregation number. In addition, there is something to lose: the charged head groups repel each other, and enlarging the aggregation number clearly increases this repulsion.

7.1. *Law of mass action*   Molecules A, and B in a dilutesolution can form a combined species AB. The work required to separate an AB into its constituents A and B at two given positions far apart is W. Show that in equilibrium the concentrations $[A]$, $[B]$, and $[AB]$ are related by

$$[A][B]/[AB] = \text{constant } e^{-W/T}.$$

7.4.1 Open aggregation: wormlike micelles

Some surfactants are constructed such that they profit from aggregation beyond that of small spheres. A spherical shape is no longer feasible beyond a diameter of about twice the tail length, for the tails could not reach the center if the sphere were larger than this. However, there are two other shapes that permit arbitrarily large K: one-dimensional **wormlike micelles** and two-dimensional **bilayers**. We consider each in turn.

In a solution of wormlike micelles, the aggregation number K may vary over a wide range to form shorter or longer worms. Each of these species must be in equilibrium with the others. These constraints control the relative number of micelles of given size, and they control how the average size depends on the overall concentration C, as we now show.

Wormlike micelles, like spherical micelles, have a minimal aggregation number K_0, below which aggregation is not worth its energetic cost. Any micelle of length aggregation number $K > K_0$ may give free surfactant molecules to the solvent, or may take them, with some cost or benefit in free energy. That is, there is an equilibrium between the free surfactants (micelles with $K = 1$) the K micelles and the $K + 1$-micelles: $(1) + (K) \leftrightarrow (K + 1)$. The law of mass action mentioned above describes the resulting relationship between the concentration of free surfactant molecules $[1]$ and the concentrations of the micelles $[K]$ and $[K + 1]$:

$$[K][1]/[K + 1] = 1/V_0 \exp(-W/T), \qquad (7.1)$$

where W is the work to carry a surfactant from the $K + 1$ micelle to the solution and V_0 is the reference volume per surfactant encountered above.

For long, wormlike micelles with $K \gg K_0$ the work to remove a surfactant is sometimes entirely due to local structure in the micelle and the solvent adjacent to the molecule being removed. This is the case if the micelles are dilute, rigid rods, or flexible chains without self interaction. Then the work W is independent of K. In this case determining the concentrations $[K]$ in terms of the total concentration C is straightforward. First, we note that the right hand side of Eq. (7.1) has dimensions of concentration. We denote it as C_0. Then the equation can be written

$$[K+1]/[K] = [1]/C_0.$$

The equation describes a geometric falloff with increasing micelle size: $[K] = C_0([1]/C_0)^K$. For very small free surfactant concentration $[1]$ the right side is much smaller than 1. Thus the K concentrations fall off rapidly with increasing K. But as $[1]$ approaches C_0 the falloff becomes slower and slower, so that large micelles become more abundant. The surfactant concentration in the form of K-micelles is $K[K]$, so that the overall concentration C is $\sum_K K[K]$. It is clear that the total concentration C becomes larger as the free surfactant concentration $[1]$ increases. Indeed, if $[1] = C_0$, the concentration of K-micelles no longer decreases with increasing K, and C becomes infinite. Since C must remain finite, $[1]$ must always remain smaller than C_0. This C_0 plays a role similar to that of the critical micelle concentration for spherical micelle-formers. We shall call C_0 the limiting free surfactant concentration or limiting monomer concentration.

To describe how $[K]$ depends on C explicitly, it is convenient to define $f(x) \equiv \sum_{k=0}^{\infty} x^k$. The series sums to $f(x) = 1/(1-x)$. Its derivative $f'(x) = x^{-1}\sum_{k=0}^{\infty} kx^k = 1/(1-x)^2$. We may express the concentration C in terms of f'. Under the conditions where long worms are prevalent, aggregates smaller than K_0 are negligible and we can write

$$C \simeq \sum_{K=K_0}^{\infty} KC_0([1]/C_0)^K = C_0 x^{K_0} \sum_{K=K_0}^{\infty} Kx^{K-K_0}, \qquad (7.2)$$

where $x \equiv [1]/C_0$. Switching to a new sum index $K' \equiv K - K_0$, this amounts to

$$C \simeq C_0 x^{K_0} \sum_{K'=0}^{\infty} (K'+K_0)x^{K'} = C_0 x^{K_0}(xf'(x) + K_0 f(x)).$$

We focus on the regime where large micelles are dominant, so that $\delta \equiv 1 - x \ll 1$. Then $f(x) \to \delta^{-1}$, $f'(x) \to \delta^{-2}$, and $C \to C_0/\delta^2$. In the same way we can find the average micelle size $\langle K \rangle \equiv \sum_K K[K]/\sum_K [K]$. The numerator is C, while the denominator is $C_0 f(x) \to C_0/\delta$. Combining, $\langle K \rangle \to 1/\delta$.

Combining this expression for $\langle K \rangle$ with the one for C above, we conclude that

$$\langle K \rangle \to (C/C_0)^{1/2}. \qquad (7.3)$$

In terms of the concentration of *micelles* $C_p \equiv C/\langle K \rangle$, this says $\langle K \rangle \to C_p/C_0$. Thus the micelle size can be controlled and predicted over a wide range of concentration based on fundamental principles.

This prediction of how concentration affects micelle size can readily be tested by scattering experiments. For example, static scattering measures the **weight-averaged** micelle size $\langle K \rangle_w \equiv \sum K^2[K]/(\sum K[K])$. All such averages scale with the same $\frac{1}{2}$ power of overall concentration $C^\dagger$. Thus

---

† We have seen that C only affects the concentrations $[K]$ via a scale factor. That is, $[K] = p(C)g(K/q(C))$. When $q(C) \gg 1$, the dominant $[K]$'s must occur for $K \gg 1$. Thus

$$\sum_K K^n[K] \simeq \int dK\, K^n p(C)g(K/q(C)).$$

Defining the variable $\tilde{K} \equiv K/q(C)$, this gives

$$\sum_K K^n[K] \simeq q(C)^{n+1} p(C) \int d\tilde{K}\, \tilde{K}^n g(\tilde{K}).$$

The integral is now a constant, independent of C. (The constant is finite, since our distributions $g(x)$ fall off faster than any power for large x.) Now, any moment ratio M_{mn} of the $[K]$ distribution has the form

$$M_{mn} = \sum_K K^n[K]/\sum_K K^m[K].$$

Our scaling law for the sums shows that $M_{mn} \simeq q(C)^{n-m}$. That is, all moments behave as though they had dimensions of $q(C)^{n-m}$. The averages discussed in the text are moments of this form: $\langle K \rangle = M_{0,1}$ and $\langle K \rangle_w = M_{1,2}$. They both grow with the same power of $q(C)$, so that their ratio remains finite as either goes to infinity.

one may verify the square root law by measuring scattering intensity versus concentration.

If the limiting monomer concentration C_0 is small, we can create solutions with very long worms. As these grow longer, they curve more and more, eventually taking the form of random walks like the polymers studied in Chapters 3 and 4. In our study of these polymers we learned that the interaction of the chain with itself affects its size in a qualitative way. Repulsive interactions lead to the fractal properties of self-avoiding walks rather than those of simple random walks. Repulsive interactions between parts of a wormlike micelle must ultimately have the same effect. In particular, the work W needed to add a monomer chain now depends on the chain size K. Expressed in terms of the equilibrium constant for K-mers assembled from monomers, the K dependence has the form

$$\frac{[K]}{[1]^K} = \frac{1}{\tilde{C}_0^{K-1}} K^{\gamma-1} \tag{7.4}$$

The $K^{\gamma-1}$ factor arises from the self repulsion, and the constant $\tilde{C}_0$ plays the role of C_0. Here γ is an universal exponent like the fractal dimension D characteristic of any self-repelling polymer [6]. The value of γ is about 1.17 [6]. This factor alters the dependence of $\langle K \rangle$ on concentration, as we may deduce by repeating the reasoning above. We first note that $\langle K \rangle \sim 1/\delta$ as before. However, the overall concentration C has a different scaling. For large $\langle K \rangle$ we may follow the scaling argument of the footnote to infer

$$C \rightarrow \text{constant } \delta^{-1-\gamma}$$

Combining, we find a new dependence of $\langle K \rangle$ on C:

$$\langle K \rangle \rightarrow \text{constant } C^{1/(1+\gamma)}. \tag{7.5}$$

Now $\langle K \rangle$ varies more weakly with concentration: the power is reduced by about eight percent owing to the repulsion. Micelle interaction also influences $\langle K \rangle$, as discussed below.

Wormlike micelles have internal degrees of freedom, as the above discussion implies. A given section of the micelle bends due to thermal fluctuations. Longer sections bend through larger angles. Thus there is some micelle length such that the mean squared bending angle is of order unity— e.g., one radian. This length is called a **persistence length** and denoted ℓ_p. We encountered the notion of persistence length in the discussion of charged polymers in Chapter 3. The persistence length has been measured by two methods in cetyl pyridinium bromide, a close cousin of CTAB in water with 0.8 M sodium bromide salt added [7]. At a concentration of 6 mM surfactant and a temperature of 35°C, the persistence length was found to be about 20 nm. The measured cylinder radius of the micelles was 2 nm. In addition to this prosaic type of internal freedom, wormlike micelles can *branch*. That is, three sections of wormlike micelle can join together. The junction region requires distortion from the normal cylindrical arrangement of the surfactants; thus the energy of a surfactant in the junction region is larger than it is elsewhere in the micelle. Nevertheless, there is a nonzero fraction of junctions and branches in equilibrium with the ordinary micelles. In favorable cases this fraction is large enough to be observable. When the micelles are above their overlap concentration, the branches act to crosslink different micelles to form a network structure. Such structures have been observed and quantitatively explained [8]. Certain surfactants *prefer* the type of packing found at a branch junction. These naturally form branches more easily.

7.4.2 Open aggregation: two-dimensional micelles

Sometimes surfactants have a preferred aggregation in the form of bilayers of surfactant. That is, the surfactants form micelles in the form of two-dimensional surfaces, as shown in Fig. 7.2. In principle we can extend the same reasoning used above to predict how the micelle mass and size depend on concentration. But in practice this is less worthwhile than for the wormlike micelles treated above. One reason is that the energy to form a two-dimensional micelle is not so straightforward as for a one-dimensional micelle. Now the role of the endcap energy is played by the perimeter or boundary energy. To know this, we must know the shape of the micelle: flat, bowl-like, or saddle-like. This additional complication makes two-dimensional aggregation much different from one-dimensional aggregation. Less can be predicted with generality. In any case, it appears that when two-dimensional micelles become large, the resulting properties can be predicted without knowing their size or size distribution in detail [9].

7.4.3 Aggregation kinetics

It takes time for micelles to form and to adjust to their environment. Generally these adjustments require surfactant molecules to pass through energetically costly configurations. This leads to barrier-dominated time dependence of the kind discussed in the Chapter 2. One can detect the time for these readjustments by a so-called T-jump experiment: one suddenly gives the solution a small pulse of heat and then monitors a property that reflects the state of the micelles, such as light scattering [5]. Typically, the solution relaxes exponentially to its new state. One generally observes two characteristic relaxation times. To illustrate these, let us imagine giving a 10° heat pulse to a solution of SDS above its critical micelle concentration. The new temperature alters the free energies of the variously sized micelles. Typically the mean aggregation number is reduced. Surfactants must leave micelles on the average to attain the new equilibrium distribution of sizes. This requires a time of the order of microseconds for small surfactants like SDS [5]. This time is called τ_1. But this exchange of surfactants is not sufficient to attain full equilibrium. The total number of micelles must in general change as well. Typically on heating, the number of micelles must increase by some factor larger than 1[†]. Thus new micelles must nucleate, and grow from aggregation number 1 to the preferred aggregation number of about 64. The necessary alteration of the system to assemble so many

[†] Since virtually all of the surfactants live in the micelles and their average size is reduced, more micelles are required to accommodate the given amount of surfactant.

molecules is great and the energy barrier is high. Thus the micelle relaxation time—denoted τ_2—is many times longer than the surfactant relaxation time τ_1. For SDS in typical conditions it is on the scale of milliseconds. It may be much longer in very pure solution where there are few impurity molecules to aid the nucleation of micelles [5].

These two characteristic times tell only part of the story of the dynamics of micelles. In the rheology section below we shall encounter further forms of relaxation that become important for interacting micelles.

7.5 Micelle interaction

When micelles form, they are often in such a dilute state that their interaction is negligible. But as their concentration increases, interactions must become important. Spherical micelles interact like a colloidal suspension: as the volume fraction of spheres rises above 10–20%, enhanced osmotic pressure and viscosity become apparent. One-dimensional, wormlike micelles reach interpenetrating, semidilute conditions, just like polymer solutions. Their osmotic pressure, cooperative diffusion, and correlation lengths can be understood by adapting the methods developed for polymers and discussed in Chapter 4.

The semidilute environment gives rise to a new effect in wormlike micelles not seen with ordinary polymers. It alters the distribution of worm lengths $[K]$. As discussed in Chapter 4, the effects of self-repulsion extend only to a correlation distance ξ, since the solution is uniform beyond such distances. These local repulsions also influence the concentrations $[K]$. We first consider micelles whose geometric size is ξ or less. Such micelles have a monomer number no greater than some g. These "blob micelles" feel the full self repulsion, so that $[g]$ follows Eq. (7.4) above:

$$[g] = \tilde{C}_0([1]/\tilde{C}_0)^g g^{\gamma-1}. \qquad (7.6)$$

Typical micelles in this semidilute solution have sizes $K \gg g$. We may infer their concentrations $[K]$ by considering the equilibrium $(K) + (g) \leftrightarrow (K + g)$. The work to assemble the $K + g$ micelle from its K and g pieces includes the work done against mutual repulsion. But this repulsion is only effective within a distance ξ, which is independent of K. Thus the work itself must be independent of K. For this reason we can find the equilibrium constant using

$$\frac{[K+g]}{[K][g]} \simeq \frac{[g+g]}{[g][g]}.$$

The micelles on the right are subject to the effects of repulsion and thus obey Eq. (7.6)

$$\frac{[g+g]}{[g][g]} \simeq \tilde{C}_0^{-1} 2^{\gamma-1} g^{1-\gamma}.$$

The repulsion suppresses the connection of the two pieces by a factor $g^{1-\gamma}$.

Repeating the reasoning used above for non-repelling micelles, we can deduce $[K]$ whenever K is an integer multiple of g:

$$[K] = \frac{[K]}{[K-g]}\frac{[K-g]}{[K-2g]}\cdots[g] = ([2g]/[g])^{(K/g)-1}[g].$$

Using Eq. (7.6), this can be written as,

$$[K] \simeq ([g]g^{1-\gamma}/\tilde{C}_0)^{(K/g)-1}[g]$$
$$= \tilde{C}_0 g^{\gamma-1}(([g]g^{1-\gamma}/\tilde{C}_0)^{1/g})^K = \tilde{C}_0 g^{\gamma-1}x^K. \qquad (7.7)$$

Here x is defined as the contents of the large $(\ldots)$. The repulsion has added a new feature: the g-dependent prefactor. Again $[K]$ falls off exponentially with the controlling parameter x, as it did without repulsion. The only change is to replace C_0 by $\tilde{C}_0 g^{\gamma-1}$. This $[K]$ varies smoothly with x, so it is approximately valid even when K is not an integer multiple of g. Repeating the reasoning after Eq. (7.2), we find

$$\langle K \rangle \to (Cg^{1-\gamma}/\tilde{C}_0)^{1/2}.$$

This $\langle K \rangle$ has a new concentration dependence, since g depends on concentration. Since g is essentially the micelle mass whose geometric size is ξ, we may use the semidilute polymer in good solvent from Chapter 4 to infer $g \sim \xi^D \sim C^{D/(D-3)}$. Thus [7]

$$\langle K \rangle \sim C^y, \quad \text{where } y = \tfrac{1}{2}[1 + (\gamma - 1)D/(3 - D)] \simeq 0.6. \qquad (7.8)$$

The self-interaction has changed the growth exponent from 0.5 to 0.6. This subtle change in how micelles grow with concentration will prove important when we consider the viscosity of these solutions below.

7.2. *Worm interpenetration* As noted above, the growth of wormlike micelles with concentration leads to interpenetration.

(a) How does the overlap ratio ϕ/ϕ^* increase with concentration, assuming (i) self-avoidance effects are negligible so that $D = 2$, and (ii) self-avoidance effects are important, so that $D = \frac{5}{3}$?

(b) Suppose that the cylindrical micelles have a radius of 2 nm and that each surfactant occupies a volume of 0.4 nm^3. Suppose that the persistence length is 20 nm, so that the persistent segments are long and slender. For significant entanglement effects to occur, one should have $\phi/\phi^* > 10$. But if the volume fraction of worms exceeds 10–20%, their packing leads to nematic order as treated in Problem 5.10. Then the resemblance to a polymer liquid diminishes. If nematic order to be avoided, $\phi < 0.1$, so that ϕ^* must be less than 0.01. Find the condition on the characteristic concentrations C_0 or $\tilde{C}_0$ that permits this much entanglement. What is the corresponding volume fraction of free monomers?

Wormlike micelles often have substantial rigidity, as noted above: they remain straight over a length ℓ_p that is several times their width. Thus at sufficiently high concentration, the micelles may align to form a nematic liquid crystal of the kind discussed at the end of Chapter 5. To understand what concentration is necessary, it is useful to think of the micelles as being

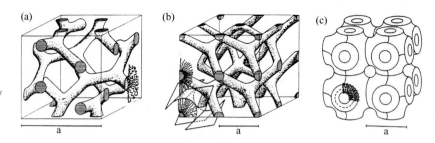

(a)   (b)   (c)

a   a   a

Fig. 7.3
Sketch of three periodic bicontinuous micellar structures discovered in concentrated surfactant solutions, courtesy John Seddon [12] with permission from Elsevier. (c) is known as the plumber's nightmare [11].

cut up into straight rods of length ℓ_p. Then one may adopt the reasoning used in Chapter 5 section on organized states.

Surfactants whose shape favors branch junctions can at sufficiently high concentration form structures that are branched *everywhere*. Such structures resemble a regular lattice of connected cylinders. The simplest such structure is a simple cubic lattice resembling a tangle of pipes, known as the **plumber's nightmare**, as shown in Fig. 7.3. For example, 60 volume percent glycerol monoolein (a C_{18} hydrocarbon chain with polar OH and $C{=}O$ groups at one end) in water is believed to form this phase [10]. Many similar periodic network structures have been discovered. These are called **Luzzati phases** [11].

7.5.1 Energy of two-dimensional micelles

As we have seen, surfactants often aggregate in the form of bilayers, so that the micelles are two-dimensional surfaces. To understand their form and their interactions, we must understand their fluctuations. Like random-walk polymers, surfactant interfaces have important thermal fluctuations. The first step in accounting for these fluctuations is to understand the energy cost of deforming the layer. First we consider the energy of stretching the layer and changing its area. For ordinary interfaces, this is the dominant form of energy, as we saw in Chapter 6. But this stretching energy is unimportant for the micelles we are considering here. Since the micelles are in equilibrium with excess surfactant, the number of surfactants per unit area saturates to a value that is set by the solution conditions (a typical density for common surfactants like SDS is 0.25 nm^2 per polar head). At such high densities, the energy to alter the surface density appreciably is large, and it is reasonable to suppose that the amount of surfactant per unit area is fixed. Thus the only important fluctuations are those involving curvature of the surface.

Our bilayer consists of two monolayers or **leaflets**. It will be useful to consider each leaflet individually. At each point on a smooth surface there are two perpendicular directions in which a straight line on the surface makes a planar arc in space. Their radii of curvature R_1 and R_2 orientation define the state of curvature at that point, as shown in Fig. 7.4. The **principal curvatures** C_1 and C_2 are defined as the inverses of these radii. Clearly zero curvatures mean a flat surface. It is convenient to define two scalar quantities that specify the two curvatures. The **mean curvature** is defined as $\frac{1}{2}(C_1 + C_2)$, and we shall denote it by c. The **Gaussian curvature** is $C_1 C_2$. We note that these two curvatures have different dimensions.

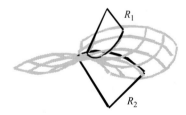

R_1

R_2

Fig. 7.4
A small section of a curved surface, showing the two principal radii of curvature R_1 and R_2. The two principal curvatures have opposite signs, so that the Gaussian curvature is negative.

We may describe how the surface deforms in thermal equilibrium once
we know the energy. Nonzero curvature costs energy. If the curvature is
gentle—so that R_1 and R_2 are much smaller than the molecular thickness
of the layer—then we can express the energy as a Taylor expansion:

$$E = \int ds[AC_1 + BC_2 + DC_1^2 + FC_2^2 + \bar{\kappa}C_1C_2 + \cdots].$$

We can simplify this general form if the surface is isotropic:

$$E = \int ds[2\kappa[c - c_0]^2 + \bar{\kappa}C_1C_2]. \tag{7.9}$$

Here κ, $\bar{\kappa}$, and c_0 are material constants of the interface. The constant κ is
called the **bending modulus** or **bending stiffness**. The κ energy is minimal
when the mean curvature has the value c_0; this c_0 is known as the **spontaneous curvature**. The term in $\bar{\kappa}$ is not relevant for our purposes. Thermal
fluctuations do not affect this energy. The reason is a geometric fact called
the **Gauss–Bonnet Theorem**. It says that $\int dsC_1C_2$ over any smooth surface with a fixed boundary is unaffected by continuous deformation of the
surface. That is, the integral is a **topological invariant**. If the deformation
increases the Gaussian curvature in one place, it must decrease it in another
so as to compensate [12]. Thus compensation is illustrated in Fig. 7.5 and
discussed in the Appendix.

Knowing the energetic cost of bending, we can now analyze the fluctuating shapes that occur in equilibrium. First we consider sections of the
surface that are small enough to be nearly flat. This must be true for small
enough sections; otherwise we would not be able to characterize our micelle
as a smooth, two-dimensional surface. For such a nearly flat surface, there
is a convenient way to describe the fluctuating shapes. We represent the
surface by giving its local height h above some reference plane (x, y), that
makes only a small angle with the surface everywhere. The smallness of
these angles means that $\partial h/\partial x$ and $\partial h/\partial y$ are much smaller than unity. This
$h(x, y)$ scheme is known as the **Monge representation**.

The energy of Eq. (7.9) can be readily expressed in the Monge representation. Indeed, the mean curvature is simply $\frac{1}{2}\nabla^2 h$, so that $E = \frac{1}{2}\int dx\, dy\, \kappa|\nabla^2 h|^{2\dagger}$. Noting that the energy density is invariant under translation, we may simplify by expressing h as a sum of plane waves. We define
$\tilde{h}(q)$ via

$$h(x, y) = L^{-2} \sum_q \tilde{h}(q)e^{iq\cdot r}. \tag{7.10}$$

The wave vectors q are a complete set of waves needed to represent functions
on this square region. Then

$$E = \frac{1}{2}\kappa L^{-2} \sum_q (q^2)^2|\tilde{h}_q|^2. \tag{7.11}$$

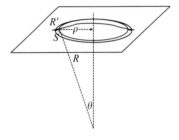

Fig. 7.5
Bump construction on a flat surface used
to demonstrate the Gauss–Bonnet theorem.
A flat surface was deformed to make the
circular bump shown, with positive
Gaussian curvature at the middle. In order
to rejoin the flat surface smoothly, the
surface must have negative Gaussian
curvature at the rim. The Appendix shows
that the Gaussian curvature averages
to zero.

† This is the energy we encountered in
Problem 5.4, with ψ corresponding to h.

The energy is a sum of normal plane-wave modes. Each fluctuates independently. Because their energy is quadratic, each has an average energy of $\frac{1}{2}T$ by virtue of the equipartition theorem of Chapter 2. From this fact, we can infer the average squared amplitude of the surface: $\langle|\tilde{h}_q|^2\rangle = TL^2/(\kappa(q^2)^2)$.

Using the behavior of $\langle|\tilde{h}_q|^2\rangle$ we can understand how the typical heights grow with lateral size of the surface. We consider a square of micellar surface of length L. Using Eq. (7.10), the mean square height relative to its average is given by

$$\langle h^2\rangle_L = L^{-4}\sum_q\langle|\tilde{h}_q|^2\rangle = \left(\frac{T}{\kappa}\right)L^{-2}\sum_q\left(\frac{1}{q^2}\right)^2.$$

The summand falls off rapidly with increasing q; evidently it is dominated by the lowest q's in the sum: $\sum_q 1/(q^2)^2 \simeq 1/q_{min}^4$, where q_{min} is the smallest nonzero q in the sum. For our patch of size L, this q_{min} is $2\pi/L$. Thus

$$\langle h^2\rangle_L \simeq (T/\kappa)L^2. \tag{7.12}$$

The height of the patch is scales linearly with its width.

A second important consequence is to see how the *orientation* of the surface fluctuates on the patch. The normal to the surface has a vertical component given by $(1-|\nabla h|^2)^{1/2} = (1-\sum_q q^2|h_q|^2)^{1/2}$. The typical slope μ of the surface is given by $\mu^2 = \langle|\vec{\nabla}h|^2\rangle = L^{-4}\sum_q q^2\langle|\tilde{h}(q)|^2\rangle$. Since the thermal average of $|\tilde{h}(q)|^2$ falls off as $(q^2)^{-2}$, we infer

$$\mu^2 = L^{-4}\sum_q q^2\langle|\tilde{h}(q)|^2\rangle = \frac{T}{\kappa}L^{-2}\sum_q q^{-2}.$$

This sum falls off only slowly with increasing q; it can be approximated as an integral, using $\sum_q \rightarrow (L/2\pi)^2\int d^2q$:

$$\mu^2 = \frac{T}{\kappa}L^{-2}\left(\frac{L}{2\pi}\right)^2\int_{q_{min}}^{q_{max}}\frac{d^2q}{q^2} \simeq \frac{T}{\kappa}\log\left(\frac{q_{max}}{q_{min}}\right). \tag{7.13}$$

We recall that q_{min} is of order L^{-1}. As for q_{max}, $1/q_{max}$ is the smallest wavelength at which significant fluctuations occur. This smallest wavelength is roughly the thickness b of the membrane. Waves with length shorter than the thickness are not adequately treated by the thin-membrane formula (7.9). Such waves cost more energy than this formula suggests, and their effect can be neglected. We conclude that the typical local slope of a patch of surface of size L is of order $(T/\kappa \log(L/b))^{1/2}$—it increases logarithmically with the patch size. If the patch size is small, the slope is smaller than unity. (Otherwise there would be no scale on which the surface was smooth.) As one enlarges the patch, there comes point, $L \equiv \ell_p$ where the typical slope becomes unity. For patches larger than this, our Monge approximation breaks down. Still, we expect the slopes to continue to grow with L. This means that for patches much larger than ℓ_p, two arbitrarily chosen points on the membrane will typically have completely independent orientations. Thus the orientations become uncorrelated over distances longer than ℓ_p. Thus ℓ_p represents a **persistence length** like the persistence length encountered above for wormlike micelles. The orientation at a given point *persists* out to distances of order ℓ_p. Since $1 = T/\kappa \log(\ell_p/b)$, we conclude

$$\ell_p \simeq b \exp(\kappa/T). \tag{7.14}$$

The persistence length increases exponentially if we increase the bending stiffness or reduce the temperature. The significance of the persistence length was first discussed by DeGennes and Taupin in the 1980s [13].

7.5.2   Energy to confine a fluctuating membrane

These membrane fluctuations have an energetic effect as well as the structural effects described above. Often these surfactant membranes are spaced closely enough so that they interfere with each other. In such situations a given membrane cannot fluctuate freely without hitting other membranes. The fluctuations are increasingly suppressed as the density of membranes grows. This suppression requires work. Thus the fluctuations induce a repulsive interaction between the membranes.

To understand the interaction quantitatively, we consider the simplified case of a single large fluctuating membrane confined between two parallel walls spaced by a distance d much smaller than the persistence length ℓ_p discussed above. Naturally constraints from the walls occur where the membrane comes in contact with a wall. At such a point, the membrane is obliged to curve away from the wall rather than fluctuating freely. A finite fraction of the possible states of curvature are excluded at the point of contact. The probability that the unconstrained system would happen to satisfy this constraint is thus some finite fraction. As we have often seen above, imposing such a constraint requires work of order T.

In a large membrane there are many points of contact, each of which carries a free energy cost of order T. To find the net work needed to bring the walls to the distance d, we need to know the number of contacts per unit area. Figure 7.6 illustrates the situation. Let us consider a patch of size L around some contact point whose width is a few times the thickness b of the membrane. As we saw above, typical places on the patch have a slope μ of order $(T/\kappa \log(L/b))^{1/2\dagger}$. Since the patch size is much smaller than the persistence length ℓ_p, this slope is much smaller than unity. As we saw above, the typical slope depends only weakly on the size of the patch. Given such a slope, we must anticipate that in a distance x of order d/μ the membrane will encounter the opposite wall, leading to another free energy penalty of order T. The density of such contacts is roughly x^{-2}, or μ^2/d^2. This leads us to expect an energy per unit area of order $T(T/\kappa)d^{-2}$. One can calculate this energy explicitly using the Monge representation. The

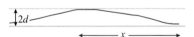

Fig. 7.6
Section of a fluctuating membrane trapped between two walls. The spacing d and the contact separation x are as defined in the text.

† To see this more simply, we note that the energy of our patch of width L is roughly $\kappa L^2 c^2$, where c is the mean curvature. Noting that this energy is of order T, we infer that $(cL)^2 \simeq (T/\kappa)$. But cL is just the difference in slope accumulated as one traverses the patch. Thus the typical slope μ near the patch is of order $(T/\kappa)^{1/2}$, as claimed.

energy ΔE per membrane of area L^2 is given by [14]

$$\Delta E = (3/\pi^2)\,(T^2/\kappa)\,(L/d)^2. \tag{7.15}$$

Interactions between adjacent fluctuating membranes evidently have the same functional form. The energy falls off as the inverse square of the distance, like the van der Waals interaction between surfaces. This universal repulsion caused by fluctuating membranes was first identified by Wolfgang Helfrich [14].

The Helfrich repulsion evidently becomes important when two-dimensional micelles become concentrated so that the distance between them is smaller than about the persistence length ℓ_p. If the micelles are charged, this adds a further source of repulsion. At such concentrations the micelles align into parallel layers or **lamellae** in order to avoid each other. The mutually oriented lamellae give an overall anisotropy to the fluid: it becomes a **smectic liquid crystal**. We encountered such liquid crystals in Chapter 5. This type of order shows up readily in scattering: the periodic alternation of solvent and surfactant in space creates diffraction peaks like those in Fig. 2.7. Knowledge of the power-law form of the Helfrich repulsion allows one to understand the distinctive power-law falloff of the scattering intensity away from the peak. This is explained in the paper from which Fig. 2.7 is taken.

7.6 Mixing immiscible liquids: microemulsions

Up till now we have studied the behavior of surfactants in a single solvent species A such as oil. In this section we ask what happens when we add a second immiscible species B such as water. Before adding any B, our surfactants are assumed to be well above the critical micelle concentration. We now specify the nature of the micelles. We shall suppose that the micelles are spherical and that the surfactants pack together with a fixed area per surfactant. Thus in adding a given amount of surfactant, we have added a specific amount of interfacial area, Σ. It will prove convenient to measure the amount of surfactant in terms of this area. We anticipate that adding solvent B may cause some B to enter the micelles. A micelle acquiring B solvent will grow in size and thus its (mean) curvature c will decrease, as indicated in Fig. 7.7. We suppose that the energy of the surfactant layer is described by a bending energy κ and a spontaneous curvature c_0, as in the last section. $E = 2\kappa \int ds(c - c_0)^2$. The curvature c_i of the micelles is generally bigger than c_0; we shall suppose that it is *much* bigger.

We now add a drop of B solvent and ask what happens. If the surfactant were absent, this droplet would keep its integrity and not mix (appreciably) with A. But with the micelles present, things are different. Let us transfer a volume V of B solvent into the centers of the micelles, and note the work required. The B solvent is now in the form of spheres surrounded by a surfactant layer. Since the area per surfactant is fixed, the total area of these layers remains at its initial value Σ. Thus the typical radius R of the water spheres is given by $V/\Sigma = \frac{4}{3}\pi R^3/(4\pi R^2) = R/3$. The curvature c of the surfactant layer has now decreased to $1/R^\dagger$.

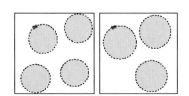

Fig. 7.7
A micellar solution swollen with water. Decreasing the curvature of the droplets with fixed total area necessarily decreases the number of droplets and increases the enclosed volume. Here the mean radius has increased by $\left(\frac{4}{3}\right)^{1/2}$, so that the area per droplet has increased by a factor $\frac{4}{3}$. Thus the surface area in the original four droplets is only enough to cover three. The volume of water has changed by a factor $\left(\frac{3}{4}\right) \times \left(\frac{4}{3}\right)^{3/2}$, i.e., $\left(\frac{4}{3}\right)^{1/2}$.

† We shall suppose that the volume V is large enough so that R is much larger than the original micelle size. Then the thickness of the surfactant layer itself is negligible compared to R.

7.3. *Size variability of micelles*   In our example we supposed that the solvent containing micelles all had the same radius R. To examine this supposition, consider the energetic cost of changing this radius. Assume that the micelle radius $R = 1/c_0$. For a given bending stiffness κ, how much may R decrease from $1/c_0$ before the bending energy E_κ increases by T?

Our surfactant interface has reduced its curvature towards the spontaneous curvature c_0. This has reduced the energy of our system. $E_\kappa(V) = 2\kappa \Sigma (\Sigma/(3V) - c_0)^2$. As long as the micelles have a radius much smaller than $1/c_0$, this energy decreases as the amount V of included solvent increases:

$$E_\kappa(V) \simeq 2\kappa \Sigma^3/(9V^2).$$

This makes a negative contribution to the work required. But there are other contributions to consider. One such contribution is the interaction energy of the transferred B molecules. Most of the B molecules transferred remain surrounded by B, so that they feel no change in interaction energy. Only the ones adjacent to the surfactant layer experience a change; the associated work is simply proportional to the area Σ, and is independent of V. We shall return to this contribution below. There is a further form of energy to be noted. Adding the B solvent reduces the number of micelles. This number N_m is evidently $\Sigma/(4\pi R^2)$, or $\Sigma^3/(36\pi V^2)$. This reduction in the number of micelles gives the solution fewer degrees of freedom. As we saw in the section on mixing principles, this requires a work $T N_m \log(v) + \text{const}$ where v is the volume fraction per micelle[†]. In terms of the solvent volume V, this amounts to a free energy cost E_T:

$$E_T \simeq (T\Sigma^3/36\pi V^{-2}) (\log v + \text{const}).$$

[†] In more detail, each small increment of micelle number δN_m costs a work $\delta W = T\delta N_m \log(\Omega/N_m)$, where Ω is the volume of the system in units of the micelle volume. To find the total work, one adds (integrates) the work δW for each increment of micelle number. The resulting integral simplifies to $T N_m \log(v) + \text{const}$ in the regime where the volume fraction of micelles is small.

The two opposing contributions to the work have similar dependences on the volume V; both vary as V^{-2}. E_T is proportional to T while E_κ is proportional to κ. If κ is sufficiently larger than T, the gain of curvature energy dominates, and the micelles swell with solvent. Concretely, if the volume fraction v is 10^{-3} or more, the micelles swell if $\kappa \gtrsim T/3.6$. Well-defined surfactant interfaces have bending moduli greater than T, as noted above. Thus there is normally a strong tendency to swell. The swelling continues until the micelles grow comparable to their preferred size $1/c_0$. They do not grow larger than this size, since then both the bending energy and the reduction of the micelle number oppose further swelling.

This simple example shows how surfactants lead to the dispersion of one solvent into another immiscible one. This equilibrium dispersed state is known as a **microemulsion**. One only needs surfactants that disperse in the host solvent and prefer to be less curved than they are in the form of bare micelles. Our analysis suggests that under these conditions a large volume fraction of solvent can be dispersed, with an indefinitely small amount of surfactant, simply by selecting a surfactant with a large κ and a small c_0. The limitations on ability to disperse arise from the interaction between the swollen micelles. Large micelles interact with each other in the same

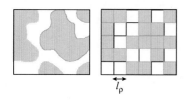

Fig. 7.8

Left: Schematic sketch of a symmetric, bicontinuous microemulsion. Right: cellular model of the microemulsion according to DeGennes and Taupin [13]. The persistence length ℓ_p is shown.

[†] A well-known Winsor microemulsion consists of SDS, brine, and toluene, with a small amount of butanol. The SDS has a concentration of 6–10 g/l, the butanol has a weight fraction of 3%, the brine contains 6.5 weight percent salt, and the toluene and water are in equal volumes within a factor 2.

ways as the large colloidal particles discussed in Chapter 5. The larger they become, the larger their attractive interaction. Beyond some given size the attraction leads the micelles to precipitate into their own micelle-rich phase rather than dispersing [15].

In our example we assumed that κ was large enough that the thermal fluctuations of the surfactant interface would be small. This meant that the micelles had to be smaller than the persistence length ℓ_p. If R becomes larger than ℓ_p then the micelles fluctuate strongly in shape. If the distance between micelles is comparable to R, the micelles can even merge to make larger irregular objects. A limiting case of these fluctuations is the situation where $c_0 = 0$, the volumes of A and B solvent are equal, and the radius $R \simeq \ell_p$. Then the distinction between the inside and the outside of the droplets disappears. The microemulsion can then be viewed as a collection of cells of size ℓ_p, each containing randomly A or B solvent, with a surfactant membrane between cells of different solvent, as shown in Fig. 7.8. Such **bicontinuous** microemulsions were first discussed in these terms by DeGennes and Taupin [13]. Since that time, the so-called Winsor III microemulsions [16] known to chemical engineers were identified as bicontinuous microemulsions [17][†] (*cf.* Fig. 1.2).

The larger the typical size of the microemulsion droplets, the smaller the surfactant fraction becomes. One may ask how small the fraction of surfactant can become. Is there a limit to how little surfactant is needed in order to mix oil and water in an equilibrium dispersion? Many Winsor microemulsions achieve a surfactant volume fraction of as low as 10^{-4} or 10^{-5}. The typical size $\hat{R}$ of the oil or water volumes appears in scattering experiments as a peak at some $q^* \simeq 1/\hat{R}$, as shown in Fig. 7.10. Such measurements indicate typical sizes of the order of 100 nm [15]. But it is difficult in practice to increase $\hat{R}$ much beyond this scale, or to reduce the amount of surfactant below this level. To increase $\hat{R}$ further requires that the bending stiffness κ be increased so that the persistence length ℓ_p increases. But in addition, the spontaneous curvature c_0 must decrease so that $c_0 \lesssim 1/\hat{R}$. Thus $1/c_0$ must become much larger than the surfactant layer thickness b. However, slight changes in the A or B solvents can make small changes in c_0 and thus reduce $\hat{R}$ significantly.

There is a great incentive to create stable microemulsions in order to mix immiscible fluids. Accordingly, the structure and phase stability of microemulsions is extensively studied. The simple scheme of random water and oil volumes sketched above has given way to much more systematic and powerful analyses [19]. A major focus of modern theories is to explain the occurrence and the coexistence of phases in the oil–water–surfactant systems. Figure 7.9 shows a typical phase diagram, indicating the multiple phases that can occur and the involuted boundaries separating them.

All of this discussion presupposes that the solvent B enters the bare micelles. In order to enter, the first B molecules must create a surfactant-B interface and pay any associated interfacial energy $\alpha \Sigma$, as noted above. If this energy price is higher than the gain in curvature energy discussed above, the energy to absorb the next B is unfavorable. For a favorable balance,

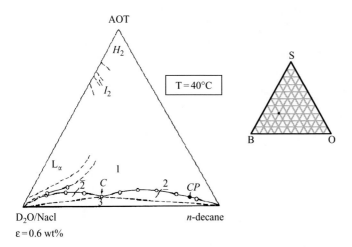

Fig. 7.9
Gibbs phase diagram for AOT, decane oil, saltwater system from [20, figure 2], © American Institute of Physics, courtesy R. Strey. The saltwater or brine has a 0.6% weight fraction percent of salt and is fully deuterated. The temperature is 40 °C. Each point on the diagram represents a particular composition. Top vertex indicates pure AOT surfactant, and the other vertices are pure brine and pure oil. Percentage of each component is the fractional distance from the opposite edge. As an illustration, the dot on the legend at the right indicates the composition with (weight) ratio 20 : 50 : 30 of oil to brine to surfactant. The region marked H_2 consists of densely packed cylinders of AOT. The phase I_2 is another surfactant-rich phase. L_α labels a lamellar phase. The regions indicated by 2 have coexisting compositions. The triangular bottom region has three coexisting compositions. One of these is the microemulsion composition indicated by C. In the region marked 3 this C composition coexists with nearly pure oil and nearly pure water. The two eyebrow-shaped lines are the upper boundary of the two phase regions. On the right-hand boundary, the black dot marked "cp" indicates a critical point where the two coexisting compositions coincide. The two open circles on either side of "cp" are two other coexisting phases.

we need $\alpha \Sigma \lesssim \kappa \Sigma c_i^2$, or $\alpha \lesssim \kappa c_i^2$. The bare micelle radius $1/c_i$ is of the order of the surfactant layer thickness b. As we have seen, the bending modulus κ is ordinarily of order T. Thus a favorable swelling energy implies $\alpha \lesssim T/b^2$. This is a rather strong requirement on the interfacial tension α. Typical interfaces have energy of the order T for every translational degree of freedom at the interface. For our surfactants this amounts to T for each surfactant head or tail area projected onto the interface. This area is normally smaller than the surfactant length b squared. Thus we need the B-surfactant energy to be small on the scale of typical interfacial energies in order for the micelles to take up the B solvent. This fact explains why one cannot readily make a microemulsion where solvent B is air, for example. No surfactant has a part that dissolves in air; thus the interface between surfactants and air is not small.

7.6.1 Interfacial tension

The spontaneous mixing seen in microemulsions leads to very low surface tensions. To understand this, we consider the interface of the simple spherical microemulsion discussed in Fig. 7.11. The system has come to equilibrium, so that the micelles have their preferred radius $1/c_0$. Here the amount of surfactant is limited, so the volume fraction of micelles is moderate. The interface between the microemulsion and the excess B solvent is naturally coated with surfactant. The interfacial tension is the work required to add a unit of area to this interface. We may expand this interface by an area $4\pi R^2$ by transferring a micelle from the microemulsion to the interface. This transfer requires the surfactant to depart from its preferred curvature $1/c_0$ and become flat, with an associated energy cost $4\pi R^2 \kappa c_0^2$ or simply $4\pi \kappa$. Removing the free micelle also reduces the number of translational configurations of the system, with a free energy cost $T \log v$ as encountered above. Since κ is generally of order T or more, our energy cost is typically dominated by the curvature contribution. The associated interfacial tension

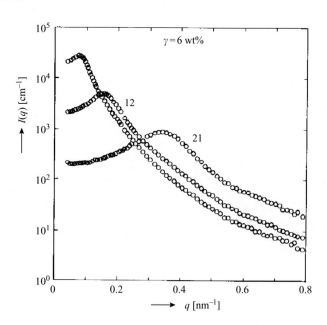

Fig. 7.10
Neutron scattering intensity for bicontinuous microemulsions, reproduced from [18, figure 17], with permission from Elsevier. The microemulsion consists of equal volumes of D_2O and n-octane. With the indicated weight percent of $C_{12}E_5$ surfactant, and closely resembles the microemulsion of Fig. 1.2. The wave vector q at the peak decreases as the amount of surfactant decreases, indicating larger droplets of oil and water.

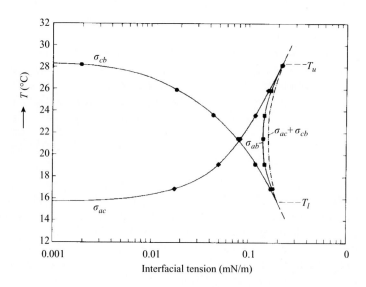

Fig. 7.11
Interfacial tension versus temperature in a bicontinuous microemulsion, from [18, figure 5b], with permission from Elsevier. Solution contains three phases, with overall composition of 58.5% water, 36.5% n-decane, and 5% C_8E_5 nonionic surfactant. As the temperature increases, the proportion of the different components changes. The three phases are denoted a for the water-rich phase, b for the oil-rich phase, and c for the microemulsion phase. The microemulsion is close in composition to that of Fig. 1.2. At the crossing point of the two curves in the middle of the figure, the microemulsion has a very low interfacial tension with both oil and water. (The very low tensions near the temperatures T_u and T_l correspond to phases that are becoming equal in composition, so a vanishing tension is to be expected.)

is thus of order κ/R^2. We recall that typical interfacial tensions are of the order T per interfacial degree of freedom, or several T per square nanometer. Our micelles have a diameter of many nanometers. Thus the interfacial tension can be orders of magnitude smaller than that of a simple-liquid interface (*cf.* Fig. 7.11).

7.6.2   Emulsions and foams

Surfactants can assist the mixing of oil and water even when they do not make a microemulsion. By assembling at the interface they can lower the interfacial tension significantly. This makes it easier to disperse e.g., water (B) in oil (A). Without the curvature effects prescribed above, the bare micelles dispersed in the A solvent need not absorb significant amounts of B solvent. (For example, the common SDS does not disperse significant amounts of oil in water.) However, by disturbing the two-phase fluid with external forces, one can achieve significant dispersion. The most common way to apply these forces is to stir the mixture, subjecting it to viscous stress. A given large droplet of B in A of radius R is distorted by the shear. This distortion is opposed by interfacial tension α that pushes the droplet back towards the shape of least area: a sphere. To simplify our thinking we may view the hydrodynamic force as exerting a tension between the two halves of the droplet. This force is the hydrodynamic stress times the area— roughly $\eta\dot{\gamma} R^2$, where η is the viscosity and $\dot{\gamma}$ is the shear rate. The opposing interfacial-tension pulls the two halves together with a force of order αR. Evidently the shear force grows faster with R than the interfacial force. For small R the interfacial force dominates and the droplet is scarcely affected by the flow. But for large R, the shear force dominates and pulls the droplet apart. Thus stirring a two-component fluid mixture at a shear rate $\dot{\gamma}$ should break it into droplets of size $R \simeq \alpha/(\eta\dot{\gamma})$. This rule, called the **Taylor formula** is roughly borne out in practice [21, 22].

Clearly anything that reduces the interfacial tension α, such as surfactants, promotes finer dispersion. The surfactants have another helpful effect. They keep the dispersed droplets, once formed, from coalescing, thus maintaining the dispersed state. Often the surfactant-coated surfaces repel each other owing to the Coulomb or steric repulsions treated in Chapter 5. Moreover, if two surfactant-coated droplets touch, they cannot merge unless many surfactant molecules reorganize. Thus it is that two adjacent surfactant-coated droplets can coalesce only on a timescale of years [23]. These effects explain the long shelf life of household emulsions like mayonnaise or cold cream.

In a foam, one of the solvents to be dispersed is air. For most purposes foams are just another type of emulsion. Often foams can be made without external stirring. Instead the foam bubbles can be made to form by starting with a saturated solution of the gas at high pressure and then suddenly lowering that pressure. The solution of gas in liquid then finds itself super-saturated, bubbles nucleate and grow, and the foam forms. This is what happens when shaving cream is released from its aerosol can or a bottle of beer is opened.

7.A   Suggested experiment: *Do not shake the soda can!*   If a can of carbonated soda is shaken rapidly before opening, there is a violent release of bubbles and gas. Even with no shaking, the opened liquid begins to bubble, because the release of pressure makes the liquid supersaturated with gas. Normally, the gas is released slowly because it cannot make a bubble without overcoming an energy barrier. The energy cost is proportional to the interfacial tension times the bubble area. Thus the gas escapes by creating bubbles at the walls (at places that cost less energy) and by expanding existing bubbles. When the can is shaken, many new bubbles are dispersed in the liquid. Then when the pressure is released, the gas may escape by expanding these preexisting bubbles: there is no barrier to overcome. Thus the release of gas is much faster. This project investigates this effect.

A 1 L can of 10 cm diameter is shaken with an amplitude of 10 cm at 10 times per second. You can assume that the center of the liquid remains stationary while the can shakes.

(a) Estimate the shear rate $\dot{\gamma}$ inside.
(b) What is the order of magnitude of the viscous stress in the liquid, assuming it has the viscosity of water?

The shearing force pushes the air at the top of the can into the fluid and breaks it up into small bubbles. The bubble size is such that the interfacial tension pushing bubbles towards a spherical shape is comparable to the viscous force tending to elongate the bubbles.

(c) For the given shear rate, what is the size of bubble for which these two forces are comparable? You may assume that the interfacial tension is the same as that between water and air.
(d) Supposing that the 5% of the original 1 L volume that was air is dispersed into these small bubbles, what is the distance between these bubbles compared to the average distance to the wall of the can?
(e) Suppose the bubbles rise as though they were Stokes spheres as discussed in Chapter 4. (Bubbles actually rise somewhat faster than solid spheres of the same size. Also a cloud of bubbles rises faster than an individual bubble.) How long will they stay in the liquid? This is the time you must wait before opening the can in order to avoid a mess.

7.7 Amphiphilic polymers

In recent decades it has become possible to make polymers with the defining characteristic of an amphiphilic molecule: two mutually immiscible parts are joined together. The simplest example is a **diblock copolymer**, in which two immiscible polymer chains are joined end to end, as described in Fig. 1.1. Melts or solutions of diblock copolymers often form micellar structures, as surfactants do. The polymeric micelles are of course larger than their small surfactant counterparts. But more importantly, the properties of polymeric micelles are far more predictable and controllable than are conventional micelles. This is because these properties are controlled by universal features of long-chain polymers. In this section we sketch how the polymer chain properties lead to precise predictions about the size and energy of a micelle. The micelles constrain the polymer chains into a state of mutual stretching that is qualitatively different from the relaxed state. This stretching gives rise to new scaling properties with molecular weight. We survey a number of situations where this stretched state is important. The theory of this stretched state of polymers has grown powerful, sophisticated and successful. A good introduction to the subject appears in two recent surveys [24, 25].

We encountered a polymeric micelle structure in Fig. 1.1. The interwoven branches of this structure show the intricate structures that polymeric amphiphiles can spontaneously make. We now describe the simplest form of structure arising from the same principles. To this end, we consider a liquid of *symmetric* diblock copolymers, in which the two joined chains are the same in every way except that they are mutually immiscible. The chain contains N monomers, with $N/2$ in each subchain or **block**. We label the two blocks A and B. We first recall the criterion for being immiscible. We denote the work to insert one isolated B monomer into a liquid of pure A polymers (or the reverse) as χT[†]. Most A and B monomers encountered in practice have χ parameters that are only a few times 10^{-2} or smaller. In our discussion below, we will make the simplifying assumption that $\chi \ll 1$. The χ parameter measures the immiscibility. As noted at the beginning of this chapter, two liquids remain miscible as long as the work to insert remains below about T. The work to insert a random coil B block into pure A is of order $N\chi T$. Unless N is sufficiently large, this work is negligible, and free A and B chains mix readily. However, as N increases, the work increases; ultimately, phase separation occurs. The same is true when A and B chains are attached to make copolymers. If $\chi N \gg 1$, the A and B blocks must phase separate, so that virtually all the A monomers are surrounded by A blocks and likewise for the B blocks. In the limit of large N any diblock polymer liquid with a positive χ must become strongly phase separated in this sense.

This strong phase separation imposes a serious constraint on the individual chains. Each junction point between the two blocks is effectively confined to the interface between A and B regions. To see why, we imagine moving a junction point from the interface into an adjacent B region, for example. This movement necessarily brings monomers from the A block of that polymer into the B region. Each monomer thus pulled into the B region requires a work of order χT. Since $\chi N \gg 1$, even a small fraction of the A block thus pulled requires a work exceeding T. This work must decrease the probability that the junction point moves spontaneously into the B region according to the Boltzmann Principle. The larger N is, the less free the junction point is to explore the B region. Below, we shall treat the junction points as being confined near the A–B interfaces.

Our liquid consists of regions of nearly pure A or pure B. Since this is polymer melt, each point in an A region is uniformly occupied by A monomers[††]. The A monomers encountered at this point have a peculiar property not shared with ordinary polymer melts. Each such monomer belongs to a block that ends at the A–B interface. The same is true for the B monomers. This constraint limits the size of the A or B regions. If a region is large, then blocks passing through the center must necessarily be elongated in order to extend to the A–B interface as required. Such elongation costs free energy and this cost makes large regions unfavorable.

7.7.1 Micelle size

The combined effects of phase separation and polymer stretching determine the size of the A and B regions in our symmetric copolymer liquid. Since there is no difference between the A and B blocks, we expect no difference in the size or shape of the A and B regions. It is thus natural to suppose that the liquid consists of planar slabs of alternating A and B. We will denote the thickness of each slab as $2h$. If we impose a specific thickness h, we can

[†] In nonsymmetric polymers, χT as conventionally defined [26] is the average of the work of inserting a B into A and the work of inserting A into B. In our symmetric case these two works are the same.

[††] The cost of altering this uniform density, by, e.g., mixing empty space into the A region, can be viewed as a mixing process with its own χ parameter. This cost is presumed arbitrarily larger than the χ for AB mixing. Thus we can neglect the effect of A–B phase separation on the individual A and B densities.

readily estimate how the free energy of the liquid depends on h for chains of a given length. It is most convenient to measure chain length in terms of the volume V that a chain displaces in the melt. Evidently V is proportional to the number of monomers N. A free chain in the melt state is a random walk, whose mean-squared end-to-end distance R^2 is also proportional to V: $R^2 = V/a$. The proportionality constant a is a characteristic length that depends on the chain structure and the packing of the monomers in the melt state, but it is independent of the chain length. We call it the **packing length**.

If we change the size h, one major effect is to alter the interfacial energy. As with any liquid interface, the energy is proportional to the area of the interface and is characterized by an interfacial energy α. The ultimate origin of this energy is the immiscibility of the A and B blocks, measured by χ. For our purposes, the connection between χ and α is unimportant. We need only observe that α arises because of a thin mixing region that is independent of chain length[†]. Thus α is independent of chain length for long enough chains. The interfacial energy per unit volume is the interfacial area per unit volume times α. The chains for a given area A of A–B interface occupy a volume $2Ah$. Thus the interfacial energy per unit volume is $\alpha A/(2Ah) = \frac{1}{2}\alpha/h$. The interfacial energy per chain, denoted U, is thus given by $U = \frac{1}{2}V\alpha/h$. Evidently this U is smallest when h is largest. The interfacial energy favors large phase-separated regions.

As noted above, there is an opposing cost to increase h. This is the cost of the required chain elongation. If the slab thickness is h, chains must stretch to lengths of order h. Indeed, any monomer at the center of a slab must belong to a chain whose end-to-end distance is atleast h. Any monomer picked at random lies at a distance of order h from the nearest interface. Thus to increase h is to increase the end-to-end distances of the chains in proportion to h. The free energy S required for this stretching is a basic property of random-walk molecules. As discussed in Chapter 3, the work needed to stretch a chain to an end-to-end distance h perpendicular to the slab is given by

$$S(h) = \frac{1}{2}T\frac{h^2}{R^2/3} \simeq T\frac{ah^2}{V}.$$

Since most of the chains extend to a distance of order h, the chains have an average stretching energy $\langle S \rangle$ of order $S(h)$.

If the system is arranged into slabs of a given size h, the energy per chain is $\langle S \rangle + U(h)$. If the system is free to adjust h, the relative probability of a given h is $e^{-(\langle S \rangle + U(h))/T}$. The most likely value h^* is that which minimizes $\langle S \rangle + U(h)$. Evidently

$$0 = \frac{\partial}{\partial h}[\langle S \rangle + U(h)]_{h^*}.$$

Since S varies as h^2 while U varies as h^{-1}, we have

$$0 = \frac{2\langle S \rangle - U(h)}{h}\bigg|_{h^*}. \qquad (7.16)$$

[†] The mixing region consists of chain segments of length k such that $\chi k \simeq 1$. This mixing region is the same as that at the interface between A and B homopolymers.

This equation tells us three things. First it gives the scaling of the preferred size h^* with chain size V

$$aT(h^*)^2/V \simeq \langle S \rangle \simeq U(h^*) \simeq \alpha V/h^*,$$

so that

$$h^* \sim V^{2/3}(\alpha/aT)^{1/3}. \tag{7.17}$$

We note that h^* becomes arbitrarily larger than the unperturbed size $R(\sim V^{1/2})$ in the long-chain limit. This confirms that the main distortion of the blocks resulting from the phase separation is indeed one of stretching, as implicitly assumed above.

Second Eq. (7.16) tells us how the free energy per chain F^* grows with the chain length V:

$$[\langle S \rangle + U(h)]_{h^*} \simeq S(h^*) \simeq Ta\,(h^*)^2/V. \tag{7.18}$$

Since h^* grows arbitrarily bigger than the resting size R, this energy grows indefinitely relative to T. Thus, the stable structure contains significant energy. Even when the stretching energy is little more than T per chain, the total stretching energy is large, because there are many chains in the structural unit. The number of chains in a cube of size h^* is $(h^*)^3/V \sim V\alpha/(aT)$. Both α and a have length dimensions of atomic scale. But the chain volume V is typically thousands of times the volume of an atom. Thus even when the immiscibility is weak, the number of chains in a region of size h^* is typically hundreds, and the energy associated with the structure is many times T.

This large energy means that the region size h is well defined and varies little from h^*. If h departs slightly from h^*, the free energy per chain can be found by expanding $\langle S \rangle + U(h)$ around its minimum. The free energy in a volume $(h^*)^3$ thus has the form $(h^*)^3 V\,F^*[1+(\text{const})(h-h^*)/h^*]^2$, where (const) is of order unity. Since F^* is large, as argued above, even a small departure of h from h^* costs more than T and is thus improbable. The size of the regions is very well defined and becomes more so with longer chain length. Likewise, significant distortions of the *shape* of the A or B regions creates stretching energy of order $\langle S \rangle$ and is thus strongly suppressed.

Finally the equilibrium condition for h^* of Eq. (7.16) gives the *ratio* of stretching energy $\langle S \rangle$ to interfacial energy U at equilibrium. Though we do not know the coefficient relating $\langle S \rangle$ to h^2, we do know that the power is 2. Thus we know rigorously that $d\langle S \rangle/dh = 2\langle S \rangle/h$. U also varies as a known power of h, and its derivative is rigorously given by $dU/dh = -U/h$. Thus Eq. (7.16) does not have unknown coefficients. When it is satisfied, we must have $2\langle S \rangle = U$. The interfacial energy must be twice the stretching energy at equilibrium. In other terms, the stretching energy is $\frac{1}{3}$ of the total energy[†].

[†] Such "virial relations" appear whenever the free energy of a system is the sum of two parts, each of which varies as a power of some variational parameter.

7.7.2 Other copolymers

The reasoning above applies far beyond the special case of symmetric diblock copolymers treated above. If for example, the A block is shorter than the B block, the A regions need not be the same size or shape as the B regions. In order to accommodate the large B volume with minimal stretching, it is natural that the interface curve away from the majority blocks. As the A blocks become shorter and shorter relative to the B blocks, there comes a point where the layered structure is no longer the most stable. Instead, the stable structure is a lattice of parallel cylinders. With further reduction of the A blocks, another transition to a lattice of spherical A micelles occurs. Further morphologies are possible in principle. Indeed the complicated double-gyroid structure pictured in Fig. 1.1 appears to be the equilibrium structure in a narrow range of composition between the layered and the cylindrical regimes [27]. More new structures may emerge in the future.

The basic energies that give rise to these structures are the same as in the symmetric case treated above. The region size is still set by a balance between stretching energy and interfacial energy. The former scales as h^2 as before, while the latter continues to scale as $1/h$. Only the numerical coefficients depend on the specific geometry of the structure. Ignoring these prefactors, we may repeat the arguments above and arrive at the same conclusions. The region size h scales as $V^{2/3}$, the energy per region is large and proportional to V, and the stretching and interfacial energies are in the ratio $1:2$. With care, these arguments can be extended to

- polymers with a range of molecular weights,
- mixtures of block copolymers and homopolymers,
- multiblock polymers in which several alternating blocks of A and B occur on each chain,
- star polymers in which several A and B blocks emerge from a common node,
- triblock copolymers with three distinct immiscible species.

It is not surprising that a wealth of self-forming structures can be created using all these variations. Since these structures arise from the simple properties of immiscibility and chain randomness, the mathematical problem of predicting the structures from the molecules is well-defined and conceptually simple. For complicated structures, the practical calculation of $\langle S \rangle$ becomes technically complicated. It is not simple in practice to find the energy of the mutually and inhomogeneously stretched chains throughout each region. Still, numerical means have been devised to accomplish this task, and the numerical predictions match successfully with experiments [24]. The usefulness of these structures is also broad. Many different types of polymers can be accommodated and their structures predicted, since it is easy to make polymers flexible and mutually immiscible without constraining other desirable properties.

7.7.3 Polymeric amphiphiles in solution

The distinctive state of mutual stretching described above becomes only stronger when one considers copolymers in solution. For example, one may add a solvent for B to the symmetric diblocks discussed above. If the solvent does not dissolve A, then the junction points are tethered to the A–B interface as before. As solvent enters the B regions, their volume increases, and the B blocks expand and stretch to dissolve in the solvent as much as possible. The result is strong stretching of the B blocks and great asymmetry between the A and B regions. The calculation of the region size h is modified, because the volume per chain is no longer fixed and the quality of the solvent must be considered. These changes, though, prove minor so that the same mathematical methods used for diblock melts can be modified to describe these solutions. Now in addition to the structures above, one can form free spherical or wormlike micelles in the solution. One can also attach chains to solid surfaces to form the brush structures discussed in Chapter 5. That chapter gives a hint of how the concentration profile of brushes or other micelles arises.

One may wonder how well amphiphilic polymers function as surfactants, dispersing immiscible solvents and reducing interfacial tension. In practice, these polymers are little used for these functions. Part of the reason lies in the large energy of these stretched polymer blocks. Many of the functions of surfactants require their micelles to be flexible and deformable. For example, the Helfrich repulsion between surfactant bilayers relies on this flexibility. But polymer micelles have little of this flexibility, because their deformation energy is large, as we saw above. For example, the bending modulus of a lamellar layer is of the order of the energy within a cube of thickness h as thick as the layer. As shown in this section, this energy grows rapidly with polymer molecular weight. Thus the main interest in amphiphilic polymers lies in the great range of ordered and controllable structures that they make.

7.8 Dynamics and rheology

The time scales for motion and relaxation in surfactant solutions and microemulsions are controlled by solvent hydrodynamics as in the polymer solutions treated in Chapter 4. The forces transmitted across these solutions arise from hydrodynamic stress and from the distortion of the surfactant interfaces. From these basic facts, we can understand the viscosity of most surfactant-structured fluids with compact micelles. If the volume fraction of micelles is small, the viscosity is near that of the solvent. If the volume fraction grows substantial, the viscosity grows, because then the micelles interact and distort each other, thus storing free energy. For example, in our spherical microemulsion, at volume fractions approaching the close-packing density of about 60%, the spheres come into contact and distort each other. Then a unit of shear distortion translates into a further distortion of the spheres, increasing their area and mean curvature by a factor of order unity. Their curvature is altered, and this increases the energy of each sphere by an amount of order the bending stiffness κ. The associated stress is of order the energy per unit volume, or κR^3. In a simple (nonequilibrium) emulsion the distorting work is done against the interfacial tension α. Then if the droplets (or bubbles) have a typical radius R, the distortion energy for a unit of shear of a droplet is of the order αR^2. The associated stress is then of order $\alpha R^2/R^3$ or α/R. Figure 7.12 shows the modulus of emulsions of various droplet sizes, confirming this scaling with droplet size.

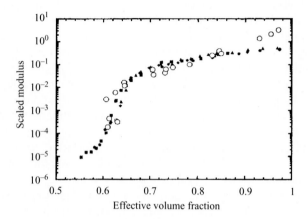

The *relaxation* of stress in these packed micelle or droplet structures can be very slow. When the droplets are concentrated enough to touch, an applied stress can only relax if the droplets can move past each other. To do this, a droplet must increase its area by a nonzero fraction. This motion thus requires an energy of order αR^2, which can be much larger than T. Thus the spontaneous relaxation of stress requires activated motion as is exponentially slow. For many purposes the system can be regarded as a solid.

7.8.1 Wormlike entanglement and relaxation

Naturally wormlike micelles experience the same entanglement constraints that polymers do, as discussed in Chapter 4. We consider worms that are most like their polymeric cousins, with negligible branching and with a persistence length ℓ_p that is much smaller than the distance ξ between micelles. On this basis, we would expect the following. (a) Entanglements should have a density of about $1/\xi^3$. (b) A small applied step strain of magnitude γ should elongate the randomly coiling segments between entanglements, thereby reducing the randomness and requiring a work per unit volume of order $(T/\xi^3)\gamma^2$. (c) To relax this applied strain, a worm should have to reptate over a distance comparable to its length L. This should require a time $\tau_{rep} \simeq \tau_\xi (L/L_\xi)^3$. Here τ_ξ is the Zimm time of an isolated worm of size ξ, and L_ξ is the length of such a segment. Some of these expectations hold true, but some are qualitatively wrong, as we now discuss. The relaxation of wormlike micelles is discussed at length by Cates and Candau [7].

The density of encounters between worms is of order $1/\xi^3$ as anticipated in (a): this fact relies only on the structural form of the worms. These encounters cannot obviously be regarded as impenetrable barriers or entanglement constraints, however. In principle, the worms can merge and pass through each other as polymers cannot. This kind of merging behavior has been proposed to explain some experiments [7]. But most wormlike micelles appear to resist merging sufficiently that merging can be neglected. If merging is negligible, the constraints amount to entanglements, and strain must elongate the worms. The energy cost again arises simply from the random-walk fluctuations common with polymers. Thus one expects a stored energy per unit volume given by (b) above. This means that the solution should have a plateau modulus of order T/ξ^3 as a polymer solution has.

The relaxation of the stress occurs by whatever routes lead to a restoration of isotropic random-walk structure in the worms. These worms have a channel for relaxation that polymers lack. They can break and reconnect. We consider a network of worms that define the entanglement constraints for a given segment of a micelle. We will take this segment to be much larger than L_ξ, but much smaller than the mean micelle length L. After a step strain, the constrained micelle may reptate to and fro along this network, but this motion in itself does not reduce the imposed anisotropy of the micelle and does not reduce the stress. Only when a micelle *end* passes through the network can it redefine the network by penetrating through it along a new path. With polymers, the relaxation must wait until the end of the chain has passed through the section of network being considered. But with micelles, there is a continual risk of breakage. Every breakage results in the creation of two ends, whose subsequent motion creates relaxed pieces of micelle. Likewise, every reconnection of two ends stops the relaxation of constraints that would otherwise have occurred.

A typical micelle of length $\bar{L}$ breaks and reconnects on a timescale τ_b that depends on the local breaking kinetics. There is a fixed rate of breakage Γ per unit length, i.e., $1/\tau_b \simeq \Gamma \bar{L}$. If this τ_b is as long as $\tau_{\rm rep}$, then micelles have time to relax much of their stress by reptation before they break. Thus breakage does not play a crucial role. On the other hand, if $\tau_b \ll \tau_{\rm rep}$, then breakage is important. We now ask how fast the stress should relax in this fast-breaking regime.

We can understand the relaxation by considering what fraction of the stress should relax in a time τ_b. In this time, a typical micelle has reptated a length ℓ given by $(\ell/\bar{L})^2 = \tau_b/\tau_{\rm rep}$. The micelle segments within a length ℓ of an end have disengaged from the network and their stress has relaxed. This motion has released a fraction $\ell/\bar{L}$ of the stress. Thus if the stress remaining after time $t + \tau_b$ is denoted $\sigma(t + \tau_b)$, we can express the decay by

$$\sigma(t + \tau_b) - \sigma(t) \simeq -\sigma(t)(\ell/\bar{L}) \simeq -\sigma(t)(\tau_b/\tau_{\rm rep})^{1/2}, \qquad (7.19)$$

or

$$d\sigma/dt \simeq -\sigma(t)(\tau_b/\tau_{\rm rep})^{1/2}/\tau_b \simeq -\sigma(t)(\tau_b\tau_{\rm rep})^{-1/2}. \qquad (7.20)$$

This suggests that the stress decays exponentially with a relaxation time $\tau \simeq (\tau_b\tau_{\rm rep})^{1/2}$. Remarkably, the relaxation time is both much shorter than the reptation time and much longer than the breaking time.

It is interesting to see how τ increases with the mean length $\bar{L}$, with everything else held constant, using $\tau_b \sim 1/\bar{L}$ and $\tau_{\rm rep} \sim \bar{L}^3$, we find $\tau \sim \bar{L}$. It increases much slower with length than the reptation time does.

This kind of breaking-dominated relaxation has been confirmed in rheological experiments [7]. Its most important effect is on the viscosity η of the fluid. As for any fluid, this viscosity is roughly the product of the step-strain modulus G_0 and the stress relaxation time. The step-strain modulus, like the osmotic pressure is roughly T/ξ^3, where ξ is the distance between chains in the semidilute solution. The relaxation time is the τ discussed above. Knowing the dependence of L, and ξ on concentration ϕ, we can readily predict how the viscosity depends on concentration. Self-repulsion effects appear important for these solutions [7], so we must use the scaling exponents D and γ introduced above. For wormlike micelles, $\bar{L} \sim \phi^y$, where y is given by Eq. (7.8). We have deduced the dependence of τ_{rep} on ξ in Eq. (7.17): $\tau_{\text{rep}} \sim \bar{L}^3 \xi^{3-3D}$. Thus for ordinary reptating polymers in these conditions,

$$\eta \sim G_0 \tau_{\text{rep}} \sim \bar{L}^3 \xi^{-3D} \sim \phi^{3D/(3-D)} \sim \phi^{3.9}. \qquad (7.21)$$

For the wormlike micelles, we combine τ_{rep} and $\tau_b \sim 1/\bar{L}$ to infer $\tau \sim (\tau_{\text{rep}} \tau_b)^{1/2} \sim (\bar{L}^2 \xi^{3-3D})^{1/2}$. Thus [29]

$$\eta \sim \xi^{-3} \bar{L} \xi^{(3-3D)/2} \sim \phi^y \phi^{(3D+3)/(6-2D)} \sim \phi^{3.72}. \qquad (7.22)$$

The micelles have virtually the same growth of viscosity with concentration as a solution of ordinary polymers. The competing effects of increasing $\bar{L}$ and the breaking kinetics nearly compensate.

As noted above, wormlike micelles can sometimes branch or join together to form a network. The stress relaxation of branched polymers is a tricky subject that depends on the state of branching, as noted in Chapter 4. The crosslinks between micelles can inhibit stress relaxation altogether. We shall not attempt to give a general account of how stress relaxes in these cases.

If the imposed shear rate is significant, the solution may be altered by the shear. Some types of alteration were discussed in Chapter 4. The chief effect is mutual alignment of polymers that reduces the degree of entanglement. In a micellar solution, new forms of alteration appear [29]. The length distribution of the worms can change, and can be different for worms with different orientations. The breaking rates can also increase.

7.8.2   Rheology of lamellar solution

The response to strain in a lamellar solution depends on the direction of the strain. We first consider the simplest case where parallel, flat lamellae are made to slide over each other, as shown in Fig. 7.13. Such a solution has little means to support stress or inhibit flow. Since most of the solution is solvent, the shear occurs mostly in the solvent. The shear rate is only slightly increased above the average shear rate, even if the surfactant layers allow no shear at all. Thus the viscosity is little higher than that of pure solvent. If we shear in a perpendicular direction, the surfactant layers themselves are obliged to shear (Fig. 7.13(b)), or to bend and compress (Fig. 7.13(c)). The layers resist such distortions strongly. Shearing a layer forces the awkwardly shaped and densely packed surfactant molecules to slide past each other. The stress required should be much higher than for similar shear in the solvent. If one bends and compresses the layers as in Fig. 7.13(c) there is

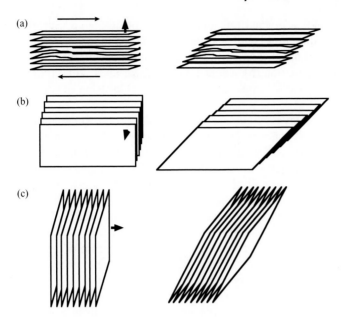

Fig. 7.13
The three principal orientations of a lamellar liquid under shear.
(a) Layer-sliding orientation,
(b) Layer-shearing orientation,
(c) Layer-stretching orientation. Shear strain is in the plane of the paper in all three pictures. The layer normal is indicated by an arrow. In (a) a fluctuation in two of the layers is shown to illustrate the resulting compression under shear.

an energy cost even with no motion; the bending cost is controlled by the bending stiffness κ; the compressional cost is controlled by the Helfrich repulsion or by Coulomb repulsion. Here steady flow is impossible without breaking or strongly distorting the layers.

If the shear is applied in a generic direction, one expects the lamellae to reorient so as to allow easier flow. That is, the lamellae should adopt orientation (a), and the viscosity should be little greater than that of the solvent. This behavior is observed sometimes when the lamellae are carefully prealigned [30]. But typically a lamellar solution is disordered, with different orientations in different places. Such solutions do not flow readily at all. It is not merely that they show a high viscosity. They have a dynamic modulus that varies roughly as the square root of frequency. This contrasts with the behavior of a viscous liquid, where the dynamic modulus is proportional to frequency. This square-root behavior may persist down to the lowest frequencies measurable, implying a stress qualitatively stronger than that of a liquid.

The reasons for this square-root behavior are ill understood. It is believed that the lamellae are forced to store elastic energy even when the shear is weak and slow. The amount of energy stored decreases as the frequency is lowered because then the lamellae have a chance to relax over larger length scales. These effects are seen most clearly when the surfactants used are amphiphilic polymers. To date the justifications for this behavior remain complicated and tentative [31].

7.8.3 Shear-induced restructuring

Shear in lamellar solutions causes mysterious forms of restructuring. In weak shear, the lamellae show a tendency to reorient so as to reduce the stress for a given strain rate, as anticipated above. Surprisingly, the alignment direction is not always in orientation (a) above. Instead, orientation (b) can occur. Sometimes orientation (a) occurs for the lowest shear rates and orientation (b) occurs at higher shear rates. This strange behavior can be understood if one looks further into the shape of the layers being sheared.

Surfactant layers fluctuate in shape; it is these fluctuations that give rise to the Helfrich repulsion discussed earlier in this chapter. These shape fluctuations persist for a certain time. If the shear occurs faster than this time, one may encounter a situation like Fig. 7.13(a) when one tries to shear in orientation (a). Instead of minimizing stress, the shear drives the bumpy lamellae into their neighbors and increases the stress. The orientation (b) avoids this situation. The shear no longer causes rippled lamellae to run into each other.

What governs the persistence time for these ripples? We first recall that the fluctuations that control the local distance between bilayers have wavelengths λ much larger than the layer spacing d. Thus to gauge the persistence time, we imagine a thought experiment involving a large scale distortion. We pinch the layers together over long distance λ so that in the pinched region the spacing is reduced by a small amount A from its equilibrium value d. We then release the pinching force and watch the spacing return to its equilibrium value. In order to restore the spacing, solvent must flow into the pinched region. The flow causes dissipation, which retards the relaxation. We denote the typical shear rate in this region as $\dot{\gamma}$ and the pinched volume by Ω. During this flow, the energy is dissipated at a rate $\dot{W} \simeq \Omega \eta (\dot{\gamma})^2$, where η is the solvent viscosity.

The return of the spacing to its equilibrium value releases compressional free energy. As we have seen, these fluctuating bilayers repel each other because compression constrains their random fluctuations. According to the Helfrich formula of Eq. (7.15) above, the energy per unit area u can be expressed in terms of the temperature T and the bending stiffness κ: $u \simeq (T/d^2)(T/\kappa)$. Thus the Helfrich energy U in the volume Ω is given by $U \simeq (\Omega/d)(T/d^2)(T/\kappa)$. Even if there is additional attraction between the layers that holds them at an equilibrium spacing, the Helfrich energy is a substantial fraction of the energy and thus the overall energy is proportional to this U.

We now express these energies in terms of our perturbed spacing $d - A$. If A increases at a rate $\dot{A}$, there must be an inward flow to supply the needed volume of fluid. The flow enters a given lamellar gap at the boundary of the pinched region, whose height is roughly d and whose circumference is roughly λ. Thus the inward current I is roughly $\lambda d v$. This current must match the increase of volume $\lambda^2 \dot{A}$, so that $v \simeq \dot{A}\lambda/d$. Finally, the shear rate $\dot{\gamma}$ is the gradient of velocity, or $\dot{\gamma} \simeq v/d \simeq \dot{A}\lambda/d^2$. Thus $\dot{W} \simeq \Omega \eta (\dot{A}\lambda/d^2)^2$. To express the perturbation in the free energy, δU, in terms of A, we imagine that d is close to an equilibrium, so that $\delta U \sim A^2$. In view of the order of magnitude of U given above, we conclude

$$\delta U \simeq (\Omega/d)(T/d^2)(T/\kappa)(A/d)^2.$$

Now we can estimate the relaxation rate $\dot{A}$ by requiring that the release rate of stored energy, balance the dissipation rate:

$$\Omega\eta(\dot{A}\lambda/d^2)^2 \simeq \dot{W} \simeq -\dot{U} \simeq -(\Omega/d)(T/d^2)(T/\kappa)\dot{A}A/d^2. \quad (7.23)$$

Simplifying this equation yields

$$\dot{A} \simeq -[(T/\eta)1/(\lambda^2 d)(T/\kappa)]A.$$

Thus A relaxes in a time $\tau \simeq 1/[\ldots]$. We can rearrange the factors to write

$$\tau \simeq (\lambda/d)^2 [6\pi\eta d^3/T] (\kappa/T)/(6\pi). \quad (7.24)$$

This type of compressional relaxation is one of the standard hydrodynamic relaxation modes of any smectic liquid crystal. It is known as the **baroclinic mode**. We recognize the [...] factor as the Zimm time for a sphere of radius d, introduced in Chapter 4. We recall that the Zimm time for a 6.1 nm radius sphere in water is 1 μs. From this time we may estimate τ for a typical lamellar solution. For simplicity we consider lamellae with a spacing of 6.1 nm. Typical bilayers have a bending stiffness κ of roughly 10 T. Thus for λ in the range of 10 d, the expected relaxation time is in the range of tens of microseconds. The persistence time for a random initial distortion of the layers is of this same order.

If one increases the shear rate to approach this inverse persistence time, the shear begins to influence the orientation of the lamellae. Increasing the shear rate further leads to more dramatic changes. The most prevalent of these is the rupturing of lamellae to form bilayer **vesicles** enclosing part of solvent. When the equilibrium state is lamellar, the vesicles contain many concentric bilayers in an **onion** or **multilamellar vesicle** morphology [33], as shown in Fig. 7.14. Such onions have been observed in a wide range of surfactant solutions from simple SDS to lipids. When they form, they generally share these features:

Fig. 7.14
Freeze-fracture micrograph of onion morphology from [32], © American Chemical Society, 1996.

- The onions form a close-packed mass, as observed in freeze-fracture electron microscopy. The micrographs show the onions pushed together so that they deform each other.
- The onions become very uniform in size after long exposure to shear. Onions often contain hundreds of bilayers and are microns in size.
- This size decreases with increasing shear rate. The selected size appears to have a baroclinic relaxation time comparable to the shear rate [33].
- If the shear is stopped, the onions can retain their integrity over months or years. Thus they can be used as controlled-release containers.
- The viscosity of an onion-phase solution is typically somewhat lower than that of the disordered lamellar phase from which it was made. This is thought to be due to a compression of the layers within each onion and a corresponding increase in the space between onions. Thus detergent liquids are often processed so that they are in the onion morphology in order to increase their concentration without excessive viscosity.

Quantitative understanding of the onion phase is currently very primitive.

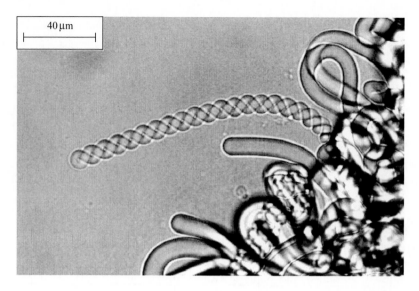

Fig. 7.15
Light micrograph of myelin structures
emerging from a drop of concentrated
$C_{12}E_5$ surfactant in water, from [34],
courtesy Mark Buchanan

A second important morphology is the **myelin** morphology pictured in
Fig. 7.15. These wormlike myelin figures consist of many concentric, cylindrical bilayers. They were first observed by biologists in the nineteenth
century [35], who likened them to the myelin structures surrounding nerve
fibers. A classic way to create myelin structures is to place a concentrated
lamellar solution under osmotic stress. For example, one can immerse a drop
or lump of concentrated SDS solution with pure water and then observe the
surface of the drop in a microscope. In the presence of this excess water, the
lamellae have an equilibrium spacing substantially larger than their initial
spacing. Thus the layers start to take in water and swell. Rather than swelling
uniformly, the solutions form bumps at their boundary, and these grow into
the micron-sized worms seen in the figure. Mysteriously, these worms are
often observed to twist upon themselves. They twist indifferently in either
direction. Quantitative understanding of myelins is even more primitive
than that of onions.

Appendix: Gauss–Bonnet Theorem

One can prove the Gauss–Bonnet Theorem elegantly using advanced methods of differential geometry [12]. One can prove it for nearly flat surfaces
by using the Monge representation introduced in Section 7.5.1. Finally,
one can show that any small deformation of a surface from s_1 to s_2 leaves
$\int ds\, C_1 C_2$ invariant. Here we explicitly verify the theorem for the case of
a small bump on a flat surface, as shown in Fig. 7.5.

The heavy solid line shows a vertical section through the bump. The
bump is small on the scale of any preexisting curvature, so that the unde-
formed surface can be considered flat. One can readily verify that the
integral of $C_1 C_2$ over the bump is zero. The integral consists of two parts.
The central part is constructed to be a section of a sphere with radius of
curvature R, so here $C_1 C_2 = R^{-2}$. The area of the spherical part is evidently

$\int R^2 d\cos\theta \, d\phi = 2\pi R^2(1 - \cos\theta)$. Thus the integral over the spherical cap of $C_1 C_2$ is just $2\pi(1 - \cos\theta)$. We may allow the bump to be shallow, so that θ is small and $(1 - \cos\theta) \simeq \frac{1}{2}\theta^2$.

To complete the integration, we must include the annular rim region. We shall choose a rim shape whose cross section is an upward curving arc with radius of curvature R'. Evidently $C_1 = 1/R'$. The other principal curvature C_2 lies in the direction perpendicular to the vertical section. The two lines curve towards opposite sides of the surface as in Fig. 7.4, so that $C_1 C_2$ is negative. We can easily find C_2 at the inner edge of the rim. This inner edge is part of the sphere and has $C_2 = 1/R$. On the outer edge, the rim meets a flat plane and $C_2 = 0$. At intermediate points within the rim we can find C_2 by using the projected radius ρ. At the inner edge $C_2 = 1/R = \sin\theta/\rho$ (Since $\rho = R\sin\theta$). As one goes outward on the rim, the radial distance ρ changes negligibly, but the inclination angle α (not shown) varies from θ to 0. We may find C_2 by imagining a cone tangent to the rim at the point of interest. The cone's C_2 is the same as the rim's. At the inner boundary of the rim, we have found that $C_2 = \sin\alpha/\rho$. The curvature of the cone depends only on the cone angle α and the radial coordinate ρ. Thus throughout the rim $C_2 = \sin\alpha/\rho \simeq \alpha/\rho$.

As we move a distance s from the outside of the rim, the inclination angle α is s/R'. Thus

$$\int_{\text{rim}} C_1 C_2 = -2\pi\rho \int ds C_1 C_2 \simeq -2\pi\rho \int_0^\theta (R' d\alpha) \left(\frac{1}{R'}\right)\left(\frac{\alpha}{\rho}\right) = \frac{-2\pi\theta^2}{2}.$$

This rim integral is independent of the rim curvature R' and is exactly the negative of the integral over the spherical region[†]. Thus, the overall integral of $C_1 C_2$ is 0, as we aimed to show.

[†] The two integrals must clearly cancel even when the angle θ is not small if the Gauss–Bonnet theorem is valid. One readily verifies that for general θ the θ^2 appearing in the two integrals should be replaced by $2(1 - \cos\theta)$. Thus the two integrals indeed cancel for large θ as well as small.

References

1. J. N. Israelachvili, *Intermolecular and Surface Forces*, 2nd ed. (London: San Diego: Academic, 1992).
2. K. G. Denbigh, *The Principles of Chemical Equilibrium: With Applications in Chemistry and Chemical Engineering*, 4th ed. (Cambridge: Cambridge, 1981).
3. D. L. Ho, R. M. Briber, R. L. Jones, S. K. Kumar, and T. P. Russell, *Macromolecules* **31** 9247 (1998).
4. These drawings are provided by the ChemIDplus (http://chem.sis.nlm.nih.gov/chemidplus/) resource using molecular data from the National Library of Medicine, 8600 Rockville Pike, Bethesda, MD 20894 (http://sis.nlm.nih.gov).
5. J. Lang and R. Zana, in *Surfactant Solutions: New Methods of Investigation*, ed. R. Zana (New York: Marcel Dekker, 1987).
6. P.-G. deGennes, *Scaling Concepts in Polymer Physics* (Cornell: Ithaca, 1979), Chapter 1.
7. M. E. Cates and S. J. Candau, *J. Phys.: Condens. Matter.* **2** 6869 (1990).
8. T. Tlusty and S. A. Safran, *J. Phys.: Condens. Matter.* **12** A253 (2000).
9. S. A. Safran, *Statistical Thermodynamics of Surfaces, Interfaces, and Membranes* (Reading MA, USA: Addison-Wesley, 1994).
10. S. T. Hyde, in *Handbook of Applied Surface and Colloid Chemistry*, ed., Krister Holmberg (New York: Wiley, 2001), Chapter 16.
11. V. Luzzati, P. Mariani, and T. Gulik-Krzywicki, in *Physics of Amphiphilic Layers*, eds. J. Meunier and D. Langevin (Berlin: Springer, 1987).

12. F. David, in *Statistical Mechanics of Membranes and Surfaces*, eds. D. Nelson, T. Piran, and S. Weinberg (Singapore: World Scientific, 1989), p. 174.

13. P. G. DeGennes and C. Taupin, *J. Phys. Chem.* **86** 2294 (1982).

14. W. Helfrich, *Z. Naturforsch.* **33a** 305 (1978).

15. M. Kotlarchyk, S.-H. Chen, J. S. Huang, and M. W. Kim, *Phys. Rev. A* **29** 2054 (1984).

16. P. A. Winsor, *Solvent Properties of Amphiphilic Compounds* (London: Butterworths, 1954).

17. S. H. Chen, *Ann. Rev. Phys. Chem.* **37** 351 (1986).

18. M. Kahlweit, R. Strey, D. Haase, H. Kunieda, T. Schmeling, B. Faulhaber, M. Borkovec, H.-F. Eicke, G. Gusse, F. Eggers, Th. Funck, H. Richmann, L. Magid, O. Söderman, P. Stilbs, J. Winkler, A. Dittrich and W. Jahn, *J. Colloid Interface. Sci.* **118** 436 (1987).

19. T. Tlusty, S. A. Safran, R. Menes, and R. Strey, *Phys. Rev. Lett.* **78** 2616 (1997).

20. S. H. Chen, S. L. Chang, and R. Strey, *J. Chem. Phys.* **93** 1907 (1990).

21. G. I. Taylor, *Proc. R. Soc. A* **146** 501 (1934).

22. R. G. Larson, *The Structure and Rheology of Complex Fluids* (New York: Oxford 1999), see Chapter 9.2.2.

23. J. Bibette, D. C. Morse, T. A. Witten, and D. A. Weitz, *Phys. Rev. Lett.* **69** 2439 (1992).

24. M. W. Matsen and F. S. Bates, *Macromolecules* **29** 1091 (1996).

25. A. Halperin, M. Tirrell, and T. P. Lodge, *Adv. Poly. Sci.* **100** 31–71 (1992).

26. P. J. Flory, *J. Chem. Phys.* **9** 660 (1941); M. L. Huggins, *J. Phys. Chem.* **46** 151 (1942).

27. H. Hasegawa, H. Tanaka, K. Yamazaki, and T. Hashimoto, *Macromolecules* **20** 1651 (1987); M. W. Matsen, *Phys. Rev. Lett.* **80** 4470 (1998).

28. T. G. Mason, J. Bibette, and D. A. Weitz, *Phys. Rev. Lett.* **75** 2051 (1995).

29. M. S. Turner and M. E. Cates, *J. Phys.: Condens. Matter.* **4** 3719 (1992).

30. D. Bonn, J. Meunier, O. Greffier, A. Al-Kahwaji, and H. Kellay, *Phys. Rev. E* **58** 2115 (1998).

31. G. H. Fredrickson and F. S. Bates, *Annu. Rev. Matter. Sci.* **26** 501 (1996).

32. T. Gulik-Krzywicki, J. C. Dedieu, D. Roux, C. Degert and R. Laversanne, *Langmuir* **12**, 4668 (1996).

33. P. Panizza, D. Roux, V. Vuillaume, C. Y. D. Lu, and M. E. Cates, *Langmuir* **12** 248 (1996).

34. M. Buchanan, *Dynamics of Interfaces in Surfactant Lamellar Phases*, Ph.D. thesis, Department of physics and astronomy (Edinburgh UK: University of Edinburgh, 1999).

35. R. L. K. Virchow, *Virchows Arch. B Cell pathol.* **6** 562 (1854) (Berlin, New York: Springer-Verlag, 1968–79).

Index